AF360730

Internet of Things and Artificial Intelligence in Transportation Revolution

Internet of Things and Artificial Intelligence in Transportation Revolution

Editors

Miltiadis D. Lytras
Kwok Tai Chui
Ryan Wen Liu

MDPI • Basel • Beijing • Wuhan • Barcelona • Belgrade • Manchester • Tokyo • Cluj • Tianjin

Editors
Miltiadis D. Lytras
Deree College—The American
College of Greece
Greece

Kwok Tai Chui
The Open University of
Hong Kong
China

Ryan Wen Liu
Wuhan University of Technology
China

Editorial Office
MDPI
St. Alban-Anlage 66
4052 Basel, Switzerland

This is a reprint of articles from the Special Issue published online in the open access journal *Sensors* (ISSN 1424-8220) (available at: https://www.mdpi.com/journal/sensors/special_issues/IOTAI).

For citation purposes, cite each article independently as indicated on the article page online and as indicated below:

LastName, A.A.; LastName, B.B.; LastName, C.C. Article Title. *Journal Name* **Year**, *Volume Number*, Page Range.

ISBN 978-3-0365-0310-3 (Hbk)
ISBN 978-3-0365-0311-0 (PDF)

Contents

About the Editors . . . . **vii**

Preface to "Internet of Things and Artificial Intelligence in Transportation Revolution" . . . . **ix**

Miltiadis D. Lytras, Kwok Tai Chui and Ryan Wen Liu
Moving Towards Intelligent Transportation via Artificial Intelligence and Internet-of-Things
Reprinted from: *Sensors* **2020**, *20*, 6945, doi:10.3390/s20236945 . . . . **1**

Xinyu Zhang, Chengbo Wang, Yuanchang Liu and Xiang Chen
Decision-Making for the Autonomous Navigation of Maritime Autonomous Surface Ships
Based on Scene Division and Deep Reinforcement Learning
Reprinted from: *Sensors* **2019**, *19*, 4055, doi:10.3390/s19184055 . . . . **5**

Yonghang Jiang, Bingyi Liu, Ze Wang, Xiaoquan Yi
Start from Scratch: A Crowdsourcing-Based Data Fusion Approach to Support
Location-Aware Applications
Reprinted from: *Sensors* **2019**, *19*, 4518, doi:10.3390/s19204518 . . . . **23**

Gianmarco Baldini, Filip Geib and Raimondo Giuliani
Continuous Authentication of Automotive Vehicles Using Inertial Measurement Units
Reprinted from: *Sensors* **2019**, *19*, 5283, doi:10.3390/s19235283 . . . . **41**

Wei Wu, Ling Huang and Ronghua Du
Simultaneous Optimization of Vehicle Arrival Time and Signal Timings within a Connected
Vehicle Environment
Reprinted from: *Sensors* **2020**, *20*, 191, doi:10.3390/s20010191 . . . . **71**

Siyu Guo, Xiuguo Zhang, Yisong Zheng and Yiquan Du
An Autonomous Path Planning Model for Unmanned Ships Based on Deep
Reinforcement Learning
Reprinted from: *Sensors* **2020**, *20*, 426, doi:10.3390/s20020426 . . . . **89**

Kwok Tai Chui, Miltiadis D. Lytras and Ryan Wen Liu
A Generic Design of Driver Drowsiness and Stress Recognition Using MOGA Optimized
Deep MKL-SVM
Reprinted from: *Sensors* **2020**, *20*, 1474, doi:10.3390/s20051474 . . . . **125**

**Yuqi Guo, Bin Li, Matthew Daniel Christie, Zongzhi Li, Miguel Angel Sotelo, Yulin Ma,
Dongmei Liu and Zhixiong Li**
Hybrid Dynamic Traffic Model for Freeway Flow Analysis Using a Switched Reduced-Order
Unknown-Input State Observer
Reprinted from: *Sensors* **2020**, *20*, 1609, doi:10.3390/s20061609 . . . . **145**

Xiangyu Zhou, Zhengjiang Liu, Fengwu Wang, Yajuan Xie and Xuexi Zhang
Using Deep Learning to Forecast Maritime Vessel Flows
Reprinted from: *Sensors* **2020**, *20*, 1761, doi:10.3390/s20061761 . . . . **161**

**Kh Tohidul Islam, Ram Gopal Raj, Syed Mohammed Shamsul Islam,
Sudanthi Wijewickrema, Md Sazzad Hossain, Tayla Razmovski and Stephen O'Leary**
A Vision-Based Machine Learning Method for Barrier Access Control Using Vehicle License
Plate Authentication
Reprinted from: *Sensors* **2020**, *20*, 3578, doi:10.3390/s20123578 . . . . **179**

Gianmarco Baldini, Raimondo Giuliani and Filip Geib
On the Application of Time Frequency Convolutional Neural Networks to Road Anomalies'
Identification with Accelerometers and Gyroscopes
Reprinted from: *Sensors* **2020**, *20*, 6425, doi:10.3390/s20226425 . . . . **197**

About the Editors

Miltiadis D. Lytras Ph.D., is an expert in advanced computer science and management, and an editor, lecturer, and research consultant, with extensive experience in academia and the business sector in Europe and Asia. Prof. Lytras is Research Professor at Deree College—The American College of Greece— a Distinguished Scientist at the King Abdulaziz University, Jeddah, Kingdom of Saudi Arabia—and a Visiting Scholar at the Effat University, Jeddah, Kingdom of Saudi Arabia. Prof. Lytras is a world-class expert in the fields of cognitive computing, information systems, technology-enabled innovation, social networks, computers in human behavior, and knowledge management. In his work, Prof. Lytras seeks to bring together and exploit synergies among scholars and experts, and is committed to enhancing the quality of education for all. He has co-edited more than 80 Special Issues in International Journals. He is currently the Editor-in-Chief of International Journal on Semantic Web and Information Systems, the Editor-in-Chief of International Journal of Smart Education and Urban Society, the Associate Editors of IEEE Access, Behavior and Information Technology, and Transforming Government: People, Process and Policy.

Kwok Tai Chui Ph.D., received a B.Eng. degree in electronic and communication engineering— Business Intelligence Minor—and Ph.D. degree from City University of Hong Kong, in 2013 and Feb. 2018, respectively. He has industry experience as a Senior Data Scientist in Internet of Things (IoT) company. He joined the Department of Technology, School of Science and Technology, at The Open University of Hong Kong as Research Assistant Professor. He was the recipient of the 2nd Prize Award (Postgraduate Category) of 2014 IEEE Region 10 Student Paper Contest. He also received the Best Paper Award in IEEE The International Conference on Consumer Electronics-China, in both 2014 and 2015. He has published more than 60 research publications. He has been serving as Managing Editor in International Journal on Semantic Web and Information Systems; Topic Editor in Sensors; Associate Editors in International Journal of Energy Optimization and Engineering and International Journal of Asian Business and Information Management; and International Advisory Board in International Journal of Healthcare Information Systems and Informatics.

Ryan Wen Liu Ph.D., received a B.Sc. degree (Hons.) in information and computing science from the Department of Mathematics, Wuhan University of Technology, Wuhan, China, in 2009, and a Ph.D. degree in imaging informatics from The Chinese University of Hong Kong, Hong Kong, in 2015. He is currently an Associate Professor with the School of Navigation, Wuhan University of Technology. He was a Visiting Scholar with the Agency for Science, Technology and Research (A*STAR), Singapore. He is also an Associate Editor of the International Journal on Semantic Web and Information Systems, and a Guest Editor of Sensors, and Journal of Advanced Transportation. His research interests include computer vision, trajectory data mining, intelligent transportation systems, and computational navigation sciences.

Preface to "Internet of Things and Artificial Intelligence in Transportation Revolution"

Human beings and goods rely heavily on safe and effective transportation, which forms a basic component of the maintainence of good economic and social development. According to the World Health Organization (WHO), annual road traffic deaths and injuries have reached 1.35 million and 50 million, respectively. As part of logistical management, goods are expected to be delivered via optimal routes. Today's smart-city visions aim at improving the operation of all sectors via transforming data into valuable information. The concept of internet-of-things (IoT) provides a scalable architecture for data collection and transmission. into smart city with the convergence of advanced technologies.

Based on the statistics from Web of Artificial intelligence (AI) offers a wide range of approaches (e.g., statistical, symbolic, cybernetics, and brain simulation) and tools (e.g., statistical learning, classifier, probabilistic reasoning, and optimization), to help cities evolve Science, increasing attention, reflected by the number of research publications (including article, review, proceedings paper, and editorial), has been paid to intelligent/smart transportation since 2007. The number of research publications using field tags for topics of intelligent transportation, intelligent transport, smart transportation, or smart transport between 2007 and 2019 has increased from 180 to 1876 per year. Particularly, the average percentage increase in the number of research publications was 51.6% from 2014 to 2019.

This Special Issue is intended to provide high-quality research on recent advances in AI and IoT in the transportation revolution, more specifically, state-of-the-art theories, methodologies and systems for the design, development, deployment, and innovative use of those convergence technologies to providing insight into the theoretical and technological revolution in transportation science and engineering. Special attention is paid to collision avoidance in surface ships, indoor localization, vehicle authentication, traffic signal control, path-planning of unmanned ships, driver drowsiness and stress detection, vehicle density estimation, maritime vessel flow forecast, and vehicle license plate recognition.

Various intelligent transportation applications are closely linked to and support some sustainable development goals. All member states of United Nations agreed to build a strong partnership to meet 17 sustainable development goals (SDGs) and 169 targets in 2015. Some of these targets are related to intelligent transportation. SDG Goal 3 Target 3.6 aims to reduce the number of traffic deaths and injuries by 50%. Besides driver drowsiness and stress recognition, driver distraction recognition could help to ensure that the driver concentrates on he front view (actual road condition) to achieve the vision of the smart city as a safe city. SDG Goal 7 aims to provide clean and affordable energy. For all kinds of transportation, energy is consumed. Various studies have applied artificial intelligence and internet-of-things to enhance the cleanliness and affordability of energy. SDG Goal 8: Decent work and economic growth states that transportation forms the foundation of environmental degradation, energy consumption, and economic growth. This foundation is closely linked to SDG Goal 9, industry, innovation, and infrastructure. The visions of SDG Goal 11 Target 11.2 include expanding public transport, improving road safety, and providing access to sustainable transport systems. In SDG Goal 12 Target 12.C, fossil-fuel subsidies are expected to be phased out. For SDG Goal 14, related to oceans, seas, and marine resources, 30% of carbon dioxide emission is dissolved into the ocean.

We would like to express our sincere gratitude to the professional staff at MDPI for their qualitative work and valuable support, as well as to all the contributors and reviewers that made this edition possible. We look forward to seeing more researchers conducting research on intelligent transportation, particularly those relying on the internet of things and artificial intelligence.

Miltiadis D. Lytras, Kwok Tai Chui, Ryan Wen Liu
Editors

Editorial

Moving Towards Intelligent Transportation via Artificial Intelligence and Internet-of-Things

Miltiadis D. Lytras [1,2], Kwok Tai Chui [3,*] and Ryan Wen Liu [4]

1   King Abdulaziz University, Jeddah P.O. Box 34689, Saudi Arabia; mlytras@acg.edu
2   Effat College of Engineering, Effat University, Jeddah P.O. Box 34689, Saudi Arabia
3   Department of Technology, School of Science and Technology, The Open University of Hong Kong, Hong Kong, China
4   Hubei Key Laboratory of Inland Shipping Technology, School of Navigation, Wuhan University of Technology, Wuhan 430063, China; wenliu@whut.edu.cn
*   Correspondence: jktchui@ouhk.edu.hk; +852-2768-6883

Received: 14 November 2020; Accepted: 28 November 2020; Published: 4 December 2020

One of the key smart city visions is to bring smarter transport networks, specifically intelligent/smart transportation. It facilitates safe and effective physical movement and interaction of humans, animals, and goods. Typical global issues include sustainable energy, traffic accidents, traffic congestion, logistic management, data analysis, security, and privacy. In recent years, the internet-of-things (IoT) and artificial intelligence (AI) have taken a leading role in achieving this smart city vision. The former provides a solid infrastructure for scalable and robust data collection and transmission. The latter brings creative and innovative elements to machines for intelligent transportation applications. In this special issue, "Internet of Things and Artificial Intelligence in Transportation Revolution", ten (10) research articles have been published. These articles generate a meaningful discussion around the impacts of AI and IoT in intelligent transportation. This editorial not only summarizes the special issue articles but also shares other hot research topics.

Transportation plays an essential role in today's economic and social development. As daily road users, we need to ensure safe and effective travel. According to the World Health Organization (WHO), annual road traffic deaths and injuries reach 1.35 million and 50 million, respectively [1]. Based on Web of Science statistics, there has been increasing attention on intelligent/smart transportation since 2007, reflected by the rising number of research publications. The average percentage increase in the number of research publications is 51.6% from 2014 to 2019.

The first article, "Decision-making for the autonomous navigation of maritime autonomous surface ships based on scene division and deep reinforcement learning" [2] authored by X. Zhang, C. Wang, Y. Liu, and X. Chen, considered maritime autonomous surface ships (MASSs). Attention has been drawn to adaptive navigation and an uncertain environment. An artificial potential field-deep reinforcement learning approach was proposed. Their experiment revealed that the proposed method significantly reduced the collision rate from 2.24% to 1.16%, compared with the traditional deep reinforcement learning approach.

Y. Jiang, B. Liu, Z. Wang, and X. Yi presented an article "Start from scratch: A crowdsourcing-based data fusion approach to support location-aware applications" [3]. Multi-dimensional crowdsourcing with multi-resolution ambient map and trade coding has been applied to the indoor localization problem. Twenty-six volunteers participated in the data collection process. In total, 931 crowdsourcing data traces were collected from five types of smartphones on two floors with a total floor space of 4000 m². Results showed that half of the data points were perfect, whereas 90% of the data points were deviations by two cells.

In [4], G. Baldini, F. Geib, and R. Giuliani published an article "Continuous authentication of automotive vehicles using inertial measurement units." It was about the continuous authentication of automotive vehicles using inertial measurement units. The workflows can be summarized as (i) data synchronization; (ii) laps extraction; (iii) segmentation; (iv) normalization; (v) feature extraction; (vi) construction of machine learning models including K-nearest neighbors, decision tree, random forest, AdaBoost, and support vector machine. The accuracy ranged from 85% (decision tree) to 90% (support vector machine).

Optimization of vehicle arrival time and signal timings is vital for traffic signal control. W. Wu, L. Huang, R. Du, presented an article "Simultaneous optimization of vehicle arrival time and signal timings within a connected vehicle environment" [5]. A time-based sliding window approach was applied to solve the optimization problem. An experiment was designed with two cases, whereby the proposed algorithm significantly reduced the number of stops by 29–77.5% and the average vehicle delay by 37.8–54%. Analysis has revealed the feasibility of the proposed model in varying communication distance, the market penetration of connected vehicles, and speed guidance's compliance rate.

In [6], S. Guo, X. Zhang, Y. Zheng, and Y. Du focused on path planning of unmanned ships in their article "An Autonomous Path Planning Model for Unmanned Ships Based on Deep Reinforcement Learning." The researchers proposed an artificial potential field based deep deterministic policy gradient approach for path planning of unmanned ships in the unknown environment. It reduced the total iteration time from 339 s to 395 s, the optimal decision time from 282 s to 236 s, convergence steps from 133 to 68, and the number of collisions from 72 to 63, compared with a traditional deep deterministic policy gradient.

Another work, "A generic design of driver drowsiness and stress recognition using MOGA optimized deep MKL-SVM" authored by K. T. Chui, M. D. Lytras, and R. W. Liu [7], presented an approach based on a multiobjective genetic algorithm and multiple kernel learning based support vector machine. It could be applied to both driver drowsiness and stress recognition. The sensitivity, specificity, and area under the receiver operating characteristic curve were 99%, 98.3%, and 97.1%, respectively, for driver drowsiness recognition. On the other hand, they were 98.7%, 98.4%, and 96.9%, respectively, for driver stress recognition.

Y. Guo, B. Li, M. D. Christie, Z. Li, M. A. Sotelo, Y. Ma, and Z. Li published an article "Hybrid dynamic traffic model for freeway flow analysis using a switched reduced-order unknown-input state observer" [8]. Dynamic graph hybrid automata with a cell transmission model was proposed for vehicle density estimation. The experimental environment has been divided into 10 cells. Traffic counting sensors were installed in 7 cells. The estimated density was close to the real density in most of the cells. The performance would be degraded when traffic flow on the main road was being affected by on-ramp vehicles.

In [9], an article "Using Deep Learning to Forecast Maritime Vessel Flows" was shared by X. Zhou, Z. Liu, F. Wang, Y. Xie, X. Zhang. A bidirectional long short-term memory network with a convolutional neural network was proposed for maritime vessel flows forecast. Four cases have been studied. Results indicated the proposed algorithm achieved the lowest error rate (20–22.5%), compared with a support vector regression (50.1–51.4%), standalone convolution neural network (24–26%), and long short-term memory (21–24%).

The article "Vision-Based Machine Learning Method for Barrier Access Control Using Vehicle License Plate Authentication" published in this special issue is authored by K. T. Islam, R. G. Raj, S. M. Shamsul Islam, S. Wijewickrema, M. S. Hossain, T. Razmovski, S. A. O'Leary [10]. An artificial neural network achieved vehicle license plate recognition. Both synthetic and real data are applied for the performance evaluation of the proposed method. The proposed algorithm statistically outperformed existing works. The reported accuracy was excellent (99.7%).

Finally, G. Baldini, R. Giuliani, and F. Geib presented an article "On the application of time frequency convolutional neural networks to road anomalies identification with accelerometers and gyroscopes" [11]. It proposed a convolutional neural network approach to detect road anomalies from data collected by an inertial measurement unit. The achieved accuracy was 97.2%. The research implication relates to the benefits of management by the road infrastructure team concerning the monitoring of road surface quality. Moreover, it helps in improving the accuracy of autonomous vehicle positioning.

In 2015, all United Nations member states agreed to build strong partnerships to meet 17 sustainable development goals (SDGs) and 169 targets [12]. Among all targets, some of them are related to intelligent transportation, including SDG Goal 3 Target 3.6, SDG Goal 7, SDG Goal 8, SDG Goal 11 Target 11.2, SDG Goal 12 Target 12.C, and SDG Goal 14.

Various emerging key applications are suggested for exploration. A study [13] has applied artificial intelligence and the internet-of-things to enhance energy's cleanliness and affordability. The work [14] has shared the use case of a 6G-enabled maritime IoT system. Parallel computing and cloud computing techniques have become proper options for high-performance computing services [15,16]. The computing services can be moved locally via edge computing [17] and fog computing [18]. The ultimate vision is a new integrated eco-system of value-adding services capable of merging the human semantic [19] and social web [20]. It will have enormous capabilities for sophisticated knowledge and data creation [21] for the dynamic composition of value-adding services in various domains, e.g., transportation [2–18] and healthcare [22]. A new revolution in computing is here to stay with a focus on cyber-physical systems.

The guest editors would like to thank the contributions of all colleagues and reviewers. We are grateful for your support.

Author Contributions: M.D.L., K.T.C., and R.W.L. contributed equally to the design, implementation, and delivery of the special issue. All co-editors contributed equally in all the phases of this intellectual outcome. All authors have read and agreed to the published version of the manuscript.

Funding: This research received no external funding.

Conflicts of Interest: The authors declare no conflict of interest.

References

1. World Health Organization. *Global Status Report on Road Safety 2018*; World Health Organization: Geneva, Switzerland, 2018.
2. Zhang, X.; Wang, C.; Liu, Y.; Chen, X. Decision-making for the autonomous navigation of maritime autonomous surface ships based on scene division and deep reinforcement learning. *Sensors* **2019**, *19*, 4055. [CrossRef] [PubMed]
3. Jiang, Y.; Liu, B.; Wang, Z.; Yi, X. Start from scratch: A crowdsourcing-based data fusion approach to support location-aware applications. *Sensors* **2019**, *19*, 4518. [CrossRef] [PubMed]
4. Baldini, G.; Geib, F.; Giuliani, R. Continuous authentication of automotive vehicles using inertial measurement units. *Sensors* **2019**, *19*, 5283. [CrossRef] [PubMed]
5. Wu, W.; Huang, L.; Du, R. Simultaneous optimization of vehicle arrival time and signal timings within a connected vehicle environment. *Sensors* **2020**, *20*, 191. [CrossRef] [PubMed]
6. Guo, S.; Zhang, X.; Zheng, Y.; Du, Y. An autonomous path planning model for unmanned ships based on deep reinforcement learning. *Sensors* **2020**, *20*, 426. [CrossRef] [PubMed]
7. Chui, K.T.; Lytras, M.D.; Liu, R.W. A generic design of driver drowsiness and stress recognition using MOGA optimized deep MKL-SVM. *Sensors* **2020**, *20*, 1474. [CrossRef] [PubMed]
8. Guo, Y.; Li, B.; Christie, M.D.; Li, Z.; Sotelo, M.A.; Ma, Y.; Li, Z. Hybrid dynamic traffic model for freeway flow analysis using a switched reduced-order unknown-input state observer. *Sensors* **2020**, *20*, 1609. [CrossRef] [PubMed]
9. Zhou, X.; Liu, Z.; Wang, F.; Xie, Y.; Zhang, X. Using deep learning to forecast maritime vessel flows. *Sensors* **2020**, *20*, 1761. [CrossRef] [PubMed]
10. Islam, K.T.; Raj, R.G.; Shamsul Islam, S.M.; Wijewickrema, S.; Hossain, M.S.; Razmovski, T.; O'Leary, S. A Vision-based machine learning method for barrier access control using vehicle license plate authentication. *Sensors* **2020**, *20*, 3578. [CrossRef] [PubMed]
11. Baldini, G.; Giuliani, R.; Geib, F. On the application of time frequency convolutional neural networks to road anomalies identification with accelerometers and gyroscopes. *Sensors* **2020**, *20*, 6425. [CrossRef] [PubMed]

12. United Nations. *Transforming Our World: The 2030 Agenda for Sustainable Development*; United Nations: New York, NY, USA, 2015.

13. Chui, K.T.; Lytras, M.D.; Visvizi, A. Energy sustainability in smart cities: Artificial intelligence, smart monitoring, and optimization of energy consumption. *Energies* **2018**, *11*, 2869. [CrossRef]

14. Liu, R.W.; Nie, J.; Garg, S.; Xiong, Z.; Zhang, Y.; Hossain, M.S. Data-driven trajectory quality improvement for promoting intelligent vessel traffic services in 6G-enabled maritime IoT systems. *IEEE Internet Things J.* **2020**. [CrossRef]

15. Huang, Y.; Li, Y.; Zhang, Z.; Liu, R.W. GPU-accelerated compression and visualization of large-scale vessel trajectories in maritime IoT industries. *IEEE Internet Things J.* **2020**, *7*, 10794–10812. [CrossRef]

16. Lytras, M.D.; Visvizi, A.; Torres-Ruiz, M.; Damiani, E.; Jin, P. IEEE access special section editorial: Urban computing and well-being in smart cities: Services, applications, policymaking considerations. *IEEE Access* **2020**, *8*, 72340–72346. [CrossRef]

17. Lin, B.; Zhou, X.; Duan, J. Dimensioning and layout planning of 5G-based vehicular edge computing networks towards intelligent transportation. *IEEE Open J. Veh. Technol.* **2020**, *1*, 146–155. [CrossRef]

18. Darwish, T.S.; Bakar, K.A. Fog based intelligent transportation big data analytics in the internet of vehicles environment: Motivations, architecture, challenges, and critical issues. *IEEE Access* **2018**, *6*, 15679–15701. [CrossRef]

19. Vossen, G.; Lytras, M.D.; Koudas, N. Revisiting the (machine) Semantic Web: The missing layers for the human Semantic Web. *IEEE Trans. Knowl. Data Eng.* **2007**, *19*, 145–148. [CrossRef]

20. Zhuhadar, L.; Yang, R.; Lytras, M.D. The impact of social multimedia systems on cyberlearners. *Comput. Hum. Behav.* **2013**, *29*, 378–385. [CrossRef]

21. Naeve, A.; Yli-Luoma, P.; Kravcik, M.; Lytras, M.D. A modelling approach to study learning processes with a focus on knowledge creation. *Int. J. Technol. Enhanc. Learn.* **2018**, *1*, 1–34. [CrossRef]

22. Spruit, M.; Lytras, M.D. Applied Data Science in Patient-centric Healthcare. *Telemat. Inform.* **2018**, *35*, 2018. [CrossRef]

Article

Decision-Making for the Autonomous Navigation of Maritime Autonomous Surface Ships Based on Scene Division and Deep Reinforcement Learning

Xinyu Zhang [1,*], Chengbo Wang [1,2,*], Yuanchang Liu [3] and Xiang Chen [4]

1   Key Laboratory of Maritime Dynamic Simulation and Control of Ministry of Transportation,
    Dalian Maritime University, Dalian 116026, China
2   Marine Engineering College, Dalian Maritime University, Dalian 116026, China
3   Department of Mechanical Engineering, University College London, Torrington Place,
    London WC1E 7JE, UK; yuanchang.liu@ucl.ac.uk
4   Department of Civil Environmental and Geomatic Engineering, London WC1E 6BT, UK;
    xiang.chen.17@ucl.ac.uk
*   Correspondence: zhangxy@dlmu.edu.cn (X.Z.); wangcb_dlmu@foxmail.com (C.W.)

Received: 19 August 2019; Accepted: 17 September 2019; Published: 19 September 2019

Abstract: This research focuses on the adaptive navigation of maritime autonomous surface ships (MASSs) in an uncertain environment. To achieve intelligent obstacle avoidance of MASSs in a port, an autonomous navigation decision-making model based on hierarchical deep reinforcement learning is proposed. The model is mainly composed of two layers: the scene division layer and an autonomous navigation decision-making layer. The scene division layer mainly quantifies the sub-scenarios according to the International Regulations for Preventing Collisions at Sea (COLREG). This research divides the navigational situation of a ship into entities and attributes based on the ontology model and Protégé language. In the decision-making layer, we designed a deep Q-learning algorithm utilizing the environmental model, ship motion space, reward function, and search strategy to learn the environmental state in a quantized sub-scenario to train the navigation strategy. Finally, two sets of verification experiments of the deep reinforcement learning (DRL) and improved DRL algorithms were designed with Rizhao port as a study case. Moreover, the experimental data were analyzed in terms of the convergence trend, iterative path, and collision avoidance effect. The results indicate that the improved DRL algorithm could effectively improve the navigation safety and collision avoidance.

Keywords: decision-making; autonomous navigation; collision avoidance; scene division; deep reinforcement learning; maritime autonomous surface ships

1. Introduction

Recently, marine accidents have been frequently caused by human factors. Based on the statistics from the European Maritime Safety Agency (EMSA), in 2017, there were 3301 casualties and accidents at sea, with 61 deaths, 1018 injuries, and 122 investigations initiated. In these cases, human error behavior represented 58% of the accidents and 70% of the accidents were related to shipboard operations. In addition, the combination of collision (23.2%), contact (16.3%), and grounding/stranding (16.6%) shows that navigational casualties represent 56.1% of all casualties with ships [1]. The important purpose of maritime autonomous surface ships (MASSs) research is to reduce the incidence of marine traffic accidents and ensure safe navigation. Therefore, safe driving and safe automatic navigation have become urgent problems in the navigation field. Future shipping systems will rely less and less on people, and the efficiency of ship traffic management is getting higher and higher. It further highlights the shipping industry's need for MASSs and their technology.

At present, many foreign enterprises and institutions have completed the concept design of MASS and the port-to-port autonomous navigation test [2–4]. However, for China, from 2009 to 2017, domestic organizations, such as the First Institute of Oceanography of the State Oceanic Administration, Yun Zhou Intelligent Technology Co., Ltd. Zhuhai, China, Ling Whale Technology, Harbin Engineering University, Wuhan University of Technology, and Huazhong University of Science and Technology, have conducted research on unmanned surface vessels (USVs). There are some differences in autonomous navigation technology of MASSs compared to USVs.

1. First, the molded dimension of a MASS is larger. Research on the key technologies of USVs pays more attention to motion control. However, for MASSs, navigation brains that can make autonomous navigation decisions are needed more.
2. Second, the navigation situation of a MASS is complex and changeable, and its maneuverability is slow to respond. Therefore, it is necessary to combine scene division with adaptive autonomous navigation decision-making in order to achieve safe decision-making for local autonomous navigation.

Owing to these differences, the autonomous navigation decision-making system is the core of a MASS, and its effectiveness directly determines the safety and reliability of navigation, playing a role similar to the human "brain." During the voyage, the thinking and decision-making process is very complex. After clarifying the destinations that need to be reached and obtaining the global driving route, it is necessary to generate a reasonable, safe, and efficient abstract navigation action (such as acceleration, deceleration, and steering) based on the dynamic environmental situation around the ship. The "brain" needs to rapidly and accurately reason based on multi-source heterogeneous information such as the traffic rule knowledge, driving experience knowledge, and chart information stored in its memory unit. In this paper, we only train navigation strategies by learning relative distance and relative position data. We assumed the following perception principles.

The input information of the autonomous navigation decision-making system is multi-source heterogeneous, including real-time sensing information from multiple sensors and various a priori pieces of information. In the environmental model, the MASS sensor detects the distance and relative azimuth between the MASS and the obstacle. Figure 1 illustrates the MASS perception. In the figure, the geographical coordinate of MASS is $S_M(x_0, y_0)$; speed is v_0; ship course is φ_0; geographical coordinate of the static obstacle is $S_O(x_o, y_o)$; relative bearing of the MASS and obstacles is δ_0; $S_P(x_p, y_p)$ is the position of the target point; dis_{M-P} is the distance between MASS and target point; and dis_{M-O} is the distance between MASS and obstacle. Among these symbols, the subscripts in the symbols are as follows: "M" is for the MASS, "P" is for target point, and "O" is for obstacle.

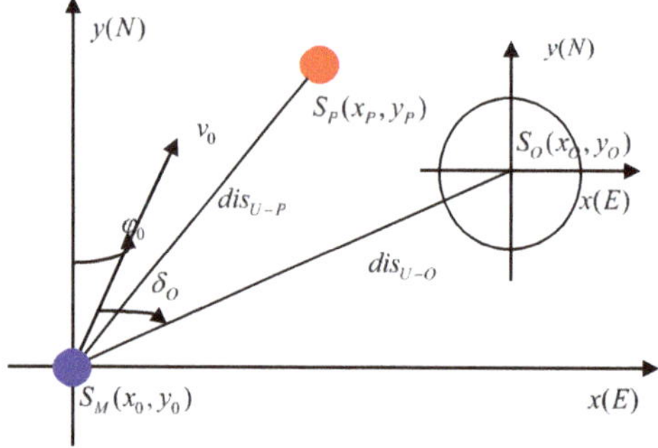

Figure 1. Schematic diagram of perception.

The current environmental status information can be expressed as $obs_t = [v_0, \varphi_0, \delta_0, dis_{M-P}, dis_{M-O}]^T$. The algorithm not only acquires the current state obs_t of the obstacle, but also obtains the historical observation state ($obs_{t-i}, i \in 1, \cdots, T_P$), where T_P is the total length of the observation memory. The data input for the final training is $X_{Perception}(t) = [obs_t \; obs_{t-1} \; \cdots \; obs_{t-T_P}]^T$. Therefore, the input of the high-level driving decision-making system at time t can be expressed as follows:

$$X_{Perception}(t) = [obs_t \; obs_{t-1} \; \cdots \; obs_{t-T_P}]^T$$

$$= \begin{bmatrix} v_t & \varphi_t & \delta_t & dis_{t,M-P} & dis_{t,M-O} \\ v_{t-1} & \varphi_{t-1} & \delta_{t-1} & dis_{t-1,M-P} & dis_{t-1,M-O} \\ \vdots & \vdots & \vdots & \vdots & \vdots \\ v_{t-T_P} & \varphi_{t-T_P} & \delta_{t-T_P} & dis_{t-T_P,M-P} & dis_{t-T_P,M-O} \end{bmatrix} \tag{1}$$

Learning from the decision-making of the officer on the voyage, this research proposes a hierarchical progressive navigation decision-making system, which mainly includes two sub-modules: a scene division module and a navigation action generation module. The main contributions of this paper are as follows:

1. We exploit ontology and the principle of divide and conquer to construct the navigation situation understanding model of a MASS, and divide the situation of MASS navigation into scenes based on the International Regulations for Preventing Collisions at Sea (COLREGS).
2. Aiming at the problem of local path planning and collision avoidance decision-making, a method of autonomous navigation decision-making for MASSs based on deep reinforcement learning is proposed, in which the reward function of multi-objective optimization is designed, which consists of safety and approaching target points.
3. An artificial potential field is added to alleviate the problem of easy-to-fall-into local iterations and slow iterations of autonomous navigation decision-making algorithms based on deep reinforcement learning.
4. Simulation results based on Python and Pygame show that the Artificial Potential Field-Deep Reinforcement Learning (APF-DRL) method has better performances than the DRL method in both autonomous navigation decision-making and algorithm iteration efficiency.

The remaining sections of the paper are organized as follows. Related works are presented in Section 2. The scene division module is presented in Section 3. The autonomous navigation decision-making module is presented in Section 4. The simulation results and algorithm improvement are presented in Section 5. The paper is concluded in Section 6.

2. Related Work

The autonomous navigation decision-making system of a MASS plays the role of the "navigation brain." The problem to be solved is to determine the best navigation strategy based on environmental information. At present, related works mainly focus on the ship's intelligent collision avoidance algorithms in specific environments.

For the study on intelligent collision avoidance and path planning of ships, the existing models mainly contain knowledge-based expert systems, fuzzy logic, artificial neural networks, intelligent algorithms (genetic algorithms, ant colony algorithms, etc.). In addition, a ship collision avoidance system based on the general structural model of the expert system has been established [5]. Moreover, a comprehensive and systematic study has been performed for the whole process of ship collision avoidance, and a mathematical model for the safe passing distance, pressing situation, and ship collision risk has been established. Fan et al. [6] combined the dynamic collision avoidance algorithm and tracking control, and as such, a dynamic collision avoidance control method in the unknown ocean environment is presented. A novel dynamic programming (DP) method was proposed to generate the optimal multiple interval motion plan for a MASS by Geng et al. [7]. The method provided the lowest collision rate overall and better sailing efficiency than the greedy approaches. Ahn et al. [8] combined fuzzy inference systems with expert systems for collision avoidance systems. They proposed a method for calculating the collision risk using a neural network. Based on the distance to closest point of approach (DCPA) and the time to closest point of approach (TCPA), the multi-layer perceptron (MLP) neural network was applied to the collision avoidance system to compensate for the fuzzy logic. Hua [9] optimized the shortest path and minimum heading of the local path and designed the surface planning of the surface unmanned submarine under the constraints of the close distance meeting model of the ship and 1972 International Collision Avoidance Rules. The target genetic algorithm realized the intelligent collision avoidance of unmanned boats through simulation. Ramos et al. [10] presented a task analysis for collision avoidance through hierarchical task analysis and used a cognitive model for categorizing the tasks, which explored how humans can be a key factor for successful collision avoidance in future MASS operations. The results provided valuable information for the design stage of a MASS. For the study on path-following and control of autonomous ships, a novel translation–rotation cascade control scheme was developed for path-following of an autonomous underactuated ship by Wang et al., and in the case of disturbance, the autonomous underactuated ship was controlled, and the trajectory point guidance was used for precise tracking and autonomous navigation [11–13].

The abovementioned models usually assume complete environmental information. However, in an unknown environment, prior knowledge of the environment is difficult to acquire. It is difficult to form a complete and accurate knowledge base, and the rule-based algorithm makes it difficult to cope with various situations. Therefore, in many practices, the system needs to have a strong adaptive ability to adjust to the uncertain environment. Recently, deep reinforcement learning combined with deep neural network models and reinforcement learning have made significant progress in the field of autonomous driving, such as unmanned surface vehicles (USV), unmanned aerial vehicles (UAV), and unmanned ground vehicles (UGV). Tai et al. [14] combined deep learning and decision-making processes into a highly compact, fully connected network with raw depth images as the input and the generated control commands as the outputs to achieve model-free obstacle avoidance behavior. Long et al. [15] proposed a novel end-to-end framework for generating effective reactive collision avoidance strategies for distributed multi-agent navigation based on deep learning. Panov et al. [16] proposed an approach for using a neural network to perform the path planning on the grid and initially realize it based on deep reinforcement learning. Bojarski et al. [17] used convolutional neural networks for end-to-end training driving behavioral data, mapping the raw pixels from a single-front camera directly to the steering commands for unmanned vehicle adaptation path planning. The performance of the model and results of learning were better than the traditional model, but the only improvement was that the model was less interpretable. Cheng et al. [18] proposed a simple deep reinforcement learning obstacle avoidance algorithm based on the deep Q learning network using a convolutional neural network to train the ship sensor image information. The interaction with the environment was included by designing the incentive function in reinforcement learning. The maximum expected value of the cumulative return was obtained, and the optimal driving strategy of the underactuated USV was derived. However, improvement in this area is needed to increase the complexity of the verification environment and dynamic obstacle environment. Compared with Cheng et al. [18], the different and better aspects of our paper are: First, we used long short-term memory (LSTM) to store historical decisions for autonomous navigation by improving the iterative effectiveness. Second, the method used in this paper learns the ship navigation state data, including relative azimuth and relative distance, to improve the accuracy and effectiveness of autonomous navigation of a MASS. Third, in order to solve the problem of easy-to-fall-into local iteration and slow iteration speed, we added a gravitational field to improve deep reinforcement learning with the target point as the potential field center. In summary, the deep reinforcement learning achieved self-adaptation to an unknown environment by self-training various experiences and using high-dimensional inputs such as raw images or environmental states.

However, few experts currently apply deep reinforcement learning to MASS intelligent navigation. Taking advantage of these, this paper uses deep reinforcement learning to solve the problem of autonomous navigation decision-making for a MASS. Learning from the decision-making of the officer on the voyage, this research proposes a hierarchical progressive navigation decision-making system, which mainly includes two sub-modules: a scene division module and an autonomous navigation action generation module.

3. Scene Division Module for a MASS

The scene division mainly organizes and describes the multi-source heterogeneous information in the driving scene with the Prolog language [19]. This research uses the ontology theory and the principle of divide and conquer to divide navigation environment into entities and attributes. Entity classes are used to describe the objects of different attributes, including chart entity, obstacle entity, ego ship, and environment. Attribute classes are used to describe the semantic properties of an object, including position attributes and orientation attributes.

Ontology is a philosophy concept, which studies the nature of existence. Ontology can be classified into four types: domain, general, application, and representation [20]. A complete marine transportation system is a closed-loop feedback system consisting of "human–ship–sea–environment."

The entity class is categorized into four sub-entity classes: chart entity, obstacle entity, MASS entity (egoship), and environmental information. Chart entity includes point entity, line entity, and area entity, where point entity refers to navigational aids and line entity refer to reporting lines. The area entity includes seapart, channel, boundary, narrow channel, divider, anchorage, junction, and segment. Obstacle entities include static obstacle and dynamic obstacle, where static obstacle entities are divided into rocks, wreck, and boundary; and dynamic obstacle entities include ships (vessel), floating ice, and marine animal. A MASS entity (egoship) is used to describe its own state information. Environmental information includes height, sounding, nature of the seabed, visibility, and disturbance.

The relationship between the MASS and the obstacles can be divided into binary relationships: MASS and static obstacles (abbreviated as ES), and MASS and dynamic obstacles (abbreviated as ED). In the azimuthal relationship, the abbreviations are as follows: HO is the head-on encounter, OT is the overtaking encounter, and CR is the crossing encounter. The MASS ontology model relationship attribute is presented in Table 1.

Table 1. MASS ontology model relationship attribute table.

ID	Object Attribute	Domain	Ranges	Comments
1	hasBehindLeftES	Egoship	StaticObstacle	
2	hasBehindES	Egoship	StaticObstacle	
3	hasBehindRightES	Egoship	StaticObstacle	
4	hasFrontLeftES	Egoship	StaticObstacle	
5	hasFrontES	Egoship	StaticObstacle	
6	hasFrontRightES	Egoship	StaticObstacle	
7	hasLeftES	Egoship	StaticObstacle	
8	hasRightES	Egoship	StaticObstacle	
9	hasFrontED-HO	Egoship	DynamicObstacle	Orientation relationship
10	hasFrontED-OT	Egoship	DynamicObstacle	
11	hasFrontED-CR	Egoship	DynamicObstacle	
12	hasBehindED-OT	Egoship	DynamicObstacle	
13	hasFrontLeftED-CR	Egoship	DynamicObstacle	
14	hasFrontLeftED-OT	Egoship	DynamicObstacle	
15	hasFrontRightED-CR	Egoship	DynamicObstacle	
16	hasFrontRightED-OT	Egoship	DynamicObstacle	
17	hasBehindLeftED-CR	Egoship	DynamicObstacle	
18	hasBehindRightED-CR	Egoship	DynamicObstacle	
19	isOnSeapart	Egoship/ObstacleEntity	Seapart	Positional relationship
20	isOnChannel	Egoship/ObstacleEntity	Channel	
21	isOnAnchorage	Egoship/ObstacleEntity	Anchorage	

ES—relationship between the MASS and static obstacles, ED—relationship between the MASS and dynamic obstacles, HO—head-on encounter, OT—overtaking encounter, CR—crossing encounter.

The scene ontology model corresponding to the relationship property of the ontology model of MASS in Table 1 is established. Figure 2 shows the ontology conceptual model of the MASS navigation scene.

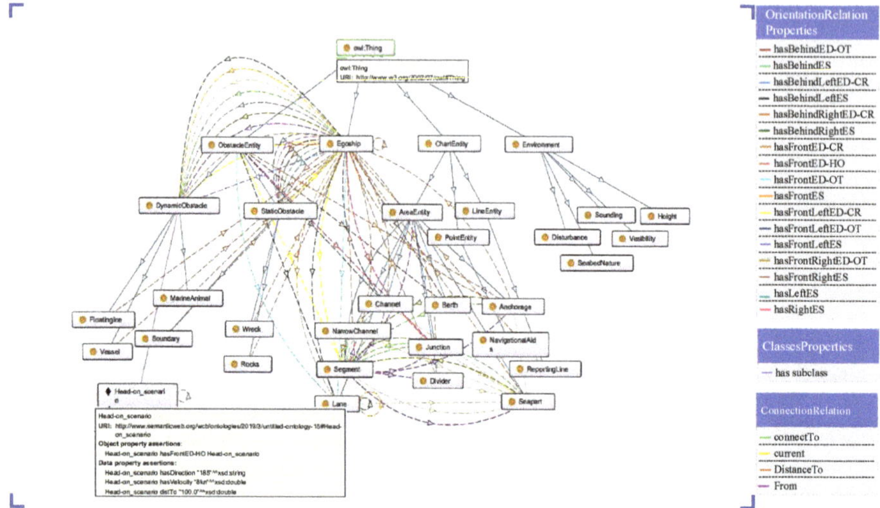

Figure 2. Ontology conceptual model diagram of the navigation scene.

Combining the COLREGS, the navigation scenes of *Egoship − StaticObstacle* and *Egoship − DynamicObstacle* are divided into six scenes: hasFront, hasFrontLeft, hasFrontRight, hasBehind, hasBehindLeft, hasBehindRight. Then, the scenes corresponding to *Egoship − DynamicObstacle* are divided into the HO sub-scenario, the CR sub-scenario, the OT sub-scenario, and the mixed sub-scenarios. However, this paper mainly analyses the scene division between MASS and dynamic obstacle. So six scenes is become hasFrontED, hasFrontLeftED, hasFrontRightED, hasBehindED, hasBehindLeftED, hasBehindRightED.

hasFrontED: $\dfrac{15\pi}{8} \sim \dfrac{\pi}{8}$, including HO, OT, and CR.

hasBehindED: $\dfrac{5\pi}{8} \sim \dfrac{11\pi}{8}$, including OT.

hasFronntLeftED: $\dfrac{3\pi}{2} \sim \dfrac{15\pi}{8}$, including CR and OT.

hasFrontRightED: $\dfrac{\pi}{8} \sim \dfrac{\pi}{2}$, including CR and OT.

hasBehindLeeftED: $\dfrac{11\pi}{8} \sim \dfrac{3\pi}{2}$, only including CR.

hasBehindRightED: $\dfrac{\pi}{2} \sim \dfrac{5\pi}{8}$, only including CR.

In summary, the visual display of the six scenes is shown in Figure 3.

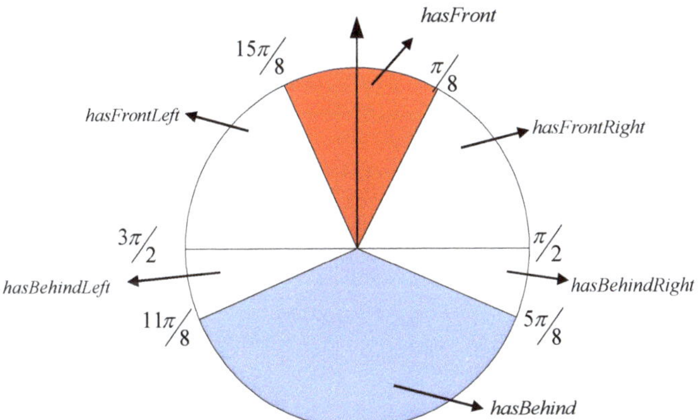

Figure 3. A quantitative map of the scene division based on the International Regulations for Preventing Collisions at Sea (COLREGS).

4. Autonomous Navigation Decision-Making Module for a MASS

Deep reinforcement learning is a combination of deep learning and reinforcement learning. In this paper, Q-learning was combined with a neural network to establish the autonomous navigation decision-making model based on deep Q-learning. The deep Q-learning algorithm uses an empirical playback algorithm whose basic core idea is to remember the historical information that the algorithm performs in this environment. In practice, the number of environmental states and behavioral states is extremely large and it is necessary to adopt a neural network for generalization. Therefore, LSTM was selected as the Q network. The core concepts of LSTM are cell state and "gate" structure. Cell state is equivalent to the path of information transmission such that information can be transmitted in sequence. This can be thought of as the "memory" of the network. LSTM network has three control gates: forget gate, input gate, and output gate. The forget gate determines which environmental states and behavioral states from the last cell state to continue to pass through the current cell. The input gate controls whether a new datum could flow into the memory and updates the cell state. The output gate decides which part of the Q value to be exported as output [21]. The three control gates weaken the short-term memory effect and regulate the predicted Q value corresponding to multiple autonomous navigation actions in the current state.

The mathematical nature of reinforcement learning can be regarded as a Markov decision process (MDP) in discrete time. A Markov decision process is defined using the following five-tuple, (S, A, P_a, R_a, γ). S represents the finite state space in which the MASS is located. A represents the behavioral decision space of MASS, i.e., a collection of all the behavioral spaces of the MASS in any state, such as left rudder, right rudder, acceleration, and deceleration. $P_a(s, s') = P(s'|s, a)$ is the conditional probability that represents the probability that the MASS will reach next state s' under state s and action a. $R_a(s, s')$ is a reward function representing the stimulus that the MASS takes from state s to state s' under action a. $\gamma \in (0, 1)$ is the discount factor of the stimulus, and the discounting at the next moment is determined according to a factor [22,23]. Figure 4 displays the schematic of the autonomous navigation decision-making of the MASS based on deep reinforcement learning. In the memory pool, the current state of the observed MASS is taken as the input of the neural network. The Q value table of the action that can be performed in the current state is the output, and the behavioral strategy corresponding to the maximum Q value is learned through training.

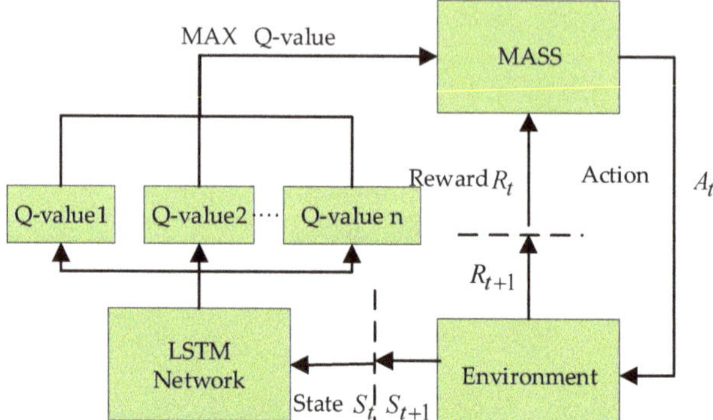

Figure 4. Schematic of the autonomous navigation decision-making of a maritime autonomous surface ship (MASS) based on deep reinforcement learning (DRL).

4.1. Representation of Behavioral Space

After setting the initial and target points, the MASS is considered as a particle in the simulation. In a real navigation process, the autonomous navigation of the MASS is a continuous state, following which, observation behavior O needs to be generalized into discrete action $\hat{A} = Generalization(A', O)$. Generally, the search action of a MASS includes the four discrete actions of up, down, left, and right. When the environment has a corner, the search behavior in the diagonal direction is increased. Centering on the mass of the MASS, the actual motion space model A is defined as eight discrete actions, *up, down, left, right, up*$_{-45°}$, *up*$_{+45°}$, *down*$_{-45°}$, *down*$_{+45°}$, namely, the matrix of Equation (2).

$$A = \begin{bmatrix} -1,1 & 0,1 & 1,1 & -1,0 & 1,0 & -1,-1 & 0,-1 & 1,-1 \end{bmatrix} \tag{2}$$

4.2. Design of the Reward Function

In the reinforcement learning system of a MASS, the activation function plays an important role in evaluating the effectiveness of the behavioral decision-making and safety of obstacle avoidance. It has a search-oriented role. The goal of reinforcement learning is to obtain the search strategy that gives the highest return value in the driverless process. The reward function consists of safety, comfort, and arrival target points. When designing the reward function, the following elements should be maximally considered [18].

(1) Approach to the target point: The search behavior of the autonomous navigation decision-making made in an uncertain environment should bring the MASS closer to the target point. A value close to the incentive function will choose the reward; otherwise it will be punished:

$$R_{distance} = -\lambda_{distance} \sqrt{(x - x_{goal})^2 + (y - y_{goal})^2} \tag{3}$$

(2) Safety: In the deep Q-learning algorithm model, the unknown environment is divided into a state space, which is divided into a safe state area and an obstacle area. The system should be selected in the local area without the obstacle to sailing. An action search strategy and "early, clear, big-amplitude" is used to avoid obstacles. Thus, in the reward function, the penalty value is added to the behavior close to the obstacle, and the reward value is increased:

$$R_{collisions} = -\lambda_{collisions} \bigvee_{i=1}^{N_{obs}} \left(\sqrt{(x - x_{obs_i})^2 + (y - y_{obs_i})^2} < Z_0 \right) \tag{4}$$

where N_{obs} is the number of obstructions that the MASS needs to avoid in the present state of the ship, $\vee$ is the symbol "OR", and (x_{obs}, y_{obs}) is the obstacle position. Z_0 is the safe encounter distance of the ship.

The safe encounter distance of the ship is related to the size of the ship (length). A large-sized ship will have a long required safety distance.

Both the environmental state set and motion state in the deep Q-learning algorithm are limited, and the actual MASS transportation process is a continuous systematic event. Thus, this study generalizes the activation function as a nonlinear hybrid function:

$$R = \begin{cases} 10, & dis_{M-P}(t) = 0 \\ 2, & s = 1 \text{ and } (dis_{M-P}(t) - dis_{M-P}(t-1)) < 0 \\ -1, & s = 0 \\ -1, & s = 1 \text{ and } (dis_{M-O}(t) - dis_{M-O}(t-1)) < 0 \\ 0, & else \end{cases} \tag{5}$$

where $s = 0$ represents the collision between the MASS and obstacle; $s = 1$ represents sailing in a safe area; $dis_{M-P}(t)$ represents the distance between the target point and the MASS at time t; $dis_{M-P}(t-1)$ represents the distance between the target point and the MASS at time $t - 1$; $dis_{M-O}(t)$ represents the distance between the obstacle and the MASS at time t; and $dis_{M-O}(t-1)$ represents the distance between the obstacle and the MASS at time $t - 1$.

4.3. Action Selection Strategy

On the one hand, the reinforcement learning system requires online trial and error to obtain the optimal search strategy, namely, exploration; on the other hand, it requires consideration of the entire route planning, so that the expectation of the algorithm to obtain rewards is at a maximum, namely, utilization. It implies that when the search behavior maximizes the action value function, the probability of selecting the action is $1 - \varepsilon + \dfrac{\varepsilon}{|A(s)|}$ and the probability of selecting other actions is $\dfrac{\varepsilon}{|A(s)|}$.

$$\pi(a|s) \leftarrow \begin{cases} 1 - \varepsilon + \dfrac{\varepsilon}{|A(s)|} & \text{if } a = \text{argmax}_a Q(s,a) \\ \dfrac{\varepsilon}{|A(s)|} & else \end{cases} \tag{6}$$

where $\pi(a|s)$ represents the navigation strategy in state s by action a. $\varepsilon \in (0, 1]$ are probabilities of exploration. $Q(s,a)$ is state-action value function of action a after scaling in state s. $A(s)$ is action-value function in state s.

4.4. Decision-Making for the Autonomous Navigation of the MASS in an Uncertain Environment

In this section, the abovementioned MASS autonomous navigation decision-making module collects the ship's own information and environmental status through the sensing layer as input for deep reinforcement learning. Through the system self-learning, the best navigation strategy is finally decided, which makes the cumulative return of MASS in the self-learning process the largest. Once the system is trained, MASS will automatically navigate to avoid obstacles and reach the destination under the command of the autonomous navigation decision-making level.

According to the designed algorithm, first, the MASS is combined with COLREGS to divide the navigation situation into the individual sub-scenarios. Second, the system takes the perceived environmental state information as an input of the current value network and then generates an action based on the current policy through training. It then performs an action to obtain the empirical data and store them in the playback memory unit. Finally, the empirical data are used to update the value function and model parameters until the error is the smallest and the cumulative return value is at a maximum. Figure 5 shows the main flowchart for the high-level driving decisions for MASS.

Algorithm 1. DRL for MASS Autonomous Navigation Decision-making

- **Input:**
Start sampling from random state s_0 and randomly select action. Sampling is terminated at T cycles or the MASS collides. The resulting sample set is S.
Each input in S must be included:
(1) Current states s_t, (2) action a, (3) return r, (4) the next state after the action s_{t+1}, and (5) the termination condition
- **Output:** weights parameter ω^* for DRL

Require: ω: a small positive number representing the allowed smallest convergence tolerance; S: the state set; $P(s', r|s, a)$: the transition probability from current state and action to next state and reward; γ: the discount factor;

1:	Initialize the optimal value function $Q(s)$, $\forall s \in S$ arbitrarily
2:	**For** episode $= 1, M$ **do**
3:	**For** $t = 1, T$ **do**
4:	repeat
5:	$\omega \leftarrow 0$
6:	for $s \in S$ do
7:	target $q \leftarrow Q(s)$
8:	$Q(s) \leftarrow \max_a[r + \gamma \sum s', rP(s', r|s, a)Q(s)]$
9:	$\omega \leftarrow \max(\omega, |q - Q(s)|)$
10:	until $\omega < 0$
11:	**end for**
12:	**end for**
13:	$\pi^*(s) \approx argmax_a[r + \gamma \sum s', rP(s', r|s, a)Q(s)]$

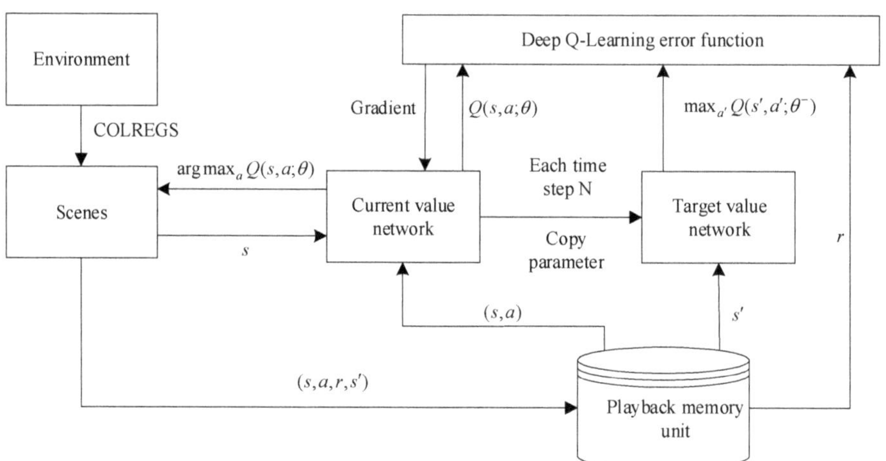

Figure 5. Main flowchart of the autonomous navigation decision-making for MASS.

5. Simulation and Evaluations

In this section, it is shown that the effectiveness of the autonomous navigation decision-making algorithm for the MASS based on deep reinforcement learning was verified by the case study. This experiment built a two-dimensional (2D) simulation environment based on Python and Pygame. Specifically, the NumPy library and sys, random, and math modules were used for the simulation. In the 2D coordinate system, each coordinate point corresponded to a state of the MASS, and each state could be mapped to each element of the environmental state set. In the simulation environment model, there were two state values for each coordinate point, which were 1 and 0, where 1 represented the navigable area, which is shown as the sea-blue area in the environmental model, and 0 represented the obstacle area, which is shown as a brown and dark gray area in the environmental model. In accordance with the simulation environment model in Figure 6, the 2D map of the state of the simulation environment was simulated, and the obstacles, such as the ship, breakwater, and shore bank were simulated in the environmental model. For the MASS, the location information for these obstacles was uncertain.

Figure 6. Simulation environment model.

5.1. Autonomous Navigation Decision-Making Based on DRL

This validation trial section was designed to combine reinforcement learning with deep learning. The goal of the navigation decision-making for the MASS consisted of two parts: the tendency toward the target and obstacle avoidance. If there were no obstacles or few obstacles in the environment, the MASS will randomly select the action to approach the target point with probability $\frac{\varepsilon}{|A(s)|}$. If the obstacle appeared within the safe encounter distance, the MASS will pass the incentive. The function interacted with the environment to avoid obstacles. Some of the model parameters in the experiment were set as: $\omega = 0.02$, $\gamma = 0.9$, and $v_0 = 8kn$.

The experiment set the initial position (128, 416) and target point (685, 36) of the MASS. As shown in Figure 7a, in the initial iteration, the MASS could not determine the temptation area in the simulation environment and fell into the "trap" sea area in the simulation port pool. As shown in Figure 7b, after 100 iterations, the system gradually planned the effective navigation path, but the collision obstacle phenomenon occurred many times in the process, and the planning navigation path fluctuated significantly. As shown in Figure 7c,d, after 200 to 500 iterations, respectively, the collision phenomenon was gradually reduced, and the planning path fluctuation slowed down. As shown in Figure 7e, all the obstacles were effectively avoided after iterating 1000 times and the planned path fluctuations were weak and gradually stabilized. As shown in Figure 7f, up until the 2000th iteration, the probability of a random search was the smallest, and the system navigated the final fixed path through the decision system to reach the target point.

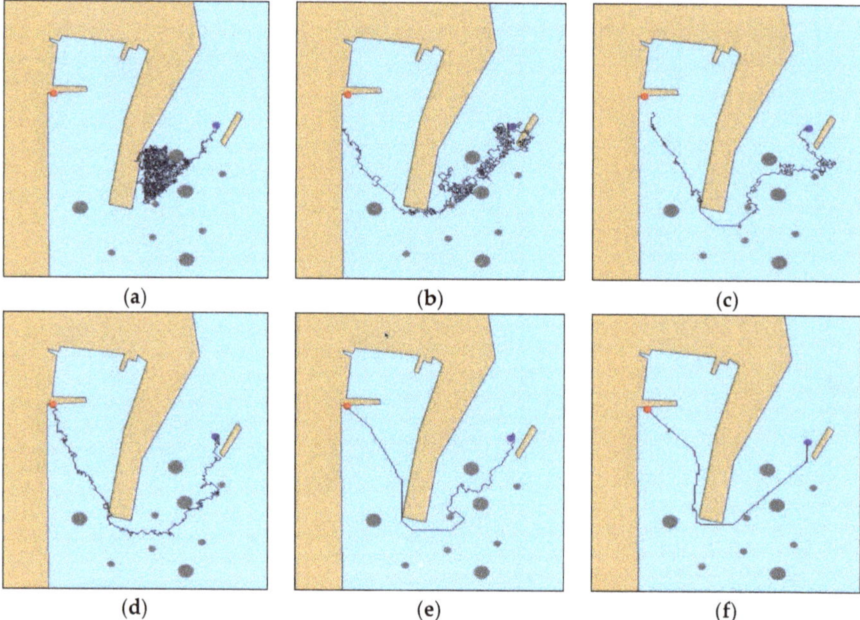

Figure 7. DRL algorithm verification experiment results: (**a**) initial iteration; (**b**) 100th iteration, (**c**) 200th iteration, (**d**) 500th iteration, (**e**) 1000th iteration, and (**f**) 2000th iteration.

5.2. Improved Deep Reinforcement Learning Autonomous Navigation Decision-making

Although DRL-based MASS navigation behavioral decision-making and path planning in an uncertain environment were realized, the algorithm iteration speed was too slow in the whole experiment, the total iteration time was up to 14 min, and it was trapped in local iterations many times. Affects the applicability and credibility of behavioral decisions. Therefore, there was a need to improve the DRL-based behavioral decision-making algorithm. Therefore, this section describes the addition of an artificial potential field (APF) to improve the DRL and establish an autonomous navigation decision-making-based APF-DRL. To this end, increasing the gravitational potential field was the initial Q value of DRL, avoiding the huge number of calculations in the complex environment, and effectively preventing the MASS from falling into the concave trap in the environment and speeding up the iteration speed of the algorithm.

The DRL algorithm had no prior knowledge of the environment, and all state value functions $V(s)$ in the initial state were equal or completely random. Each step of action a was produced in a random state, i.e., the Markov decision process (MDP) environment state transition probability was equal. For the track decision problem, the return value R was only be changed when the destination was reached or obstacles are encountered. The sparsity of the reward function resulted in an initially low decision efficiency and numerous iterations. Particularly for large-scale unknown environments, there was a large amount of invalid iterative search space.

Combining the APF method to improve the autonomous navigation decision-making based on DRL:

1. Starting point coordinate A and the target point coordinate B were determined. A gravitational potential field with the target point as the potential field center in the initialization state value function was established. The $V(s)$ table was initialized according to the position of the target point as prior environmental information, and the initialized $V(s)$ value was set as larger than or equal to 0.
2. This study conducted a four-layer layer-by-layer search of the environment map. If an obstacle was found, an operation was performed according to $v(S(t+1)) = -A$.
3. The state-action value function table was updated using the environment state value function:

$$Q(S,A) = r + \gamma V(s')$$ (7)

4. The MASS explored the environment from the starting point and only considered the state of $v(S(t+1)) \geq 0$ as an exploitable state. It adopted a variable greedy strategy and updated the state-action value each time it moved. After reaching the target point, this round of iteration ended, and the next round of exploration started from the starting point.

For autonomous navigation decision-making based DRL in complex environments, the action state space is large and iteration speed is slow. When the gravitational field of target point was added as the potential field center to improve the DRL algorithm, the autonomous navigation decision-making of unmanned ships tends to the target point more quickly and iteratively, and the navigation strategies given in each state are directional, whereas a random strategy ensures that it does not fall into a local optimal solution. The dynamic and experimental parameters of the ship were the same as those in Section 5.1.

The experiment set the initial position (128, 416) and target point (685, 36) of the MASS. In the early stages of the experimental iterations, the MASS collided with the obstacle at different time steps, and there was no collision after the initial iteration in the experiment. The system maneuvered the MASS back to the previous navigation state and re-decided the navigation path planning strategy. Compared with the experiment in Section 5.1, as shown in Figure 8a, in the initial iteration, the MASS fell into a local iteration. As shown in Figure 8b, after 100 iterations, the system first planned an effective navigation path, and the collision obstacle phenomenon occurred multiple times in the process, but the path of the experiment with the same iteration step was shorter than that in Section 5.1. As shown in Figure 8c,d, compared with the Section 5.1 iteration steps, the collision phenomenon was reduced and the path fluctuation was significantly slowed down. As displayed in Figure 8e,f, at the 1500th iteration, the system had completed the high-level navigation decision-making and acquired the optimal navigation strategy until 2000 iterations. The final authenticated publication is available online in reference [24].

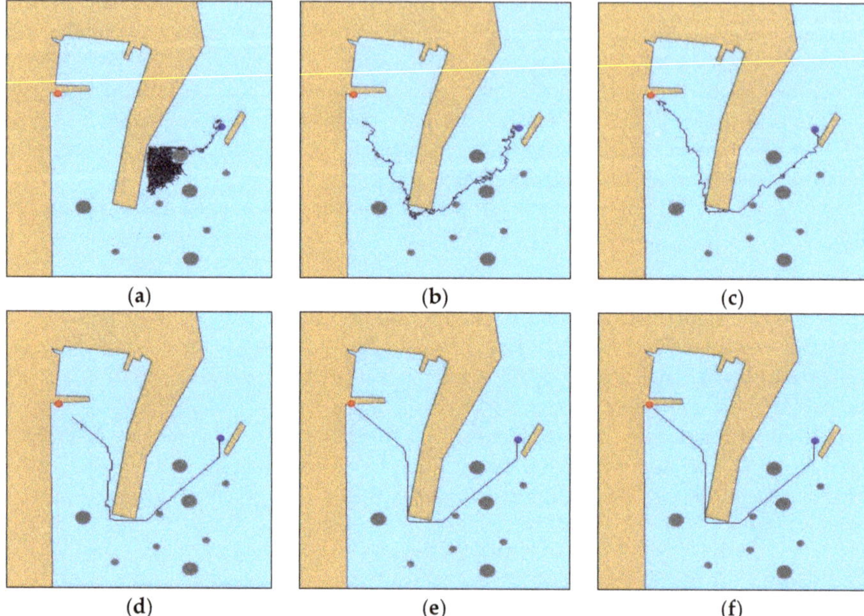

Figure 8. Improved DRL algorithm to verify the experimental results: (**a**) initial iteration; (**b**) 100th iteration, (**c**) 500th iteration, (**d**) 1000th iteration, (**e**) 1500th iteration, and (**f**) 2000th iteration.

5.3. Result

By comparing the verification experiments of the DRL algorithm in Section 5.1 and the APF-DRL algorithm in Section 5.2, the autonomous navigation decision-making-based APF-DRL had a faster iteration speed and a better decision-making ability. Plotting a graph with the training times of the model as the abscissa and the number of steps required to move from the starting point to the end of each iteration as the ordinate allows us to visually demonstrate the training speed and training effect of the two algorithms. The iterative convergence trend comparison presented is shown in Figure 9. The solid blue line represents the iterative trend of the APF-DRL algorithm, and the green dashed line represents the iterative trend of the DRL algorithm. The APF-DRL algorithm did not fall into local iteration after the 500th iteration (step > 500), while the DRL algorithm was iteratively unstable and would still fall into local iteration after the 1500th iteration. The APF-DRL iteration trend had converged by the 1500th iteration, and a fixed path was decided.

Two sets of experimental data were extracted, and the performance of the two navigation decision-making algorithms were compared and analyzed from the aspects of the number of local iterations, large fluctuation iterations (waves more than 300 times), collision rate, optimal decision iteration number, and optimal decision iteration time. As presented in Table 2, compared with the DRL algorithm, the decision algorithm that added the artificial potential field to improve the deep reinforcement learning had fewer local iterations. Moreover, the number of fluctuations and the trial and error rate were reduced. The simulation result shows that the APF-DRL algorithm had a fast convergence rate.

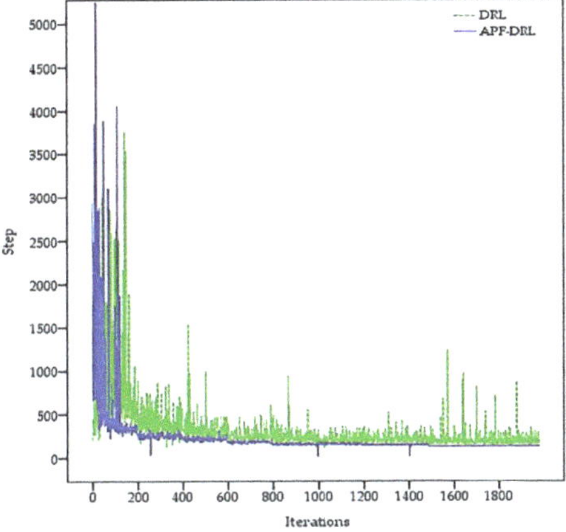

Figure 9. Iterative convergence trend comparison result.

Table 2. Experimental data comparison.

Verification Experiment	Trapped into Local Iterations (Times)	Fluctuations > 300 Iterations (Times)	Rate of Collision	Optimal Decision Iterations (Times)	The Iteration Time to Optimal Decision(s)
DRL	56	604	2.24%	2000	859
APF-DRL	29	192	1.16%	1498	442

In addition, the step size of each iteration of the two sets of experiments and iterative trends were visually analyzed. Figure 10 displays the comparison of the iterative step size scatter distribution of the two sets of experiments. The color of the scatter in the graph represents the path length of the decision. The DRL-distance was up to 4000, and the APF-DRL algorithm had a maximum iteration step size of 3000, which was much smaller than 4000. Furthermore, the distribution trend of the two sets of scattering points exhibited that the scatter points were almost concentrated on the DRL experimental iteration surface, which indicated that the improved algorithm had a shorter distance for each iteration. Thus, the experimental iterations done by the new procedure were better.

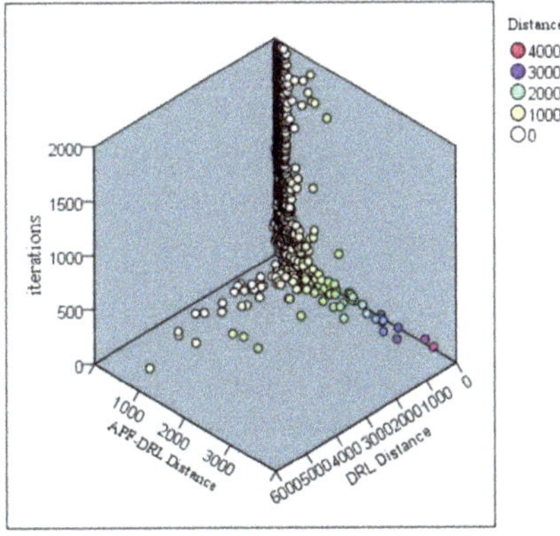

Figure 10. Experimental iteration step scatter plot.

Figure 11 shows the experimental collision avoidance parallel diagram of the two algorithms. The three parallel axes in the figure, from left to right, were DRL avoidances, APF-DRL avoidances, and epochs. The heavier the color, the worse the algorithm learning effect and the slower the convergence. For the APF-DRL algorithm, the avoidance presents a regular distribution, where as the number of iterations increased, the number of collisions decreased, and the self-learning success rate of the algorithm improved. On the contrary, DRL's collision avoidance iteration diagram is rather messy. It shows that the algorithm's self-learning ability was unstable.

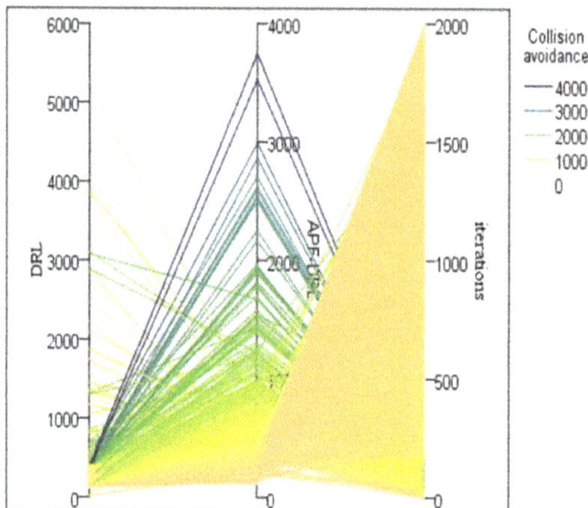

Figure 11. Experimental collision avoidance parallel diagram of the two autonomous navigation decision-making algorithms.

In summary, by comparing the convergence and decision-making effects of APF-DRL and DRL algorithms from multiple aspects, this study finds that the performance of the models and algorithms based on an artificial potential field to improve deep reinforcement learning was better.

6. Conclusions

In this paper, the autonomous navigation decision-making algorithm for a MASS based on the DRL algorithm is proposed. The model was established based on the four basic elements of deep Q-learning. However, the simulation results were not satisfactory, and the easy-to-fall-into local iterations and slow iteration convergence speed problems were apparent. To solve this problem, we added a gravitational potential field centered on the target point, and established an autonomous navigation decision-making algorithm of the MASS based on APF-DRL. In the initial stage of the interaction with the environment, the MASS had little knowledge of the environmental status information, and there were collisions and large fluctuations in the navigation path planning. As the number of iterations increased, the MASS accumulated learning experience, completed the adaptation to the environment, and finally successfully planned the path and reached the target point. In future research, the algorithm still needs significant improvements:

1. The deep Q-learning algorithm based on the Markov decision process could obtain the optimal path through the trial and error algorithm, but its convergence speed was still slower and the number of iterations was large. The first intended improvement involves enhancing the adaptive ability of DRL such that a small number of iterations can be used to learn the correct navigation behavior with only a small number of samples.

2. The strategy functions and value functions in the DRL were represented by deep neural networks, where the networks were poorly interpretable. This unexplained the security problem, which is

unacceptable in unmanned cargo ship transportation. The second intended improvement is to improve the interpretability of the model.

3. In the actual voyage, the navigation behavior of the unmanned cargo ship had a complex continuity. In this simulation experiment, only a simple generalization was performed to divide the navigation behavior of the MASS into the eight navigation actions. As such, the third intended improvement direction is to increase the ability of the model to predict and "imagine."

Author Contributions: Data curation—X.Z.; Formal analysis—C.W., Y.L., and X.C.; Methodology—C.W.; Project administration—X.Z.; Validation—C.W.; Writing—original draft—C.W.; Writing—review and editing—X.Z.

Funding: This work was supported by the National Key Research and Development Program of China (grant no. 2018YFB1601502).

Acknowledgments: Our deepest gratitude goes to the anonymous reviewers and guest editor for their careful work and thoughtful suggestions that have helped improve this paper substantially.

Conflicts of Interest: The authors declare no conflict of interest.

References

1. EMSA (European Maritime Safety Agency). *Annual Overview of Marine Casualties and Incidents*; EMSA: Lisbon, Portugal, 2018.
2. MUNIN (Maritime Unmanned Navigation through Intelligence in Network). Available online: http://www.unmanned-ship.org/munin/ (accessed on 10 June 2016).
3. Shipunov, I.S.; Voevodskiy, K.S.; Nyrkov, A.P.; Katorin, Y.F.; Gatchin, Y.A. About the Problems of Ensuring Information Security on Unmanned Ships. In Proceedings of the IEEE Conference of Russian Young Researchers in Electrical and Electronic Engineering (EIConRus 2019), Saint Petersburg/Moscow, Russia, 28–31 January 2019; pp. 339–343.
4. AAWA—The Advanced Autonomous Waterborne Applications. Available online: https://www.rolls-royce.com/media/press-releases.aspx#27-11-2017-rr-opens-first-ship-intelligence-experience-space (accessed on 27 November 2017).
5. Peng, L. AIS-Based Intelligent Collision Avoidance Expert System for Inland Vessels and Its Implementation. Master's Thesis, Wuhan University of Technology, Wuhan, China, 2010.
6. Fan, Y.; Sun, X.; Wang, G. An autonomous dynamic collision avoidance control method for unmanned surface vehicle in unknown ocean environment. *Int. J. Adv. Robot. Syst.* **2019**, *16*, 1729881419831581. [CrossRef]
7. Geng, X.; Wang, Y.; Wang, P.; Zhang, B. Motion Plan of Maritime Autonomous Surface Ships by Dynamic Programming for Collision Avoidance and Speed Optimization. *Sensors* **2019**, *19*, 434. [CrossRef] [PubMed]
8. Ahn, J.H.; Rhee, K.P.; You, Y.J. A study on the collision avoidance of a ship using neural networks and fuzzy logic. *Appl. Ocean Res.* **2012**, *37*, 162–173. [CrossRef]
9. Chen, H. Preliminary Study on Local Path Planning of Surface Unmanned Boats. Master's Thesis, Dalian Maritime University, Dalian, China, 2016.
10. Ramos, M.A.; Utne, I.B.; Mosleh, A. Collision avoidance on maritime autonomous surface ships: Operators' tasks and human failure events. *Saf. Sci.* **2019**, *116*, 33–44. [CrossRef]
11. Wang, N.; Pan, X. Path-following of autonomous underactuated ships: A translation-rotation cascade control approach. *IEEE ASME Trans. Mechatron.* **2019**. [CrossRef]
12. Wang, N.; Karimi, H.R.; Li, H.; Su, S. Accurate trajectory tracking of disturbed surface vehicles: A finite-time control approach. *IEEE ASME Trans. Mechatron.* **2019**, *24*, 1064–1074. [CrossRef]
13. Wang, N.; Karimi, H.R. Successive waypoints tracking of an underactuated surface vehicle. *IEEE Trans. Ind. Inform.* **2019**. [CrossRef]
14. Tai, L.; Li, S.; Liu, M. A deep-network solution towards model-less obstacle avoidance. In Proceedings of the IEEE/RSJ International Conference on Intelligent Robots and Systems, Aejeon, Korea, 9–14 October 2016; pp. 2759–2764.
15. Long, P.; Liu, W.; Pan, J. Deep-Learned Collision Avoidance Policy for Distributed Multiagent Navigation. *IEEE Robot. Autom. Lett.* **2017**, *2*, 656–663. [CrossRef]
16. Panov, A.I.; Yakovlev, K.S.; Suvorov, R. Grid Path Planning with Deep Reinforcement Learning: Preliminary Results. *Procedia Comput. Sci.* **2018**, *123*, 347–353. [CrossRef]

17. Bojarski, M.; Del Testa, D.; Dworakowski, D.; Firner, B.; Flepp, B.; Goyal, P.; Jackel, L.D.; Monfort, M.; Muller, U.; Zhang, J.; et al. End to end learning for self-driving cars. *arXiv* **2016**, arXiv:1604.07316.
18. Cheng, Y.; Zhang, W. Concise deep reinforcement learning obstacle avoidance for underactuated unmanned marine vessels. *Neurocomputing* **2018**, *272*, 63–73. [CrossRef]
19. Geng, X. Research on Behavior Decision-Making Approaches for Autonomous Vehicle in Urban Uncertainty Environments. Ph.D. Thesis, University of Science and Technology of China, Hefei, China, 2017.
20. Takahashi, T.; Kadobayashi, Y. Reference Ontology for Cybersecurity Operational Information. *Comput. J.* **2015**, *58*, 2297–2312. [CrossRef]
21. Yu, Z. Smartphone Sensor Fusion for Indoor Localization: A Deep LSTM Approach. 2018. Available online: http://hdl.handle.net/10415/6440 (accessed on 27 November 2017).
22. Liu, Q.; Zhai, J.-W.; Zhang, Z.-Z.; Zhang, S. A Survey on Deep Reinforcement Learning. *Chin. J. Comput.* **2018**, *41*. [CrossRef]
23. Zhao, D.; Shao, K.; Zhu, Y.; Li, D.; Chen, Y.; Wang, H.; Liu, D.-R.; Zhou, T.; Wang, C.-H. Review of deep reinforcement learning and discussions on the development of computer Go. *J. Control Theory Appl.* **2016**, *33*, 701–717. [CrossRef]
24. Wang, C.; Zhang, X.; Li, R.; Dong, P. Path Planning of Maritime Autonomous Surface Ships in Unknown Environment with Reinforcement Learning. In *Communications in Computer and Information Science*; Sun, F., Liu, H., Hu, D., Eds.; Springer: Singapore, 2019; Volume 1006, pp. 127–137.

Article

Start from Scratch: A Crowdsourcing-Based Data Fusion Approach to Support Location-Aware Applications [†]

Yonghang Jiang [1], Bingyi Liu [2,*], Ze Wang [2] and Xiaoquan Yi [2]

[1] Department of Computer Science, City University of Hong Kong, Hong Kong; yhjiang4-c@my.cityu.edu.hk
[2] School of Computer Science and Technology, Wuhan University of Technology, Wuhan 430070, China; cs_wangze@whut.edu.cn (Z.W.); yixiaoquan@whut.edu.cn (X.Y.)
[*] Correspondence: byliu@whut.edu.cn; Tel.: +86-027-87216780
[†] This paper is an extended version of our paper published in Proceedings of the IEEE International Symposium on Dynamic Spectrum Access Networks, Newark, NJ, USA, 11–14 Novermber 2019.

Received: 1 September 2019; Accepted: 15 October 2019; Published: 17 October 2019

Abstract: As one of the most important breakthroughs for modern transportation, the indoor location-based technology has been gradually penetrating into our daily lives and underlines the foundation of the Internet of Things (IoT). To improve the positioning accuracy and efficiency, crowdsourcing has been widely applied in indoor localization in recent years. However, the crowdsourced data can hardly be fused easily to enable usable applications for the reason that the data are collected by different users, in different locations, at different times, with different noises and distortions. Although different data fusing methods have been implemented in different crowdsourcing services, we find that they may not fully leverage the data collected from multiple dimensions that can potentially lead to a better fusion results. In order to address this problem, we propose a more general solution, which can fuse the multi-dimensional crowdsourced data together and align them with the consistent time and location stamps, by using the features of the sensory data only, and thus build high quality crowdsourcing services from the raw data samplings collected from the environment. Finally, we conduct extensive evaluations and experiments using different commercial devices to validate the effectiveness of the method we proposed.

Keywords: internet of things; crowdsourcing; indoor localization; data fusion

1. Introduction

With the proliferation of wireless communication and Internet of Things (IoT), location-based services have shaped and enabled a wide range of applications for our safety and convenience. While GPS dominates outdoor localization and becomes an essential element of the modern transportation, the prospect of indoor navigation has also gained attention in the last few decades as it holds important value for ubiquitous applications in location-based service (LBS) such as inventory management, logistical and supply chain management, smart home and smart building monitoring, retails and sport analytics, mall navigation, virtual reality, etc.

However, indoor location is inaccessible for GPS signals due to structural blockages and severe multiple propagation effects. Currently, there is no general solution for indoor localization, but there are several potential technologies available to provide indoor positioning and map construction solutions such as radio frequency identification (RFID), ultra wide band (UWB), micro-electromechanical systems (MEMS) multi-sensors, and wireless local area networks (WLAN). To fully leverage the data collected from multiple dimensions that can potentially lead to a better indoor positioning result, crowdsourcing has drawn significant attention in recent years. More and more crowdsourcing approaches have been presented, and the popularity of smart devices boosts their applications. The term crowdsourcing describes a new computing paradigm that distributes the work previously handled by employees to a large undefined network of individuals. Obviously, a crowd of people can coordinate and solve problems faster than a single person, and the crowd can generate data rapidly about the particular location.

Such model is an efficient way for indoor location-aware applications. The crowdsourced data, however, can hardly be fused easily to enable usable applications for the reason that the data are collected by different users, in different locations, at different times, with different noises and distortions. At first, the sensory readings are inaccurate due to the deviation of low-cost sensing devices, and different individuals can report slightly inconsistent readings on the same parameter. Since the crowdsourced data possess the advantage of their large quantity, many researchers propose to improve the accuracy through data fusion. These methods, however, may suffer from the fact that crowdsourced data traces are distorted both temporally and spatially. As a result, in this case, the fusion of multiple data sources can hardly be proceeded. The lack of precise time and location information largely limits the usability of the crowdsourcing data, and thus there is an urgent need for efficient crowdsourced data alignment to support varying applications.

As the time calibration can be achieved through time synchronization, we focus on the geographical trace alignment in this paper. Theoretically, collecting data from different locations should be based on a localization service. However, in the indoor scenarios, the indoor positioning systems are not widely deployed and used in practice. There are many existing indoor localization approaches such as the fingerprint-based methods [1–5], the ranging based methods [6], and the like. They, however, may suffer from the high dynamics and complexities of the indoor environment and thus output erroneous results. The authors in [7,8] propose the methods to combine the sensor information with the constraints imposed by the map, thereby filtering out infeasible locations and converging on the true location. However, they need special constraints of the floor plan, which may not always be available, as the floor plan cannot be acquired in some circumstances. A recent state-of-the-art approach Walkie-Markie [9] presents an indoor pathway mapping scheme. It automatically reconstructs the internal pathway maps of buildings, using the trend of WiFi signals as landmarks to calibrate the location of pathways. This method, however, we find still has room to be further improved due to several limitations. Firstly, as this system works based on WiFi-Marks, it is less capable of processing short data traces with hardly any WiFi-Marks. Moreover, it assumes that the distortions of traces are simple because they only come from incorrect stride lengths and headings. As a matter of fact, due to a variety of localization methods, the distortions could be more complicated in practice. Some other works use geomagnetic information [10,11] or WiFi information [12,13] for indoor localization. Although a high localization performance can be achieved, they focus on the feature of a single data dimension (such as WiFi), i.e., they cannot be applied to fuse other sensory data with different features from WiFi signals. Moreover, in the indoor environment, there are various signals and noises which will interfere with the feature of the single data dimension, and thus may affect their positioning accuracy.

In order to address this problem, we propose a novel solution that aligns the multi-dimensional crowdsourced data with the right time and location stamps, and thus builds high quality traces from raw data samplings. Generally, crowdsourced traces consist of multiple data dimensions such as the WiFi signal strength, the ambient temperature/humidity, the magnetic field information, and the like. Each dimension corresponds to one underlying parameter of the physical world and exhibits its unique feature on the data distribution and variation. For example, the wireless signal follows the strength loss model and the magnetic follows the magnetic field model. This field feature seams consistent as the fluctuations of gradient along a specific trace, even if the deviations varies for different devices. Figure 1 shows the gradient of access point (AP) and magnetic field in ideal models. More importantly, with only one dimension information, the trace may not be unique, e.g., the trace departing from the same AP to different directions can have a similar decrease gradient feature. However, the multi-dimensional information, e.g., signals from more than one AP and base stations, three directional components of magnetic field, etc, could give us the chance to conduct integrated fusion and align traces accurately without a priori knowledge of the environment.

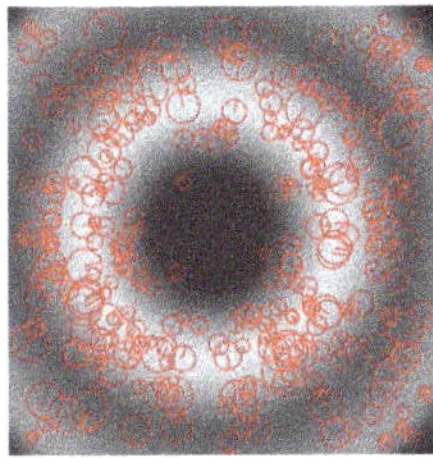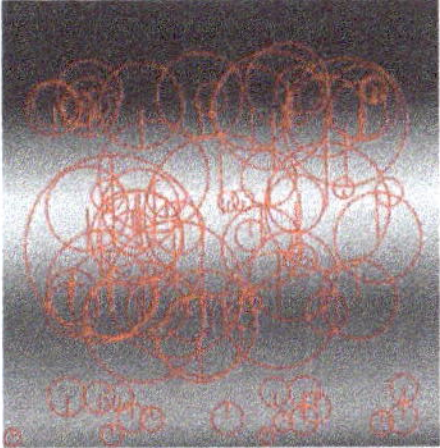

Figure 1. Gradient of AP signal and magnetic field in ideal models.

Based on these observations, we form the crowdsourced trace alignment as a multi-dimensional data consistency optimization problem and search for the more accurate solution based on the mutual correlation among the multi-dimensional data. In summary, the contributions of this work are as follows:

- We extensively study the problem of multi-dimensional crowdsourced data fusion problem and propose a novel method to fuse noisy and distorted crowdsourced data in different environments.
- We propose a new trace coding and transformation approach to conduct the trace alignment, in order to deal with a complex trace distortion issue.
- We propose an optimization scheme to find the crowdsourced trace alignment solution.
- We conduct extensive evaluations and experiments to validate the effectiveness of our method.

The rest of this paper is organized as follows. In Section 2, we discuss existing efforts related to this work. Section 3 provides a detailed statement of the crowdsourcing data alignment problem. Section 4 describes our main idea and the framework of the proposed approach. In Section 5, we discuss the 3D indoor trace collection and illustrate the evaluation results. We reach a conclusion in Section 6.

2. Related Work

There are some existing efforts on map reconstruction that are related to our work. FootSLAM [14] proposes a Bayesian estimation approach that applies foot mounted inertial sensors to construct the internal map. PlaceSLAM [15] incorporates human-reported place information. Escort [16] explores the possibility of using mobile phone sensors and opportunistic user-intersections to navigate users. UnLoc [17] uses environmental signatures as landmarks to re-calibrate users' locations, which is an unsupervised localization approach to avoid the labor intensive war-driving for building the localization feature dataset. In this paper, we further explore the distribution of multi-dimensional data and conduct integrated optimization on all dimensions for the data alignment task, where the primary version of the paper is published in [18]. A recent state-of-the-art approach Walkie-Markie [9] is an indoor pathway mapping system. It automatically reconstructs internal pathway maps of buildings without any a priori knowledge, using the trend of WiFi signals as landmarks to calibrate the location of pathways. Walkie-Markie, however, considers only one dimension of the sensory data and focuses on pathways that can hardly be applied to an open indoor space.

Moreover, there are some existing approaches on indoor localization [2,6,8] and navigation [10,12,19,20] that are related to this work. A major category of indoor localization approaches is fingerprint-based, which collects measurements at known locations in advance. The fingerprint collection process is usually called a site survey. In the localization stage, the fingerprint-based methods compare the observed measurements with the fingerprints in the database and determine the new locations according to the best matched fingerprints. Many different types of measurements can be applied as fingerprints, such as radio-frequency (RF) [1], GSM [2], FM radio signal [4], geo-magnetism [3], CIR (channel impulse response) [5], and the like. [11] puts forward a kind of fingerprint distinguishability evaluation model to find a high discernible fingerprint extracted method for indoor localization, e.g., the authors in [13] propose a novel deep learning framework for indoor localization tasks using WiFi fingerprints.   In these works, the need for a survey is a key bottleneck since it is labor-intensive and time-consuming. Especially in large-scale applications, fingerprints that store the buildings' environment characteristics must be mapped. As collecting fingerprints in thousands of buildings is quite costly, some researchers propose crowdsourcing approaches [21,22] to conduct a site survey. The authors in [23] provide an unsupervised learning-based location labeling technique for crowdsourced fingerprints, then generate indoor high-precision maps, and provide indoor localization and navigation services.

With the rich on-board sensors on mobile devices, the mobile crowdsourcing (MCS) [24,25] appears through mixing the mobile techniques and crowdsourcing, which can enable many location-aware crowdsourced tasks. Recently, the MCS sensing applications have provided benefits to many aspects of people's everyday lives. For example, TrafficInfo [26] can provide a real-time public transport information service through sharing information and sending feedback with participatory sensing. Ear-Phone [27] can generate a noise map to facilitate monitoring of environmental urban noise pollution through crowdsourcing data collection. In addition, MCS sensing applications can provide a solution to the lack of floor plans of indoor localization service. The authors in [28–30] propose the methods for reconstructing the indoor floor plans by utilizing inertial sensors (accelerometers, compasses and gyroscopes), visual and spatial information crowdsourced from the users' smart phones. Since such applications rely on users' participation, many incentive mechanisms [31,32] have been proposed to encourage users to share their sensed data. At the same time, when participating in crowdsourcing tasks, users may suffer security and privacy issues. The authors in [33,34] have studied the privacy leaks and potential threats in crowdsourcing tasks to handle such issues. Moreover, the collected data in crowdsourcing tasks are usually not well structured and noisy. Thus, some data analysis technologies are proposed to improve the quality of collected data. For instance, [35] proposes a quality-sensitive answering model to satisfy user required accuracy through guiding the crowdsourcing query engine to process and monitor human tasks that can provide an estimated accuracy for each generated result. The authors in [36] propose an anchor point-based forward-backward smoothing method to generate indoor localization databases and a quantitative framework to evaluate the quality of sensed data automatically. Conversely, there are some works that motivate users to provide high-quality data through data quality control [37,38].

Although the aforementioned methods are important for improving the positioning accuracy and supporting location-aware applications, few of which consider the solution of multi-dimensional crowdsourced trace fusion and alignment to deal with complex trace distortion issues. Moreover, the realistic indoor scenario, such as the coexistence of multi-dimensional data, has not been fully considered in the literature. Motivated by these facets, we propose a framework to find the crowdsourced trace alignment solution in this paper.

3. Problem Statements and Definitions

In common crowdsourcing applications, the collected data can be represented as an ordered set of trituples $\{(d, t, p)\}$. For each tuple, d denotes the sensory data, which may contain multiple dimensions such as wireless signal strength, magnetic field, acoustic information, etc.; t is the time stamp and p is the location stamp.

Given stamps t and p, the ground truth of d is unknown in most cases due to the existence of measurement errors and noises. Nevertheless, a large amount of noisy samplings may be collected by a crowd of N mobile devices, say $\{P_j\}$. In this work, we denote the crowdsourced data as $d + n_j$ in which n_j is the additive noise part. Consequently, we will discuss the properties of the crowdsourced data sets.

3.1. Distortion of Crowdsourced Data

For the three tuples of one record, the time stamp is relatively easy for alignment through the time synchronization. The distortion of the collected data from crowd mainly results from two sources: the volumetric aspect and geometric aspect. The inaccurate or uncalibrated on-broad sensors can report noisy readings and the geometric distortion is caused by the inaccurate localization. Firstly, we consider the noise in sensory values. Let (x, y, z) denote the coordinates of a location in which z indicates the floor number. Different from the outdoor environment, buildings can have multiple floors and the data traces are collected on different floors. To the real value $d(x, y, z)$ of a physical parameter, the observed measurement could be:

$$o(x, y, z) = d(x, y, z) + n(x, y, z). \tag{1}$$

In order to derive the real value $d(x, y, z)$ from $o(x, y, z)$, many researchers propose to use the fusion techniques. For example, some approaches assume that the noise $n(x, y, z)$ comes from different devices are independent. As a result, $n(x, y, z)$ can be regarded as a random variable following the zero-mean Gaussian distribution. In this case, averaging N different noisy measurements from different traces can achieve a good estimation of the real value,

$$\bar{o}(x, y, z) = \frac{1}{N} \sum_{j=1}^{N} o_j(x, y, z).$$

Then, it follows that

$$E\{\bar{o}(x, y, z)\} = d(x, y, z),$$

where $E\{\bar{o}(x, y, z)\}$ is the expected value of $\bar{o}(x, y, z)$. Let $\sigma^2_{\bar{o}(x,y,z)}$ and $\sigma^2_{n(x,y,z)}$ denote the variances of $\bar{o}(x, y, z)$ and $n(x, y, z)$, respectively. Then, we have:

$$\sigma_{\bar{o}(x,y,z)} = \frac{1}{\sqrt{N}} \sigma_{n(x,y,z)}.$$

As N increases, the equation above implies that the variability of the measurement at each location decreases. In typical crowdsourcing scenarios, when there are more than dozens of users, $\bar{o}(x, y, z)$ can achieve a good estimation about the noiseless value $d(x, y, z)$ and the approximation accuracy increases as the user number increases.

In practice, the observations $\{o_j(x, y, z)\}$ should be aligned to the accurate location coordinates; otherwise, the averaging process does not provide a better fusion result. For the outdoor environment with GPS services, the average-based data fusion can be applied for a wild range of crowdsourced services. In the indoor environment, however, the lack of efficient localization services becomes the main obstruction for the data alignment and indoor map construction.

Employing an indoor localization method, the result comes out with an error for each location $l = (x, y, z)$:

$$p = l + e(x, y, z) = (x + e_x, y + e_y, z + e_z).$$

May existing localization methods are fingerprint based, whose performance degrades if the sensory data are inaccurate and unstable. For example, in a localization approach using AP signal strength as the fingerprint, the change of the furniture layout or AP position can significantly affect the signal strength observed in the same location. In addition, different places can exhibit similar signal strength observations, leading to the sensing ambiguity.

3.2. Multi-Dimensional Data Alignment

In this work, we address the multi-dimensional data alignment problem that assigns proper locations to each data point by analyzing the multi-dimensional sensing traces.

Formally, each type of information is referred to as a *dimension* m_i, e.g., the signal strength of a specific AP or the geomagnetic field. Each observed trace O_j is a set $\{(o_j, t_j, p_j)\}$ ordered by t_j. With multiple on-board sensors on the mobile device, the observed value o_j contains information of multiple dimensions, e.g., $o_j = \{o_j^i\}$. Data in varying dimensions could possess different distributions. For example, the signal strength of a specific AP may exhibit a radial distribution centered at the AP's location, and the geomagnetic field is approximately along the longitudes. Based on the observation, our proposed scheme tries to find an alignment of the data traces that best fits the underlying data distributions cross all the data dimensions.

4. Framework

4.1. Mechanism Overview

As illustrated in Figure 2, our approach is an iterative method that basically consists of four coupled components: (1) a coding scheme that models the locations of a data trace; (2) a map metric that measures the quality of ambient trace data; (3) a class of transformations that can be applied to the traces during the fusion process; and (4) a solver that searches for the next transformation to improve the map-level data quality in all data dimensions.

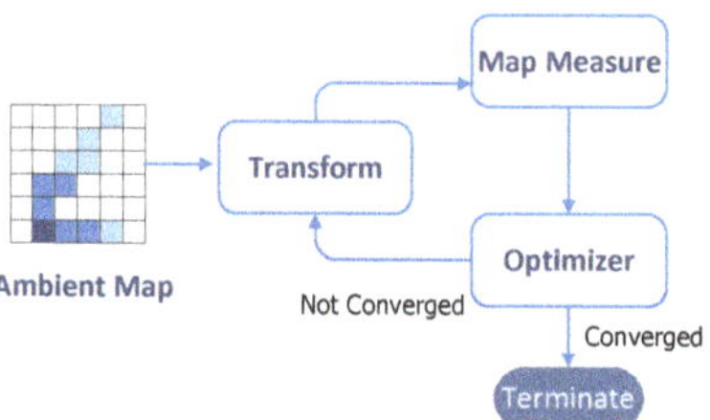

Figure 2. Mechanism overview.

For the purpose of measuring the effectiveness of the data alignment, we propose a map-level trace quality metric, $Q(\mathbf{M_i})$. We denote $\mathbf{M_i}$ as an *ambient map* of the ith data dimension. An *ambient map* is defined as the data map of a specific dimension. For example, the ambient map of Wi-Fi signal consists of all the signal strength readings at different physical locations. For the purpose of data fusion, the data map is usually embedded into grids, and data samplings within the same cell are fused together. In our analysis, we propose to use multi-resolution ambient maps for the reason that different data dimensions show different signal changing rates.

We design a family of feasible spatial transformations $T_k(\cdot)$. Each transformation $T_k(O_j)$ is an atomic operation that fine-tunes the location and shape of a trace O_j, which may also affect the quality of all the corresponding maps. A multi-dimensional data alignment solution is a sequence of transformations $\mathbf{T}$ applied to the collected traces. For a solution $\mathbf{T}$, the quality of $\mathbf{M_i}$ is $Q(\mathbf{M_i}|\mathbf{T})$. Then, the multi-dimensional data alignment problem can be expressed as follows:

$$arg \max_{\mathbf{T}} \sum_i Q(\mathbf{M_i}|\mathbf{T}). \tag{2}$$

In the following subsections, we will firstly present our basic fusion algorithm and then discuss some design issues and data structures in the proposed approach.

4.2. Basic Fusion Scheme

As we have discussed above, the crowdsourced traces are noisy and distorted. To address this problem, we propose a novel scheme that aims to find an optimal transformation **T** to align the collected data traces and thus maximizes the resulting *ambient maps'* quality.

Before introducing the detailed algorithm design, we firstly discuss how to assess the quality of an *ambient map*. The *ambient map* in a specific resolution is generated by fusing the observed values in a small region or called cells from all the data traces.

Note that the physical parameters exhibit unique features in their value distributions. If the collected data traces are well aligned with the accurate locations, all of the crowdsourced traces will follow the distribution of the physical parameter in each dimension, and thus different traces will be consistent with each other. As the ground truth of the underlying physical value is unknown in common crowdsourced applications, we assess the quality of *ambient maps* through measuring the data consistency among different traces. Then, for each pair of traces, we find their overlapped segments and calculate the correlation score between them. Because the crowdsourced data are noisy due to the device variation, during the computation of the correlation score between trace segments, we focus on the variation trend of the data series instead of the values in each single cell. In this approach, we apply the *Pearson correlation coefficients* to measure the correlation of two series. For each dimension, we have:

$$r_{uv}^i = \frac{\sum_{j=1}^N (u_j - \bar{u})(v_j - \bar{v})}{\sqrt{\sum_{j=1}^N (u_j - \bar{u})^2}\sqrt{\sum_{j=1}^N (v_j - \bar{v})^2}}. \tag{3}$$

In the equation, u_j and v_j are the values of series u and v on the jth overlapped location. r_{uv}^i is their correlation of the ith dimension. Then, the final correlation score is the summation of r_{uv}^i from all dimensions. We further extend the correlation computation to the multi-layer case, in which the ambient data map is represented in different dimensions. The details of the multi-resolution ambient map will be discussed in the following subsection. The quality of an *ambient map* is assessed by the correlation scores of all the overlapped trace segments:

$$R = \sum_{u \neq v} R_{uv} = \sum_{u \neq v} \sum_i r_{uv}^i \times l, \tag{4}$$

where R_{uv} is the all dimensional correlation of series u and v, and l is the length of the overlapped series.

In the fusion process, we move and stretch some traces, and thus make the highly correlated ambient data series be located along the same path to maximize the overall correlation score. The transformation of each trace may bring some new overlapped segments, which may contribute new correlation scores and destroy some existing overlapped segments as well. Thus, the effect of each transformation can be represented by a correlation gain function $g(T(O_j))$ that is defined as the difference of the *map correlation score* before and after the trace transformation as follows:

$$g(T(O_j)) = R' - R. \tag{5}$$

In each iteration, we choose a transformation with a maximum gain and then update the correlation score of the integrated *ambient map* R.

The number of potential transformations increases exponentially with the number of traces. In order to speedup the optimization process, we present several pruning strategies. Firstly, the traces usually start from some specific locations, for example, the entrance of the building. Since the location errors will accumulate as the trace length increases, the trace segments near the entrance can be more accurate and easier to align. Thus, we conduct an ordered search starting from such entrance points. In the first step, we only consider traces near the building entrance. Secondly, we consider some restrictions of the transformation. If the indoor maps are available, the transformations are guided by the structures of the indoor map. In most cases, it is reasonable to assume that the trace distortion can be bounded by some thresholds (as the traces can be not very long), thus we let the size of each transformation be less than certain thresholds. Thirdly, in order to further reduce the solution space, we propose to extract transformations with high potential gain. For each pair of traces, we calculate the possible highest correlation score between them after some transformation. That is, we slip one trace from the beginning of the other to the end, during which we find the overlapped segments with the highest correlation score. Then, the transformation that can align the overlapped segments to be physically coincided will be regarded as a good candidate.

4.3. Multi-Resolution Ambient Map

As mentioned above, we introduce the concept *ambient map* to model the multi-dimensional ambient data distribution.

In this paper, we introduce a multi-resolution pyramid grid structure to encode the indoor maps as shown in Figure 3. Then, an ambient map is the combination of multiple weighted data layers in different resolutions. Different dimensions of data may show different changing rates and thus their significance on different resolutions are different. For example, the indoor temperature changes slowly over spatial locations, so we assign relatively low weight to the high resolution layer of temperature map because of the fact that they contribute little information in the trace alignment. In contrast, the Wi-Fi signal strength dimension is more location-sensitive, and we will lose valuable information if this signal is mapped to a low-resolution map.

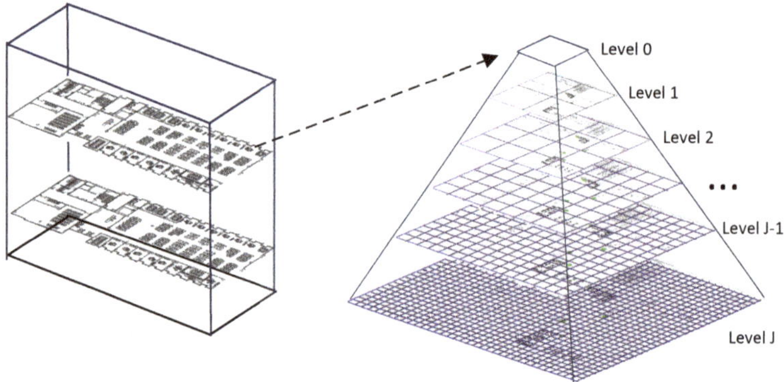

Figure 3. Illustration of the multi-resolution building map.

Formally, the indoor map of the building is represented by grids of different scale. The map for a building contains a tree of floors for the whole building and a quadtree for each floor. Moving up the pyramid, both size and resolution decrease. For level k, we set its size as $2^k \times 2^k$. Then, the map for dimension m_i of a building with the pyramid description is $\mathbf{M}_i^k(v, w, f, t)$, here k is its resolution level, (v, w) is the location of the grid, and f indicates the floor level.

4.4. Trace Coding

For a continuous trace $O = \{(o_j, t_j, p_j)\}$, the location $p_j = (x_j, y_j, z_j)$ in the earth coordinate can be projected to the grid representation (v_j, w_j, f_j) by the nearest neighbor rule as shown in Figure 4, e.g., if p_j is in the range of a grid, then it is projected to this grid. Note that the absolute location p_j of the crowdsourced data traces cannot be accurately determined in many indoor situations. Nevertheless, the shape of traces can be estimated using on-board sensors, like accelerometer and gyroscope, which provide information about the user's movements and thus help us estimate the shape of trace segments. In order to utilize such kind of raw traces without absolute locations, as well as facilitate the trace representation, storage and optimization process, we propose an 8-directional chain code to model a crowdsourced data trace. As mentioned above, we apply multi-resolution grids to divide the physical map. In this case, we can assign the start point p_0 of a trace to a start cell S, and the consequent points are projected into a series of cells, thus the whole trace can be embedded into the grid. For each pair of successive cells, we define c as the moving direction from the start cell to the end one. In this work, direction c is coded using a 3-bit code. Then, the locations of an integrated trace can be represented by a start cell and a sequence of direction codes. The observed data of each point will be associated with corresponding cells. For each cell, the observed data of dimension m_i is the mean value of all readings located in this cell. A coded trace is $CO = \{P_s, \{o_j, t_j, c_j\}\}$, here P_s is the start cell and c is a 3-bits direction. In the implementation, the coded trace of the supported highest resolution will be constructed firstly, then the lower resolution trace can be generated from the higher resolution trace. Figure 4 illustrates the direction codes and examples of coded trace in different resolutions. The direction codes of the trace in a lower level (higher resolution) is "17811231235545556", and in a higher level is "711313555".

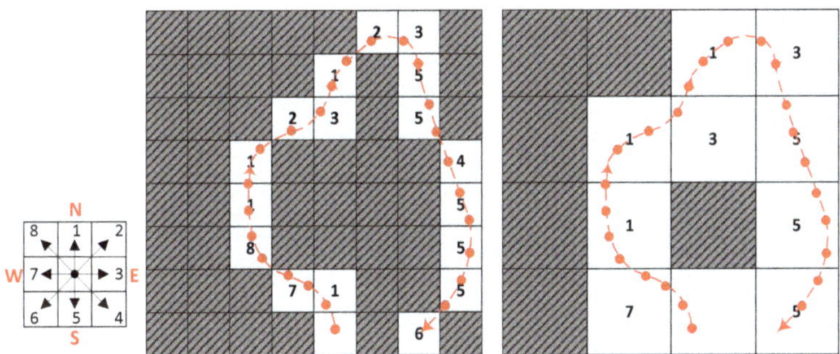

Figure 4. Trace coding projects a raw moving trace to a grid-based 8-directional chain coded presentation.

In cases that the absolute coordinates of some points in a trace can be obtained at GPS capable positions, e.g., the entry of a building or the balcony, we treat these GPS-enabled points as anchors for the trace. We call the trace with anchors a *rooted trace*. In this work, we focus on the rooted trace with only a fixed start point, which is the most common situation because it is easy to obtain the GPS location of the building entrance. The traces with anchors at the end point can be simply reversed to traces with anchors at the start point. For those traces with more than one anchor, the segments between two anchors can be easily aligned, leaving the rest of the segments with only one anchor fall into the rooted trace situation. Therefore, we define the rooted traces as the following statement:

Definition 1 (Rooted Trace). *The rooted trace contains one anchor at the start point, whose accurate location is known, i.e., P_s is known for trace $CO = \{P_s, \{o_j, t_j, c_j\}\}$.*

Sometimes, a user may turn on the tracking service after he/she enters the building, and the trace has no anchor point, i.e., only direction codes are available, We call this kind of trace the floating trace.

Definition 2 (Floating Trace). *The floating trace doesn't have any anchor point, i.e., P_s is unknown for trace $CO = \{P_s, \{o_j, t_j, c_j\}\}$.*

As we can see, giving the start point of one floating trace makes it become a rooted trace. In our method, the rooted traces are processed first, and then the floating traces will be further considered.

4.5. Trace Transformation

To align the trace for data fusion, a sequence of transformations will be applied to the trace $(x', y') = T((x, y))$. In this paper, we employ the affine transformation, as shown in Figure 5, which has the general form of

$$[x'\ y'\ 1] = [x\ y\ 1] \times T = [x\ y\ 1] \times \begin{bmatrix} t_{11} & t_{12} & 0 \\ t_{21} & t_{22} & 0 \\ t_{32} & t_{32} & 1 \end{bmatrix}. \tag{6}$$

Considering the main misalignment caused by the distance and orientation miscalculation while the user moves in the building, the rotation and scaling transformations are defined for the rooted trace and one additional translation transformation for the floating trace.

1. Rotation:

$$T_r = \begin{bmatrix} \cos\theta & \sin\theta & 0 \\ -\sin\theta & \cos\theta & 0 \\ 0 & 0 & 1 \end{bmatrix}. \tag{7}$$

2. Scaling:

$$T_s = \begin{bmatrix} c_x & 0 & 0 \\ 0 & c_y & 0 \\ 0 & 0 & 1 \end{bmatrix}. \tag{8}$$

3. Translation:

$$T_t = \begin{bmatrix} 1 & 0 & 0 \\ 0 & 1 & 0 \\ t_x & t_y & 1 \end{bmatrix}. \tag{9}$$

We can obtain:

$$\mathbf{T} = T_r \cdot T_s \cdot T_t = \begin{bmatrix} c_x \cdot \cos\theta & c_y \cdot \sin\theta & 0 \\ -c_x \cdot \sin\theta & c_y \cdot \cos\theta & 0 \\ t_x & t_y & 1 \end{bmatrix}. \tag{10}$$

It shall be noted that c_x and c_y can be treated as equal for the trajectory of dead reckoning in the scaling operation, thus it can be regarded as a similarity transformation. The reason we choose similarity transformation rather than affine transformation is that the trajectory of dead reckoning does not need shearing and flipping, which is a difference between similarity transformation and affine transformation. The other reason is that the similarity transformation has four degrees of freedom, while the affine transformation has six. Thus, the problem scale can be reduced. The trace coding discretizes the trace presentation and simplifies the transformation, since the rotation and scaling operations are only applied to the direction codes of a trace, despite its start point. In addition, we use nearest neighbor interpolation to generate the trace after transformation. The translation is even simpler by just changing the start point P_s, leaving the rest of the trace unchanged.

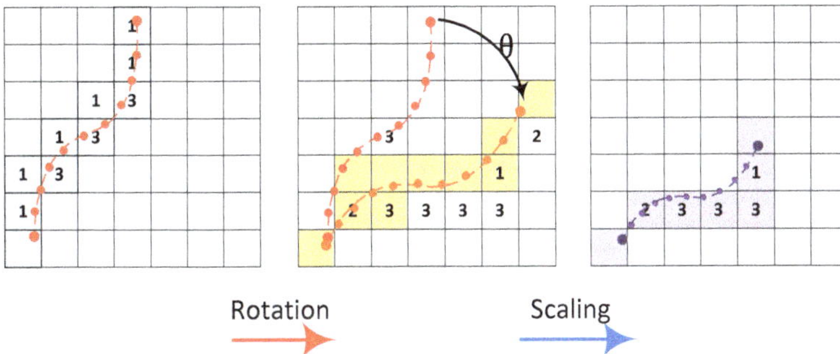

Rotation   Scaling

Figure 5. Illustration of different trace transformation operations.

We observe from our collected trace data that, in most cases, the deviation of trace directions is within 20 degrees, and the offsets of the coordinates are within 3 m, so we restrict the size of transformations to each trace. In addition, to avoid an infinite search space, we assign discrete values to the three transformation operations. For example, we select eight rotation angles between anticlockwise 20 degrees and clockwise 20 degrees. The scaling of traces is set to 10 different levels from 2% to 10% expansion or reduction.

Another design issue is the granularity of the trace transformation. Integrated data traces sometimes exhibit long length and complex formation. With the three types of transformation operations, it is efficient to adjust traces with a relatively simple structure like a straight path. It, however, is difficult to fine-tune some local formations in a complex trace. To address this issue, we propose to segment the data traces and conduct transformations one by one. Each entry in a coded trace denotes the direction of the next cell, so the difference between consecutive items indicates angle of the direction change of the trace. One unit difference of the code numbers corresponds to the path direction change of 45 degrees and, in this work, we define the direction change equal to or more than 90 degrees (two unit difference) as a significant change of the path way. Then, we divide the trace according to positions that experience significant direction changes.

Defining these transformations, our fusion algorithm also needs anther preparation. Because of a variety of localization methods, the noises and distortions of crowdsourced traces are of great complexity. In order to align the traces and recover their real shapes, we need to separate the traces into segments according to the shapes of traces, and transform all the segments to find the optimal transformations. Intuitively, with our trace coding method, we only need to check each of the coded traces, and cut it when the its code number changes. However, the fact is more complex. The code of the first trace shown in Figure 5 is "113131311". If we segment this trace by the code number, it should be cut into seven segments, most of which are only one grid length, so that they cannot be transformed. In fact, in this case, cutting into two segments should be appropriate. As trace coding eliminates many direction features of the traces, we prefer to do segmentation on the non-code traces. For the raw traces, we can use Biezer curve least square fitting [39] to do the segmentation. Then, we code all segments as described before. Of course, during this process, though being transformed, the segments of a trace should keep being linked together.

5. Experiments and Evaluation

5.1. Experiment Setup

We implement a crowdsourcing data collection application in five different types of commodity smart phones, including SamSung e120s, SamSung e120k, HTC one 3D, HTC one X and HTC one. We conduct the indoor data trace in a large lab building. The data traces are mainly collected on two floors with a total floor space of 4000 m^2. Our data traces span different indoor infrastructures, such as the corridor, working place, exhibition hall, office, meeting room, and the like. Figure 6 illustrates the floor plan of these two floors.

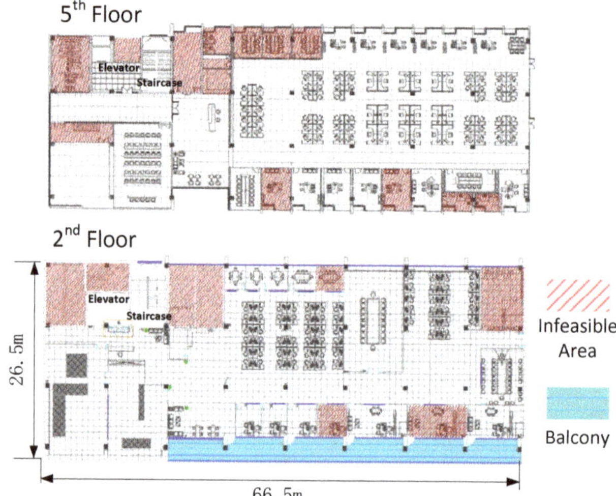

Figure 6. The ground truth of the floor plan for the 5th floor and the 2nd floor.

In total, 26 volunteers with heights from 160 cm to 185 cm join this experiment, who carry smart phones of different models with our indoor trace collection application during their daily work. The data collection lasts for 12 days and we get 931 traces in all. Three kinds of ambient information are collected, including the signal strength of all sensible Wi-Fi APs, the signal strength of the phone base station and the magnetic field. There are 11 APs deployed on these two floors.

5.2. Evaluation

In order to test and evaluate our data fusion method, we need a method to know the traces of the crowdsourced data to simulate the distorted results of the inaccurate indoor localization services. We describe the design and implementation of our crowdsourcing approach to obtain indoor multi-dimensional traces without any pre-knowledge of the building using the on-board sensors in off-the-shelf smart phones. The captured acceleration values are in the coordinate system fixed to the smart phone, and here we refer it as a phone coordinate system. Considering a user could hold the phone in any position, we firstly convert the acceleration from the phone coordinate system to the earth coordinate system, i.e., north, east, gravity.

With the on-board accelerometer of a smart phone, a user's movement can be sensed and used for activity recognition. In our work, we recognize the following most common activities: walking, climbing stairs and taking an elevator. We use these events, climbing stairs and taking an elevator, to decide the beginning of detecting walking trace. Knowing the beginning of walking, we need a detecting algorithm to detect the walking trace by using only smart phones. We implement the dead reckoning method described in Montage [40] on both Android and IOS smart phones to simulate localization services.

The method we recognize for the three activities is based on the following observation. As shown in Figure 7, accelerations generated by common walking changes positively and negatively regularly. Each step consists of a continuous sequence of positive accelerations and a followed sequence of negative accelerations that have similar amplitudes. The recognizing step method is included by the trace detecting algorithm mentioned before. In fact, climbing stairs is a kind of walking, and its acceleration springs profile is similar to that of walking, as shown in Figure 7. As presented in

Figure 7a,b, the acceleration springs' profiles of climbing upstairs and going downstairs are very similar. Their main difference is climbing upstairs starting with a peak and going downstairs beginning with a valley. As shown in Figure 7c, according to the changing amplitude levels, the acceleration data can be separated into several segments, in which the segments have higher amplitude level corresponding to steps of climbing processes, and the lower amplitude parts stand for walking on platforms.

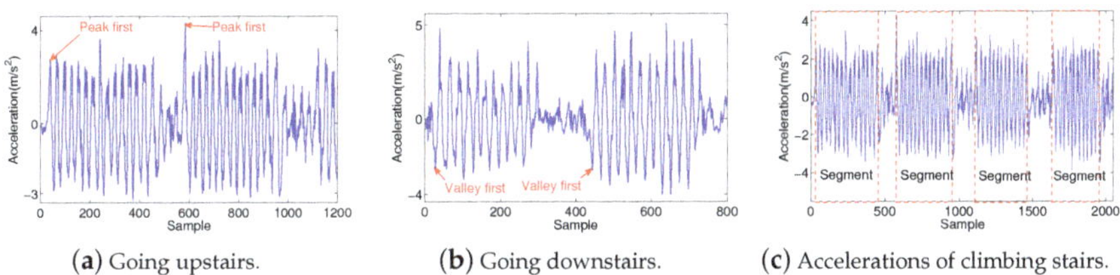

(**a**) Going upstairs.   (**b**) Going downstairs.   (**c**) Accelerations of climbing stairs.

Figure 7. Accelerations of climbing stairs.

Figure 8 demonstrates the acceleration profiles of taking an elevator. As shown in Figure 8a, when taking an elevator upstairs, we find that the trend of acceleration increases significantly at the beginning, then it returns to nearly zero and lasts a few samples; finally, it decreases and returns to about zero. The amplitude and duration time is quite similar between the increase–return process and the decrease–return process. Taking an elevator downstairs is the opposite of the upstairs process, as shown in Figure 8b. The profiles of taking an elevator to different floors are very similar. The only difference is the duration of the steady moving process, as shown in Figure 8c. These significantly distinguishing patterns can be distinguished by a simple Deterministic Finite Automaton (DFA) based algorithm: at the beginning, we use a low-pass filter to extract the acceleration whose spectrum is lower than 5 Hz. Then, use the step detecting algorithm in Montage [40] to detect and count steps. If we find that the difference between average amplitude of 8–15 successive steps and the following 2–5 steps is more than 1, we suppose that the user is climbing stairs. In addition, if one cannot detect any steps, we begin to detect the elevator patterns. Specifically, build a simple DFA to find the three segments of the acceleration, and compare the durations and maximum amplitudes of the first and the third segment; if their differences are lower than 5%, and the standard deviation of the second segment is lower than 0.3 and the average value is lower than 0.2, we suppose the user is taking an elevator.

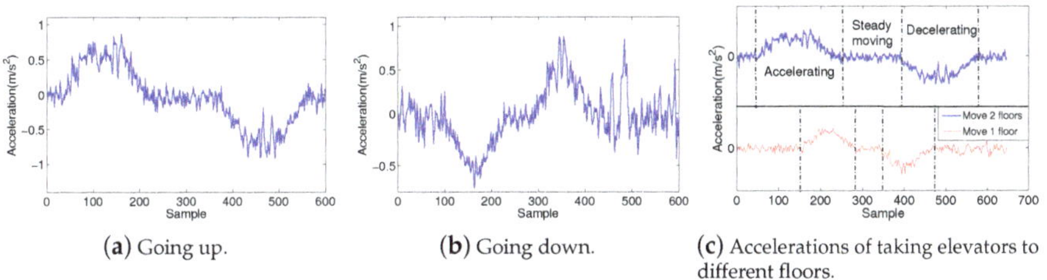

(**a**) Going up.   (**b**) Going down.   (**c**) Accelerations of taking elevators to different floors.

Figure 8. Accelerations of taking elevators.

Figure 9 shows different dimensions of ambient data from the users, including the vertical component of the geomagnetic field, the horizontal component of the geomagnetic field, signal strength of AP1, and signal strength of AP2. The color of each cell indicates the ambient value of each cell, and we can clearly find the data distribution of each dimension. In this experiment, we examine the features of the collected multi-dimensional traces. With the on-board accelerometer and gyroscope sensor readings, we utilize the dead reckoning method [40] to track the user's movement. Three pieces of data traces are collected from three different users. For these three users, they start from the same point, that is, the entrance of the floor, and share a common pathway. Then, they walk to their own seats.

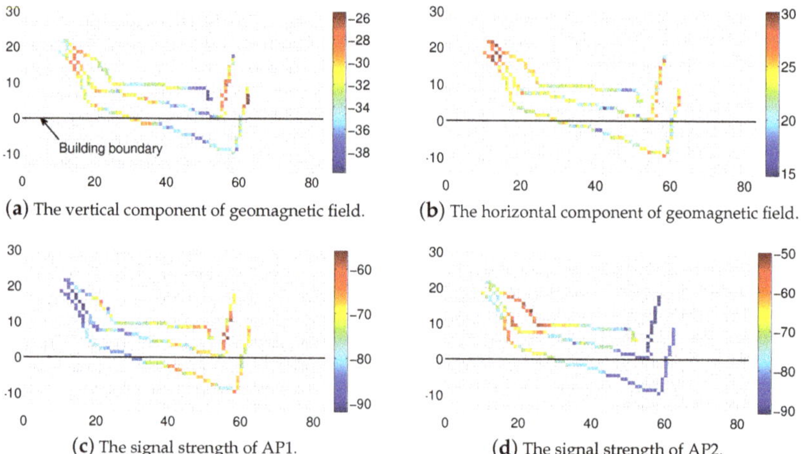

(**a**) The vertical component of geomagnetic field.

(**b**) The horizontal component of geomagnetic field.

(**c**) The signal strength of AP1.

(**d**) The signal strength of AP2.

Figure 9. The ambient data map on different dimensions.

We can see that the traces experience significant distortion on their locations. The recorded coordinates of some points are outside of the building boundary. In addition, it is easy to find that three traces exhibit nearly the same data change trend on the first half of the traces. Similar trends can be observed in the other three dimensions as well. It seems very likely that the three users share a common path on the first half of the trace that meets the ground truth. Thus, we are confident enough to conduct transformation on the three traces and move them together in the first half. In fact, it is hard to determine the concrete transformations just based on the information in the snapshot. In the integrated optimization, we need more information to make decisions such as the moving direction, scaling size, and the like.

Figure 10 shows the fusion results of the traces in Figure 9. The solid black line denotes the real route of three users. The optimized traces match the groudtruth, especially in the first half part. The latter half, however, still deviates from the real path. Through careful examination of the trace data, we find that the result is due to the lack of enough information for generating the optimization decision. It shall be noted that the accumulative error may deteriorate the dead reckoning performance, especially when the trajectory is long and has many turns. Although the end point of the trajectory has huge offset, each short segment (for indoor cases, the segments are usually not long) is not very erroneous (despite the initial offset). Otherwise, after segmentation, each segment's shape is always quite similar to the ground truth. Thus, we can use a 'small' similarity transformation to cancel the error. Actually, the length of the traces will not obviously affect the accuracy, except when all the paths are very short. In Figure 11, we show an integrated data map constructed by 11 different users after data alignment. We can find that the combined traces precisely sketch the structure of the underlying indoor map.

Finally, we summarize the location deviation in traces before and after the optimization, and the results are illustrated in Figure 12. In this experiment, we choose a 0.8 m × 0.8 m grid as the distance unit that equals the size of the floor tile. Nearly 50% of the trace points are aligned to their right locations and around 90% of points show deviations within two cells. Compared with the unaligned traces, the optimized traces achieve significant improvement on the accuracy.

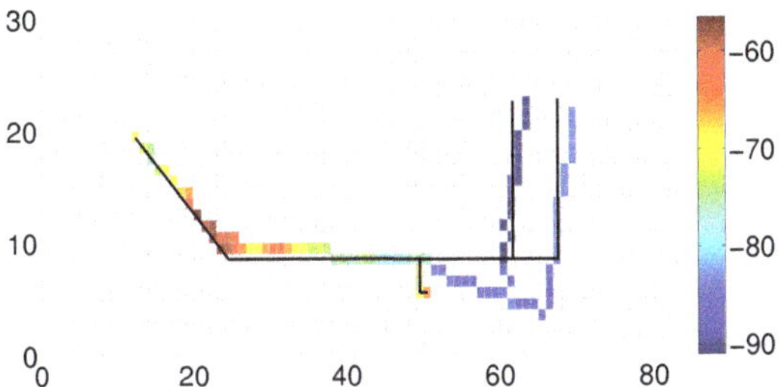

Figure 10. An example of the trace alignment result.

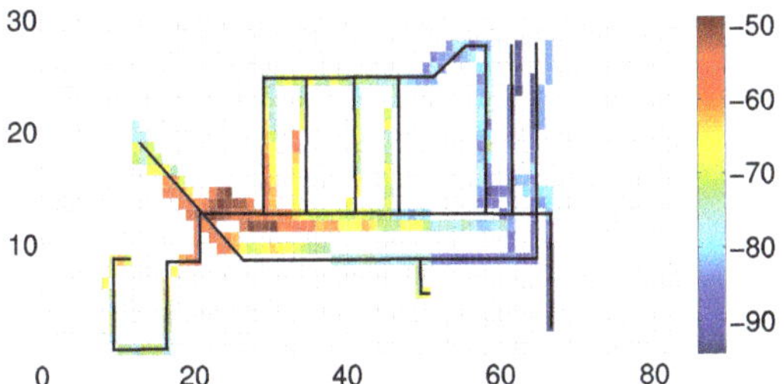

Figure 11. Map reconstruction with the aligned data traces.

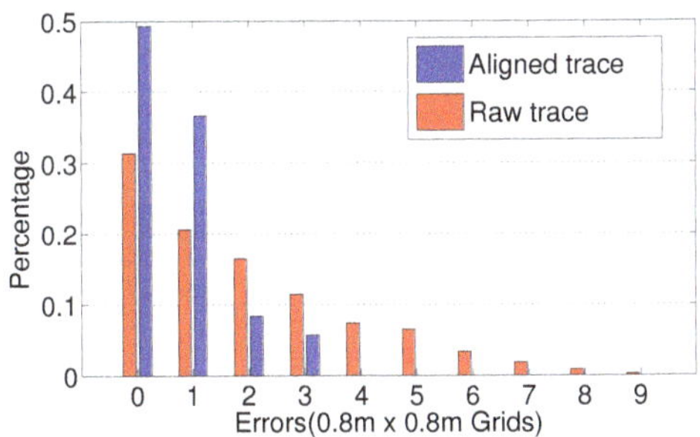

Figure 12. Location deviation of crowdsourced traces.

6. Conclusions

In this work, we extensively studied the multi-dimensional crowdsourced data fusion problem and proposed a novel method to fuse the noisy and distorted crowdsourced data from different types of sensory data together in different situations. We proposed a new trace coding and transformation approach to conduct the trace alignment, in order to deal with complex trace distortion issues. Moreover, we proposed an effective fusion scheme to find the crowdsourced trace alignment solution. Finally, we implemented the proposed approaches on different commercial devices and conducted experiments to verify the effectiveness of our method. Since our system is an initial and general framework, and the similar works are limited so far, we did not compare it with the dedicated systems in this paper. In our future work, we will spend more time to make a comparison to some similar works to demonstrate our merits and limitations, and improve our current work.

Author Contributions: The work described in this article is the collaborative development of all authors. Y.J. and B.L. contributed to the idea of the data fusion and alignment model, performed the experiments, and participated in the writing of the paper. Z.W. and X.Y. made contributions to data measurement and analysis.

Funding: This work was supported in part by the National Natural Science Foundation of China (61802288), the Major Technological Innovation Projects in Hubei Province (2019AAA024).

Acknowledgments: The authors are grateful for all of the valuable suggestions received during the course of this work.

Conflicts of Interest: The authors declare no conflict of interest.

References

1. Bahl, P.; Padmanabhan. V. RADAR: An in-building RF-based user location and tracking system. In Proceedings of the IEEE INFOCOM, Tel Aviv, Israel, 26–30 March 2000; pp. 775–784.
2. Varshavsky, A.; de Lara, E.; Hightower, J.; LaMarca, A.; Otsason, V. GSM indoor localization. *Pervasive Mob. Comput.* **2007**, *3*, 698–720. [CrossRef]
3. Chung, J.; Donahoe, M.; Schmandt, C.; Kim, I.; Razavai, P.; Wiseman, M. Indoor location sensing using geo-magnetism. In Proceedings of the 9th International Conference on Mobile Systems, Applications, and Services, Bethesda, MD, USA, 28 June–1 July 2011; pp. 141–154.
4. Matic, A.; Papliatseyeu, A.; Osmani, V; Mayora-Ibarra, O. Tuning to your position: FM radio based indoor localization with spontaneous recalibration. In Proceedings of the Pervasive Computing and Communications, Mannheim, Germany, 29 March–2 April 2010; pp. 153–161.
5. Jin, Y.; Soh, W.; Wong, W. Indoor localization with channel impulse response based fingerprint and nonparametric regression. *IEEE Trans. Wirel. Commun.* **2010**, *9*, 1120–1127. [CrossRef]
6. Liu, H.; Gan, Y.; Yang, J.; Sidhom, S.; Wang, Y.; Chen, Y.; Ye, F. Push the limit of WiFi based localization for smartphones. In Proceedings of the 18th Annual International Conference on Mobile Computing and Networking, Istanbul, Turkey, 22–26 August 2012; pp. 305–316.
7. Rai, A.; Chintalapudi, K.; Padmanabhan, V.; Sen, R. Zee: Zero-effort crowdsourcing for indoor localization. In Proceedings of the 18th Annual International Conference on Mobile Computing and Networking, Istanbul, Turkey, 22–26 August 2012; pp. 293–304.
8. Yang, Z.; Wu, C.; Liu, Y. Locating in fingerprint space: Wireless indoor localization with little human intervention. In Proceedings of the 18th Annual International Conference on Mobile Computing and Networking, Istanbul, Turkey, 22–26 August 2012; pp. 269–280.
9. Shen, G.; Chen, Z.; Zhang, P.; Moscibroda, T.; Zhang, Y. Walkie-Markie: Indoor pathway mapping made easy. In Proceedings of the 10th USENIX conference on Networked Systems Design and Implementation, Lombard, IL, USA, 2–5 April 2013; pp. 85–98.
10. Liu, D.; Guo, S.; Yang, Y.; Shi, Y.; Chen, M. Geomagnetism-Based Indoor Navigation by Offloading Strategy in NB-IoT. *IEEE Internet Things J.* **2019**, *6*, 4047–4084. [CrossRef]
11. Shao, W.; Zhao, F.; Wang, C.; Luo, H.; Zahid. T.; Wang, Q.; Li, D. Location Fingerprint Extraction for Magnetic Field Magnitude Based Indoor Positioning. *J. Sens.* **2016**, *2016*, 1945695. [CrossRef]
12. Zhuang, Y.; Syed, Z.; Li, Y.; El-sheimy, N. Evaluation of Two WiFi Positioning Systems Based on Autonomous Crowdsourcing of Handheld Devices for Indoor Navigation. *IEEE Trans. Mob. Comput.* **2016**, *15*, 1982–1995. [CrossRef]
13. Wang, F.; Feng, J.; Zhao, Y.; Zhang, X.; Zhang, S.; Han, J. Joint Activity Recognition and Indoor Localization With WiFi Fingerprints. *IEEE Access* **2019**, *7*, 80058–80068. [CrossRef]
14. Robertson, P.; Angermann, M.; Krach, B. Simultaneous Localization and Mapping for Pedestrians using only Foot-Mounted Inertial Sensors. In Proceedings of the 11th International Conference on Ubiquitous Computin, Orlando, FL, USA, 30 September–3 October 2009; pp. 83–96.
15. Robertson, P.; Angermann, M.; Khider, M. Improving simultaneous localization and mapping for pedestrian navigation and automatic mapping of buildings by using online human-based feature labeling. In Proceedings of the IEEE/ION Position, Location and Navigation Symposium, Indian Wells, CA, USA, 4–6 May 2010; pp. 365–374.
16. Ionut, C.; Xuan, B.; Martin, A.; Romit, C. Did you see Bob?: Human localization using mobile phones. In Proceedings of the Sixteenth Annual International Conference on Mobile Computing and Networking, Chicago, IL, USA, 20–24 September 2010; pp. 149–160.
17. Wang, H.; Sen, S.; Elgohary, A.; Farid, M.; Youssef, M.; Choudhury, R. No need to war-drive: Unsupervised indoor localization. In Proceedings of the 10th International Conference on Mobile Systems, Low Wood Bay, Lake District, UK, 25–29 June 2012; pp. 197–210.

18. Jiang, Y.; Liu, Y.; Li, Z. Data Fusion and Alignment for Location-Aware Crowdsourcing Applications. In Proceedings of the IEEE International Symposium on Dynamic Spectrum Access Networks, Newark, NJ, USA, 11–14 November 2019.

19. Lin, W.; Su, X.; Zhang, X.; Liu, W.; Li, B. Spee-Navi: An Indoor Navigation System via Speech Crowdsourcing. In Proceedings of the First, International Workshop on Human-centered Sensing, Networking, and Systems, Delft, The Netherlands, 5 November 2017; pp. 43–48.

20. Teng, X.; Guo, D.; Guo, Y.; Zhou, X.; Liu, Z. CloudNavi: Toward Ubiquitous Indoor Navigation Service with 3D Point Clouds. *ACM Trans. Sens. Netw.* **2019**, *15*, 1:1–1:28. [CrossRef]

21. Ricardo, S.; Marília, B.; Ricardo, L.; Hugo, G. Fingerprints and Floor Plans Construction for Indoor Localisation Based on Crowdsourcing. *Sensors* **2019**, *19*, 919.

22. Yang, J.; Zhao, X.; Li, Z. Crowdsourcing Indoor Positioning by Light-Weight Automatic Fingerprint Updating via Ensemble Learning. *IEEE Access* **2019**, *7*, 26255–26267. [CrossRef]

23. Jung, S.; Lee, S.; Han, D. A Crowdsourcing-based Global Indoor Positioning and Navigation System. *Pervasive Mob. Comput.* **2016**, *31*, 94–106. [CrossRef]

24. Wang, Y.; Jia, X.; Jin, Q.; Ma, J. Mobile crowdsourcing: Framework, challenges, and solutions. *Concurr. Comput. Pract. Exp.* **2017**, *29*, e3789. [CrossRef]

25. Zhou, P.; Zheng, Y.; Li, M. How long to wait?: Predicting bus arrival time with mobile phone based participatory sensing. In Proceedings of the 10th International Conference on Mobile Systems, Applications, and Services, Low Wood Bay, Lake District, UK, 25–29 June 2012; pp. 379–392.

26. Farkas, K.; Nagy, A.; Toms, T.; Szab, R. Participatory sensing based real-time public transport information service. In Proceedings of the 2014 IEEE International Conference on Pervasive Computing and Communication Workshops, Budapest, Hungary, 24–28 March 2014; pp. 141–144.

27. Rana, R.;Chou, C.; Bulusu, N.; Kanhere, S.; Hu, W. Ear-Phone: A context-aware noise mapping using smart phones. *Pervasive Mob. Comput.* **2015**, *17*, 1–22. [CrossRef]

28. Alzantot, M.; Youssef, M. CrowdInside: Automatic construction of indoor floor plans. In Proceedings of the SIGSPATIAL 2012 International Conference on Advances in Geographic Information Systems (formerly known as GIS), Redondo Beach, CA, USA, 7–9 November 2012; pp. 99–108.

29. Gao, R.; Zhao, M.; Ye, T.; Ye, F.; Wang, Y.; Bian, K.; Wang, T.; Li, X. Jigsaw: Indoor floor plan reconstruction via mobile crowdsensing. In Proceedings of the 20th Annual International Conference on Mobile Computing and Networking, Maui, HI, USA, 7–11 September 2014; pp. 249–260.

30. Chen, S.; Li, M.; Ren, K.; Qiao, C. Crowd Map: Accurate Reconstruction of Indoor Floor Plans from Crowdsourced Sensor-Rich Videos. In Proceedings of the 35th IEEE International Conference on Distributed Computing Systems, Columbus, OH, USA, 29 June–2 July 2015; pp. 1–10.

31. Wang, Y.; Car, Z.; Tong, X.; Gao, Y.; Yin, G. Truthful incentive mechanism with location privacy-preserving for mobile crowdsourcing systems. *Comput. Netw.* **2018**, *135*, 32–43. [CrossRef]

32. Wang, Z.; Hu, J.; Wang, Q.; Lv, R.; Wei, J.; Chen, H.; Niu, X. Task-Bundling-Based Incentive for Location-Dependent Mobile Crowdsourcing. *IEEE Commun. Mag.* **2019**, *57*, 54–59. [CrossRef]

33. Meng, Y.; Zhang, W.; Zhu, H.; Shen, X. Securing consumer IoT in the smart home: Architecture, challenges, and countermeasures. *IEEE Wirel. Commun.* **2018**, *25*, 53–59. [CrossRef]

34. Shu, J.; Liu, X.; Jia, X.; Yang, K.; Deng, R. Anonymous privacy-preserving task matching in crowdsourcing. *IEEE Internet Things J.* **2018**, *5*, 3068–3078. [CrossRef]

35. Liu, X.; Lu, M.; Ooi, B.; Shen, Y.; Wu, S.; Zhang, M.; Cdas: A crowdsourcing data analytics system. *Proc. VLDB Endow.* **2012**, *5*, 1040–1051. [CrossRef]

36. Zhang, P.; Chen, R.; Li, Y.; Niu, X.; Wang, L.; Li, M.; Pan, Y. A localization database establishment method based on crowdsourcing inertial sensor data and quality assessment criteria. *IEEE Internet Things J.* **2018**, *5*, 4764–4777. [CrossRef]

37. Daniel, F.; Kucherbaev, P.; Cappiello, C.; Benatallah, B.; Allahbakhsh, M. Quality control in crowdsourcing: A survey of quality attributes, assessment techniques, and assurance actions. *ACM Comput. Surv.* **2018**, *51*, 7:1–7:40. [CrossRef]

38. Hu, Q.; Wang, S.; Ma, P.; Cheng, X.; Lv, W.; Bie, R. Quality Control in Crowdsourcing Using Sequential Zero-Determinant Strategies. *IEEE Trans. Knowl. Data Eng.* **2019**. [CrossRef]

39. Khan, M. Approximation of Data Using Cubic Biezier Curve Least Square Fitting. Available online: http://130.235.212.213/trac/raw-attachment/wiki/BezierPaths/cubicbezierleastsquarefit.pdf (accessed on 14 October 2019).

40. Zhang, L.; Liu, K.; Jiang, Y.; Li, X.; Liu, Y.; Yang, P.; Li, Z. Montage: Combine Frames with Movement Continuity for Realtime Multi-User Tracking. *IEEE Trans. Mob. Comput.* **2017**, *16*, 1019–1031. [CrossRef]

Article

Continuous Authentication of Automotive Vehicles Using Inertial Measurement Units

Gianmarco Baldini [1,*], Filip Geib [1,2] and Raimondo Giuliani [1]

[1] European Commission, Joint Research Centre, 21027 Ispra, Italy; filip9geib@gmail.com (F.G.); raimondo.giuliani@ec.europa.eu (R.G.)
[2] Faculty of Electrical Engineering, Czech Technical University in Prague, 160 00 Praha, Czech Republic
* Correspondence: gianmarco.baldini@ec.europa.eu; Tel.: +39-0332-78-6618

Received: 14 October 2019; Accepted: 25 November 2019; Published: 30 November 2019

Abstract: The concept of Continuous Authentication is to authenticate an entity on the basis of a digital output generated in a continuous way by the entity itself. This concept has recently been applied in the literature for the continuous authentication of persons on the basis of intrinsic features extracted from the analysis of the digital output generated by wearable sensors worn by the subjects during their daily routine. This paper investigates the application of this concept to the continuous authentication of automotive vehicles, which is a novel concept in the literature and which could be used where conventional solutions based on cryptographic means could not be used. In this case, the Continuous Authentication concept is implemented using the digital output from Inertial Measurement Units (IMUs) mounted on the vehicle, while it is driving on a specific road path. Different analytical approaches based on the extraction of statistical features from the time domain representation or the use of frequency domain coefficients are compared and the results are presented for various conditions and road segments. The results show that it is possible to authenticate vehicles from the Inertial Measurement Unit (IMU) recordings with great accuracy for different road segments.

Keywords: security; authentication; Inertial Measurement Units; road transportation

---

1. Introduction

Authentication is an important function in security as it ensures that the identity of an entity is recognized by a system whose services the entity would like to access. Authentication is usually combined with other security functions like authorization where the entity, once authenticated, is authorized to access specific services or data. The verification of the identity can be implemented using different means including information known by the entity (e.g., login/password) based on what the entity is (e.g., biometric features). This information can be provided only once at the beginning of the interaction between the entity and the system or more than once either periodically or in a continuous way. The latter methods are considered more robust than one time login against external attacks because an attacker, who aims to impersonate the real entity, has to provide authentication information more than once, which is considerably more difficult [1,2]. On the other side, the request for an additional login/password after the initial login may negatively impact the usability of the human-machine interface [3]. In this context, the concept of continuous authentication has been mostly proposed in the literature for human authentication. It makes use of the physiological and behavioral biometrics using built-in sensors and accessories. For example, the research results in the literature have proven that persons can be uniquely identified by the way they use mobile devices [1,4], they type on a keyboard [5] or perform different activities. Details on the literature results are provided in the following paragraphs.

Continuous authentication can be performed in the application scenarios where the entity must access services to the system for a considerable amount of time and where the biometrics information can be collected from the authentication system. For example, smart city application scenarios could benefit from this approach [6]. Various techniques to implement continuous authentication for persons have been proposed and validated by the research community in recent times. While continuous authentication has been researched for human beings in recent years, there has been limited attention to other entities especially in the transportation domain (e.g., a study on the identification of transport modes rather than transport vehicles is Reference [7]). Details on the existing related work are presented in Section 2.

Our Contribution

This paper proposes a proof of concept on the application of the continuous authentication technique (used in the literature to identify human beings) to the road transportation domain and more specifically to the problem of device identification where the device is actually a vehicle (i.e., a car in this paper). This paper investigates the possibility of authenticating a car on the basis of its "biometrics" derived from the way the car responds to the irregularities (e.g., potholes, bumpers) of the road surface. The rationale is that the components (e.g., bumpers) of different cars provide a different response to the same segment of the road surface. Such responses can be evaluated by using the digital output provided by Inertial Measurement Units (IMU) and its components (e.g., accelerometers, gyroscope) installed in a car in a similar way that continuous authentication is implemented for human beings using wearable sensor or smartphones (which also contain IMUs). This paper presents the evaluation results for the application of techniques derived from the literature on continuous authentication. In particular, two different approaches are compared both from an identification accuracy and an execution time point of view. The results presented in this paper show that each vehicle can be uniquely identified with high accuracy by the analysis of the data collected by Inertial Measurement Unit (IMU) accelerometers and gyroscopes while the car is driving. To the knowledge of the authors, it is the first time that the concept of continuous authentication is applied to vehicles (i.e., car).

A potential scenario for the use of continuous authentication of vehicles (or cars as the two terms will be used with the same meaning in the rest of this paper) is based on Intelligent Transport Systems (ITS) applications, where the authentication of the vehicle is needed a number of times during the vehicle trip like "insurance as you drive" or electronic tolling. The continuous authentication can complement (as multi-factor) authentication based on cryptographic means or replace it when the set-up of a Public Key Infrastructure (PKI) is too costly or complex to achieve. Then, the motivation of the paper is to propose a different form of device (i.e., the car) identification and authentication when a cryptographic mean cannot be applied or in case of the compromise of the cryptographic system (e.g., PKI) and the presence of an invalid cryptographic information (e.g., certificate) in the vehicle. In other words, this approach can be applied in all the cases where a cryptographic mean is not available, either because it is not implemented or because it has been compromised. The authors agree that, in the majority of the cases, a cryptographic solution is the primary solution because of its proven effectiveness. In addition, it can be noted that there are also examples of regulations for commercial vehicles were data is manipulated and tampered even if the cryptography authentication is not compromised [8]. Because the form of authentication described in this paper is based on the data analysis, it can mitigate such issues. The proposed approach is based on the consideration that modern vehicles are increasingly equipped with a variety of sensors including IMU, which are used for a variety of purposes but mostly to improve transportation comfort and driveability [9]. Many authors have also proposed the use of smartphones mounted in the vehicle for easier accessibility to the data generated by the built-in IMU [10].

The deployment scenario is similar to the continuous authentication of a human-using behavioral metric. In an initial training phase, data samples are acquired through remote communication by an entity external (e.g., a cloud application) to the vehicle. Then, discriminating features are identified and extracted from the data. In a subsequent phase, an authentication system (either based on the same cloud application or connected to it) uses a smaller data subset and compares it against the training model for identification. The two key performance metrics are (a) the identification accuracy and (b) the processing time for authentication, which is proportional to the amount of the data, which must be processed. Because IMU data ca have hundreds of samples per second, a dimensionality reduction is necessary, while preserving the identification accuracy. In this paper, we used professional IMUs to record the data (while driving) originating from accelerometers and gyroscopes with high sample rates. In the subsequent analysis, the sampling rate is decreased to the rate commonly used by mass market smartphones.

In this initial study, the focus is to investigate the capability of authenticating the vehicle (i.e., car) in similar conditions and on the same road surface. In a practical application of this concept, different variables can influence the classification accuracy including the driver, different loads in vehicle (number of passengers), aging of the vehicle (e.g., tyre consumption) and different environmental conditions (e.g., rain, sunny weather). Each of these variables can influence the application of the method, but it is important to investigate the impact of each variable independently starting from a baseline. We also note that there is already an extensive literature on the identification of the driver and his/her driving style based on data collected by accelerometers (see References [7,11]) and it was not the intention to repeat such specific analysis here. Then, the focus of this paper was to establish a baseline where the most effective continuous authentication methods could be identified in a quantitative way. Future developments will evaluate the impact of each of the variables (see also Conclusions section).

The structure of this paper is as follows: Section 2 provides an overview of the related work on continuous authentication in all domains, authentication of vehicles and on the use of IMU to collect behavioral statistics in ITS. Section 3 describes the materials used in the experimental work. Section 4 describes the adopted methodology to collect the data and to perform the authentication. Section 5 provides the results of the implementation of the authentication where the impact of different hyper-parameters was quantitatively evaluated. Section 6 summarizes and discusses the key findings from the results and identify potential future developments.

2. Related Work

A very recent (2019) survey [12] reviews the techniques for continuous authentication using Internet of Things devices. Another review provides an overview of the state of art on continuous authentication using mobile devices [1]. Both surveys are specific to the problem of continuous authentication of human beings on the basis of behavioral bio-metrics rather than explicit and single phase authentication mechanisms like the password, a Personal Identification Number (PIN), or a secret pattern on the display of the smartphone. Both reviews highlight the weaknesses of the traditional approaches and the advantages of the continuous authentication approach. For example, passwords or PINs can be forgotten or systems may be vulnerable immediately after the initial login. In addition, some authentication mechanisms require the implementation of sophisticated Public Key Infrastructure (PKI)s, which are also vulnerable to attacks. To overcome these issues, the biometrics and security research communities have developed techniques for continuous authentication on mobile devices [1]. Most of these methods make use of physiological and behavioral biometrics, using built-in sensors and accessories such as the gyroscope, touch screen, accelerometer, orientation sensor and pressure sensor, to monitor and authenticate the identity of the user.

In this review of the state-of-the-art, we focus specifically on continuous authentication based on accelerometers and gyroscopes as these components are also used in this paper for vehicle authentication. This type of continuous authentication is based on the concept of identifying the gait of a human being while s(he) is performing a specific activity like walking, running or lifting an object. The data needed to perform the continuous authentication are often measured using the built-in components of a smartphone like accelerometers. Once the raw data are collected and measured, discriminative features are extracted, which are then fed into a classifier to distinguish users. This is a similar approach used in this paper. Various approaches are used in the literature and it is quite common to use either (or both) time domain and frequency domain representations of the recordings of the accelerometers and gyroscopes data. For example, the authors in Reference [13] extract statistical features from sensor readings, which are then used for user classification using a Support Vector Machine (SVM) machine learning algorithm. The statistical features used in [13] are mean, standard deviation, kurtosis, skewness, entropy and so on, both in the time domain and in the frequency domain. Then, from the methodology point of view, the approach used in Reference [13] is very similar to this paper with the fundamental difference that this paper focuses on vehicle authentication, which is a novel approach in the literature. In a similar way, the authors in Reference [14] have used the accelerometer data from a smartphone attached to a pocket of a person while walking. Each step was the gait, which could be used for authentication. An important element in the methodology was the segmentation of the accelerometer recordings to identify the repeated steps. Then, the identified segments were fed to a SVM classifier. The issue that persons are walking at different speeds was mitigated using Dynamic Time Warping (DTW). The appropriate segmentation of the recordings and the different speeds of the entity is also a problem addressed in this paper. A similar approach was also used by the authors in Reference [15], where the main coefficients of the frequency domain transform (based on FFT) of accelerometers data of walking persons were used for person authentication. Instead of using the Fast Fourier Transform (FFT), the authors have applied the Wavelet transform in Reference [16] in the context of a similar scenario of the previous papers: person authentications through mobile phone built-in accelerometers. Another recent paper, which presents continuous authentication of human beings using accelerometers recordings is Reference [17] which focuses on passive and continuous user authentication by analyzing and recognizing the unique characteristics of the physical activity patterns of human beings. A number of statistical features were extracted from the sensors recording—mean, standard deviation, skewness, kurtosis, energy and entropy in the time domain and frequency domain. Then, three machine learning algorithms (e.g., SVM, Random Forests and Decision Trees) were used for classification and authentication. As we have described in this section, there is an extensive literature on continuous authentication of human beings using accelerometers and gyroscopes and extraction of statistical features. The novel approach presented in this paper is to apply the techniques described above for the continuous authentication of vehicles which has never been investigated in the literature.

Authentication of vehicles is mostly based on cryptography means [18] and it has been applied to various interactions between vehicles and specific applications like the vehicle to the electric grid infrastructures in Reference [19] or for C-ITS applications [20]. In many cases, the authentication and authorization process is standardized as in the case of C-ITS domain [21–23]. Various cryptographic architectures are proposed for vehicular networks including PKI or Identity-based cryptography.

The use of IMU and their digital output in ITS applications has received considerable attention in recent years thanks to the increasing computing power of smartphones and their built-in IMU. On the other side, no papers have used such data to implement the continuous authentication of vehicles. The focus has been mostly on the estimation of the driver style (e.g., aggressive) using data from accelerometers in Reference [24] or (in a similar way) the driving behavior as in Reference [25]. A paper that is mostly similar to ours is Reference [7] where the transport mode (e.g., vehicle, bicycle) is identified on the basis of the digital output from IMUs, while this paper focuses on the identification of the specific vehicle in a specific transport mode.

Then, we can conclude that the idea of using the concepts of continuous authentication of human beings to vehicle is rather novel and it will be comprehensively explored in the rest of this paper.

3. Materials

The goal of the experiment was to obtain cars' driving characteristics under comparable driving conditions. We planned an Experimental Lap (EL) (in the rest of this paper, the term EL and lap will be used with the same meaning) inside the Joint Research Centre (JRC) area with conventional road rules and realistic traffic. This lap had a length of 2065 m and average driving time of 182.5 s (see Figure 1). To collect the measurements, every car had to drive this lap for at least twenty times in a row with just a tiny stop between each other (from 2 to 10 s). We distinguish and label the three types of points on the EL as follows—Orientation Point (OP) (e.g., any kind of left or right turn that we later used to separate the smaller parts of the EL), Speed Bump (SB) (e.g., four big speed bumps spread through the EL) and Road Feature (RF) (any roughness on the surface of the road, such as speed bump or an asphalt joint). Examples of those points with their descriptions and Global Navigation Satellite Systems (GNSS) positions are in Table 1. The disposition of the points is also marked in the map of the EL (see Figure 1), the photo documentation of the most significant ones is shown in Figure 2 and a description of the most significant points with the related latitude and longitude is provided in Table 1. The EL starts at $OP01$ and continues in clockwise.

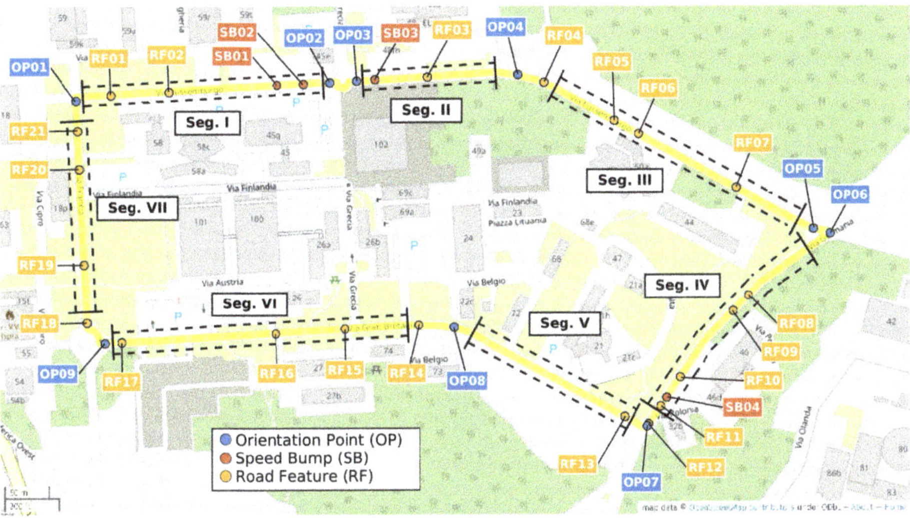

Figure 1. Map of the Experimental Lap (EL) with marked points and the most important segments.

Table 1. Description and GNSS position of points recognised in the EL.

Name	Description	GPS Lat.	GPS Lon.
SB01	sequence of ten rumble strips in the right driving lane	45.811832	8.627199
SB02	speed bump before the roundabout OP02	45.811845	8.627569
SB03	speed bump after the roundabout OP03	45.811883	8.628583
SB04	speed bump in front of the Visitor's center	45.808842	8.632676
RF06	small road fix only in the center of the road	45.811370	8.632282
RF15	long road patch through both driving lanes	45.809502	8.628149

For the IMU, we used the microelectromechanical system based motion tracker supplied by Xsens with the model number $MTi - 100 - 2A8G4$ (Xsens Technologies B.V., Enschede, The Netherlands). This sensor device is designed to measure the three axis acceleration and rate of turn in 2000 Hz sample rate and three axis magnetic field in 100 Hz sample rate. The IMU was mounted using a strong double sided foam tape at the same spot and orientation for every car. For details see Figure 3. We decided to place the sensor on the top of car's dashboard in the middle of the car. No measuring tools were used and the alignment was fully empirical. The x-axis of the IMU was always pointing towards the driving direction and z-axis in the vertical direction. The car position was also recorded using the GNSS receiver with the antenna mounted on the car's roof. The set used for this experiment consisted of twelve different cars. The specifications of the vehicles are listed in Table 2 and the photo documentation of those cars is in Figure 4. The same driver and co-driver were present in every car during the experimental data collection (i.e., driving the car) to minimise the potential bias of the driver. We highlight that half of the used vehicles were of the same model (Panda Active). The choice of such datasets with a predominance of vehicles of the same model and brand was intentional. While, this aspect is not quite relevant in the continuous authentication of human beings (because each human being is unique), this aspect is quite relevant in device authentication as discussed in the related literature [26], where it is discussed how intra-model classification is more difficult than inter-model classification (because components of the same model are the same or similar). To clarify—intra-model classification is when the devices are of the same model and brand (e.g., mobile phones of brand/model Sony Experia or Fiat Panda as in this scenario), while inter-model classification is when devices are of a different brand and/or models. We wanted to target the most difficult problem of intra-model classification rather than inter-model. This is why we have used a majority of vehicles of the same brand and mostly of the same model.

Figure 2. Photo documentation of the most significant points detected on the EL.

Table 2. Order and specifications of used cars.

Car	Manufacturer	Model	Gen.	Version
1	Fiat Automobiles	Panda	2nd	Active
2	Fiat Automobiles	Panda	2nd	Active
3	Fiat Automobiles	Panda	2nd	Active
4	Fiat Automobiles	Panda	2nd	Active
5	Fiat Automobiles	Panda	2nd	Active
6	Fiat Automobiles	Panda	2nd	Active
7	Fiat Automobiles	Punto	2nd	3-door
8	Fiat Automobiles	Doblo	1st	Facelift
9	Fiat Automobiles	Tipo	3rd	Hatchback
10	Mitsubishi	Colt	6th	CZ3
11	Škoda Auto	Octavia	3rd	Estate
12	Mazda Motor Corp.	Mazda3	4th	Hatchback

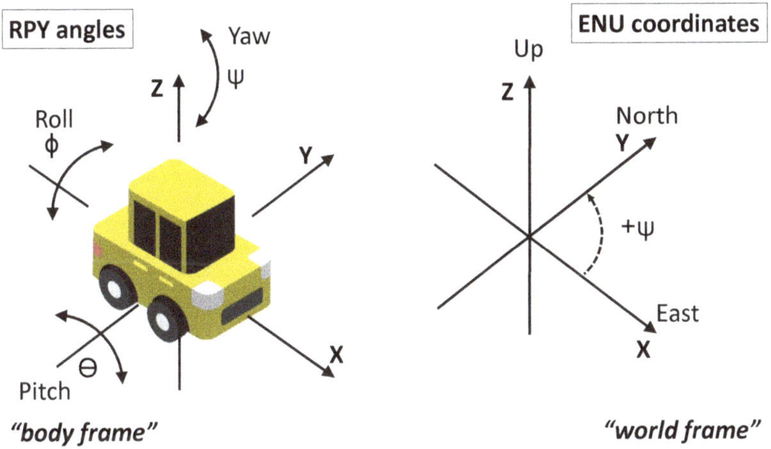

Figure 3. Alignment and orientation of IMU used during measurements.

Figure 4. Photo documentation of the cars set used in this paper.

4. Methodology

4.1. Workflow

A pictorial description of the methodology workflow is provided in Figure 5, where each step is described in detail in the following subsections.

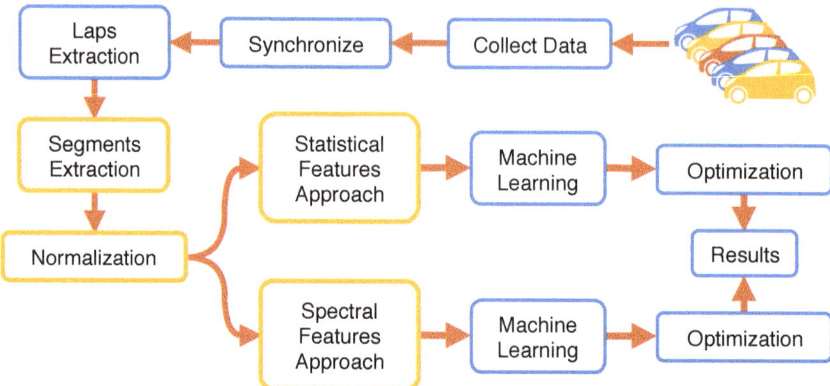

Figure 5. Methodology workflow.

4.2. Data Collection

The first step is to collect the data from each vehicle. As described in the Section 3, the same driver was used for all the recordings. The driving was performed in different environmental conditions (e.g., sunny days, rainy days) always on the same loop for 20 times. The data were recorded from the Xsens IMU with a sample rate of 2000 Hz, which is much higher than the sample rate available in commercial smartphones. For this reason, the sample rate was decimated to a lower sample rate as described in the Section 5. The car was driven at different speeds in the EL because of the varying traffic conditions.

4.3. Synchronization and Laps Extraction

The second step is to synchronize the data recordings and to extract the identified laps. To facilitate this step, each lap driving was performed in the following way—the car was stationary on *OP*01 at the beginning of each lap with a running engine for 10 s. This initial Background Noise Window (BNW) segment was used for laps synchronization and extraction. To compute an overall motion of the car the Kalman filter is used to estimate three dimensional angular velocity from the gyroscope sampled to 20 Hz. The variance computed from the BNW is used as a filter parameter for the noise reduction. Then the computed angular velocity is normalized to only one dimension and the one-dimensional convolution is applied. The convolution matrix is normalized to 20 elements long (due to the 20 Hz down-sample frequency). The start and end of each lap is calculated using the value of the angular velocity, when it is less than the threshold for the specific time defined above. The output of this procedure is a time table with starting and ending times of each lap. A description of the pseudo algorithm is shown in Algorithm 1. The visualization of the synchronization and laps extraction for the first two laps is shown in Figure 6.

Algorithm 1: Detect laps and compensate different alignment for one car.

Function `veloMean`(*data*):

$$delta = \sum_{(a,b)\in\{(x,y),(x,z),(y,z)\}} abs(a - b);$$

  return mean(norm(*data*), *delta*);

Function `bound2lap`(*bound*):

  start = detectStarts(*bound*);

  ends = detectEnds(*bound*);

  return timetable(*start*, *ends*);

Input:

  data = IMU raw data at 2000 *Hz*

  bnw = BNW part of IMU raw data

Main:

velo = downsample(*data*, 20*Hz*);

vari = variation(*bnw*);

velo = kalmanFilter(*velo*,*vari*);

velo = veloMean(*velo*);

$convoMat = [\ (\frac{1}{20})_1, (\frac{1}{20})_2, \ldots, (\frac{1}{20})_{20}\ ];$

convo = convolution(*velo*, *convoMat*);

thres = mean(*velo*) + std(*velo*);

bound = *convo* < *thres*;

lapsTimetable = bound2lap(*bound*);

quater = findQuaternionRot(*bnw*, *vari*);

quater = mean(*quater*);

data = quaternionRot(*data*, *quater*);

return *data*, *velo*, *lapsTimetable*;

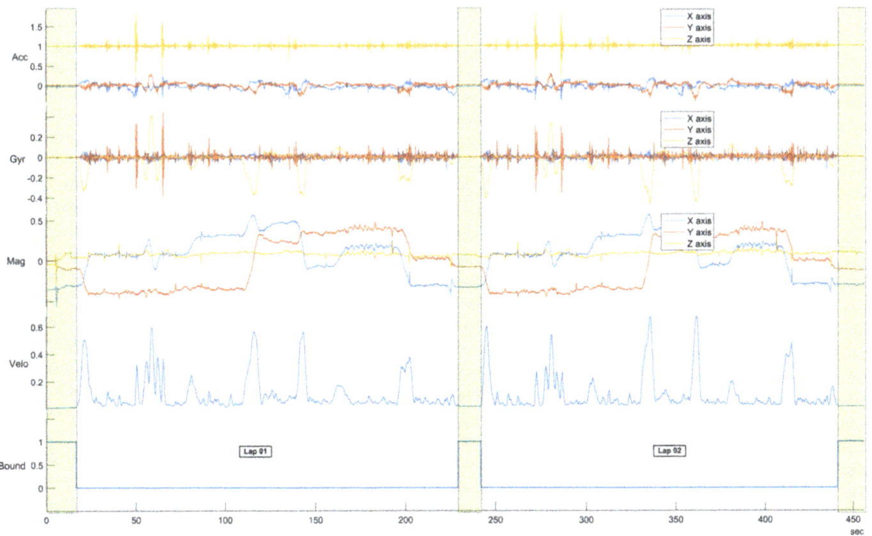

Figure 6. Detailed representation of the first two laps after synchronization. This plot shows that the proposed synchronization and laps extraction approach does not lose data between the laps (area highlighted with a green colour).

4.4. Segments Extraction

In the following step, the different road segments are extracted from each lap. The goal is to divide the entire lap in specific segments on which the classification process for vehicle continuous authentication is performed. The issue is already known in literature for continuous authentication of persons [13,14] and it is obviously based on the consideration that the speed of the car will never be exactly the same across laps in a similar way to a person walking or moving at different speeds. A possible solution to the different cars' speed problem across laps is to re-sample the laps records to have the same number of data points. The result of such a squeezing effect is similar to the case in which every car has the same speed pattern in every lap. It is important to divide the lap into many smaller parts similar for every car and to squeeze them separately. For detecting those parts we took advantage of the shape of the EL with many curves. Every one of them and also the readings between them are considered as a separate parts.

To detect turns, the z-axis of angular velocity is used. It is negative during the right turn, positive during the left turn and very close to zero when the car drives straight. One dimensional convolution is applied. The turns are detected separately in every lap of current car to minimize the potential speed biases between those laps. The threshold is set on the sum of median and one fourth of standard deviation of the velocity in the current lap. Every velocity value between the negative threshold and the positive threshold is considered as a turn. This method also detects many smaller turns which are invalid for the next usage. The second level of detection is applied to erase those small turns. For every turn, the area between its velocity and threshold is computed. If this area is smaller than the length of this threshold, it is marked as a false turn and erased. The output of this procedure is a time table with the starting times of each detected segment for each lap. We were able to detect sixteen of them. For more information, see Figure 7. For the description of the pseudo algorithm, see Algorithm 2.

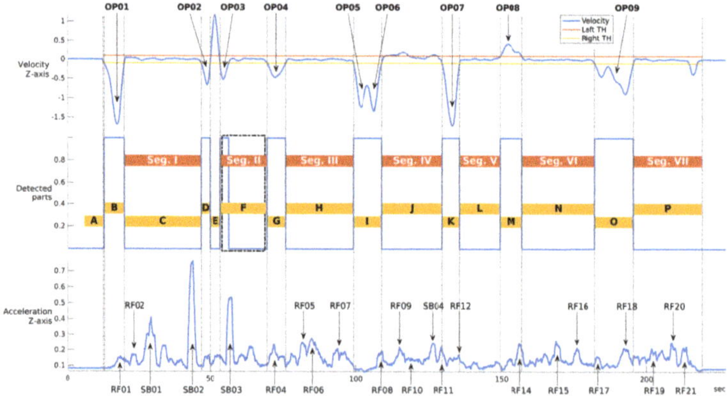

Figure 7. An example of the segments identified in a lap of the first car. The first plot shows the convoluted z-axis of the axial velocity with its positive (right turn) and negative (left turn) threshold. The Operational Point (OP) positions from Figure 1 are marked above in form of text arrows. The detected turns are shown at the second subplot (1 means turn and 0 means straight drive). The vertical gray lines separates different segments which are individually marked by yellow rectangles with alphabetical letters. Those letters represents segment names and stands for: A-Start, B-StartTurn, C-FastFirstBump, D-PreRound, E-RoundOne, F-SecodBump, G-RightCurve, H-WindowOne, I-CrossOne, J-VisitBump, K-CrossTwo, L-WindowTwo, M-LeftCurve, N-WindowThree, O-RoundTwo, P-WindowFour. The orange rectangles identify the seven segments used for the machine learning classification. The last subplot shows the z-axis of the acceleration with the identified EL road landmarks from the Figure.

Algorithm 2: Detect laps parts for one car.

Input:
   | *data* = compen. *data* from Algorithm 1
   | *velo* = z-axis of *velo* from Algorithm 1
   | *laps* = *lapsTimetable* from Algorithm 1
Main:

bound = Null · [size(*velo*)];
velo = downsample(*velo*, 20*Hz*);
convoMat = [$(\frac{1}{20})_1, (\frac{1}{20})_2, \ldots, (\frac{1}{20})_{20}$];
velo = convolution(*velo*, *convoMat*);
for *lap* **in** *laps* **do**
   | *tmp* = *velo.lap*;
   | *thres* = median(*tmp*) + $\frac{1}{4}$ std(*tmp*);
   | *thresRA* = max(*tmp*) - *thres*;
   | *turns* = -*thres* < *tmp* < *thres*;
   | *turns* = removeRA(*turns*, *thresRA*);
   | **for** *turn* **in** *turns* **do**
   | | *area* = curvesArea(*turn*, ± *thres*);
   | | *thresArea* = size(*turn*) · ± *thres*;
   | | **if** *area* < *thresArea* **then**
   | | | *turns* = removeTurn(*tun*);
   | | **end**
   | **end**
   | *turns* = removeTime(*tuns*, *last*10*s*);
   | *bound.lap* = *turns*;
end
partsTimetable = bound2part(*bound*, *laps*);
return *partsTimetable*;

4.5. Segments Records Re-Sampling

For every segment, its longest occurrence is found between records of every car and its every lap. This maximal segment is used as a reference and re-sampled directly to the desired frequency. Then the algorithm executes throughout the rest of the segment occurrences (for all cars and laps) and re-sample them with individually computed sample rates so that it will match the length of the already re-sampled reference segment. Because the longest segment was chosen as the reference, the computed sample rate are always higher than the desired sample rate. For the description of the pseudo algorithm, see Algorithm 3. The sample rate of our measurements (i.e., 2000 Hz) determined the lower and upper frequency boundaries of re-sampling approach. We found that frequencies below 50 Hz were not enough to hold usable data quality for classification. On the contrary, frequencies above 500 Hz were too high for down-sampling only and required up-sampling for some particular segments. In addition, we note that a sample rate above 500 Hz is unrealistic for most of the mobile phones available in the market.

Algorithm 3: Squeeze separately data parts of all cars to lowest desired frequency.

Input:

    $freq$ = lowest desired output frequency in *Hz*;   *cars* = list of cars

    data = compensated *data* from Algorithm 1 merged for all cars

    partsTimetable = *partsTimetable* from Algorithm 2 merged for all cars

    lapsTimetable = *lapsTimetable* from Algorithm 1 merged for all cars

Main:

```
for part in partsTimetable do
    maxPartSize = findGlobalMaximalPartSize(partsTimetable,part);
    for car in cars do
        for lap in lapsTimetable do
            freqResa = (freq · maxPartSize) / size(data.car.lap.part);
            data.car.lap.part = resample(data.car.lap.part,freqResa,'MEAN');
        end
    end
    minPartSize = findGlobalMinimalPartSize(partsTimetable,part);
    for car in cars do
        for lap in lapsTimetable do
            if minPartSize < size(data.car.lap.part) then
                data.car.lap.part[from minPartSize + 1 to end] = null;
            end
        end
    end
end
return data;
```

4.6. Description of the Classification Techniques

After the alignment and synchronization, the classification of the vehicles is performed using two different techniques. The first technique is based on the extraction of statistical features and it is presented in Section 4.6.1. The second technique is based on the application of the FFT transform and the analysis is conducted in the spectral domain representation and it is described in Section 4.6.2.

4.6.1. Statistical Features Approach

To perform dimensionality reduction, statistical features are applied to each segment. Such statistical features are commonly applied to time series classification and in particular for continuous authentication as described in the related work in References [13,17]. The list of applied statistical features is shown in Table 3.

Table 3. Statistical Features used for dimensionality reduction.

Feature Id	Feature Description
1	Variance
2	Skewness
3	Kurtosis
4	Shannon Entropy
5	Wavelet Spectral Shannon Entropy
6	Wavelet Spectral Log Entropy
7	Permutation Entropy with Embedding dimension 3
8	Permutation Entropy with Embedding dimension 4
9	Aproximate Entropy
10	Distribution Entropy

A brief description of the statistical features is provided here. Variance, Skewness and Kurtosis are moments of different levels of a specific quantitative measure of the shape of a function. Variance (second central moment) is the expectation of the squared deviation of a random variable from its mean. Skewness (third moment) is a measure of the asymmetry of the probability distribution of a real-valued random variable about its mean. Kurtosis (fourth moment) is a measure of the tailedness of the probability distribution of a real-valued random variable. The entropy state function is simply the amount of information (in the Shannon sense) that would be needed to specify the full microstate of the system and it can be used as a statistical measure of randomness. The Wavelet Spectral Shannon Entropy is the Shannon entropy calculated in the spectral representation of the IMU recording transformed using the wavelet transform. The Wavelet Spectral Log Entropy is the entropy calculated using a logarithmic scale on the spectral representation of the IMU recording transformed using the wavelet transform. The Permutation Entropy (PE) was initially proposed in Reference [27] and it has been applied in many domains from healthcare, to analyze EEG recordings [28], to the detection of mechanical faults [29]. PE [27] is a complexity measure for a time series x of T elements and an embedding dimension $D \geq 2$. The time series is embedded to a D-dimensional space $X_t = \{x(t), x(t+1), ..., x(t+D-1)\}$, with t ranging from 1 to $T - D + 1$. Given an ordinal set $R_D = \{r1, r2, ..., rD\}$, where $r1 < r2 < ... < rD$, there are $D!$ permutations π_i, with i ranging from 1 to $D!$. Then, X_t is mapped to π_i, so that $\{x(t), x(t+1), ..., x(t+D-1)\} \mapsto \{r1, r2, ..., rD\}$ and $\{x(t) \leq x(t+1) \leq ... \leq x(t+D-1)\}$. Probabilities p_i of each π_i are calculated as the number of occurrences of each π_i out of the size of the dimensional space $T - D + 1$. The permutation entropy of order D is then calculated with the formula $PE = -\sum_{i=1}^{D!} p_i \times \log p_i$, while the normalised permutation entropy is $NPE = PE/\log(D!)$. As the embedding dimension $D \geq 2$ is an important factor, two values of it has been used in our evaluation. Higher values are not considered because of the limitation on the size of the segments, which are imposed by the equations described above. Lower values are also not significant. Approximate entropy (ApEn) is a recently developed statistic quantifying regularity and complexity, which appears to have potential application to a wide variety of relatively short (greater than 100 points) and noisy time-series data (as in this case, because IMU recordings can be noisy). Distribution entropy (DistEn) is also a recent entropy measure, which has been developed based upon the probability density of vector-to-vector distances in state space. Further details on Distribution Entropy are presented in Reference [30].

It is not known a priori which features are more relevant for classification and a feature selection process is adopted. An additional reason to reduce the number of features used for classification is related to the curse of dimensionality in machine learning. Because a small dataset (20 paths for 12 vehicles) is used in this analysis, a limited number of features should be used for classification to avoid the curse of the dimensionality [31]. In this paper, we adopt the rule of thumb described in Reference [32] which states that there should be at least 5 samples for dimension. In this case, we adopt a number of features in the range of 4 to 6 (20 samples for each car divided by 5 is equal to 4). Experimental evaluation also shows that the use of all the features decreases the accuracy (e.g., 0.818 using all features against 0.832 with the 4 best features with sample rate of 250 Hz and $Gyro_y$). Because of its simplicity and time efficiency, in this paper, we adopt the ReliefF algorithm, which is part of the family of wrapper feature selection algorithms. The ReliefF algorithm [33] is based on the estimate of the quality of attributes according to how well their values distinguish between instances that are near to each other. The estimate can be implemented using the K-Nearest Neighbors (KNN) algorithm. Further details are presented in Reference [33].

4.6.2. Spectral Domain Approach

The other technique is based on the transformation of the signal from the IMU recording in the frequency domain using the FFT. Then, the frequency domain representation is divided in segments (i.e., frequency bands) with a number N_{freq}, which is one of the hyper-parameters identified in the Section 5. The Root Mean Square (RMS) of each frequency band is calculated to obtain N_{freq}

features. This approach is similar to the continuous authentication approaches for human beings, which are described in the related work in References [13,17]. The rational for this approach in vehicle authentication is that the main components of a vehicle (e.g., tyres, absorbers and wheels) will have smoother or harder reactions to road irregularities depending on the vehicle model. The smoother or harder response corresponds respectively to a lower occupation of the frequency bands or a higher occupation of the frequency bands. This can be seen from Figure 8, which shows the amplitude frequency representation with $N_{freq} = 6$ of each vehicle for the same segment I using a sample rate of 500 Hz and $Gyro_y$. It can be seen from the Figure that the first 6 vehicles (all Pandas) have a similar distribution of the amplitude coefficients, while the other vehicles have a different distribution. In more detail, the amplitude of the second frequency component is only slighter less than the first frequency component, while there is wide gap between the amplitude of the first and second frequency components in the vehicle number 12, which is of a completely different brand and model and more importantly of a different class—sport car. On the other side, the frequency response can depend by many different factors including the speed of the vehicle and how it can change from one lap to another. Then, the goal is to identify the optimal hyper-parameters, which can optimize the vehicle classification regardless of the vehicle speed. Note that the frequency representation of a time series is complex and it is not known a priori which components of the time series representation are useful for classification. As shown in the results Section 5, the magnitude component of the frequency domain representation is much more relevant for classification than the phase component. Because of these reasons, the amplitude component of the FFT transform is used. It could be argued that other spectral representations apart from the one based on FFT could also provide a good classification accuracy. This paper has also used a wavelet based representation but the results are slightly worse than the frequency domain representation as shown in Section 5.

4.7. Machine Learning

The authors of this paper used different machine learning algorithms to produce the results shown in Section 5. Considering that it is a small subset of data, only shallow machine learning algorithms have been used. A brief description of the algorithms is provided here.

- K-Nearest Neighbors (KNN) classifies a data sample (called a query point) based on the labels of the near data samples. Different functions can be used to determine how near or distant are the nodes. The most common function is the euclidean distance but other distance metrics can be used like the Mahalanobis or Minkowski distance. The advantage of the KNN is that it is computationally efficient and it does not need high computational power in the training phase, while the classification phase could be more computational intensive than other algorithms. Apart from the distance metric, the K parameter must also be optimized.
- Decision Trees, where the algorithm iterates through the input data by using the features properties to reach a specific category, which is more similar to the labeled data. The implementation of decision trees is usually very simple and fast if the data is well structured. An additional advantage is that they performs well even with high dimensional datasets. The disadvantages are the long training time and that the orders of the features in tree nodes have adverse effect on performance. In this paper, the maximum number of splits is the hyper-parameter, which must be optimized. Because individual decision trees tender to overfit, ensemble methods are also used. In particular, we have used the Random Forest algorithm (bagged decision trees) and the Ada Boost algorithm as well. The Ada Boost algorithm has been used in combination with the Decision Tree.

- SVM is a supervised algorithm, which learns to classify the data points (e.g., originating from the observables), from the labeled training samples (e.g., the reference fingerprints). SVM separates the labeled set in two areas on a multi-dimensional surface by using a separating function, which can be of different types—linear, Radial Basis Function (RBF), polynomial, sigmoidal are the most common. Because the multi-dimensional surface is divided in two areas, SVM is a binary classifier and it can be directly used to distinguish between two mobile phones or for validation (to validate that the claimed identity of a mobile phone). The extension of SVM to multi-classifier identification has been proposed by various authors. In this paper the OneVsOne approach is used. The RBF kernel is used and the RBF scaling factor γ must be optimized together with the C factor of the SVM algorithm.

Two main classification metrics are used—(a) the confusion matrices where each row of the matrix represents the instances in a predicted class while each column represents the instances in an actual class and (b) the identification accuracy, which is the sum of the true positive plus the true negatives divided for the total number of samples.

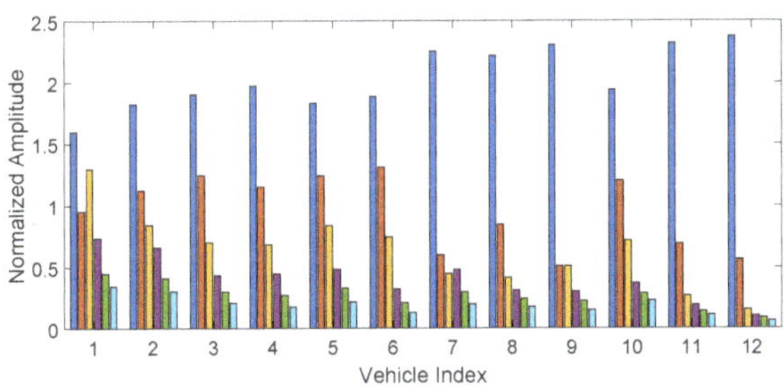

Figure 8. Amplitude of the components ($N_{freq} = 6$) in the frequency representation for each of the 12 vehicles using Gyroscope Y and the 500 Hz sample rate.

5. Results

5.1. Hyper-Parameters to Optimize

This section describes the results of the methodology proposed in the previous sections using the two main different techniques—the first technique based on the use of statistical features and the second technique based on the frequency domain (i.e., spectral approach). Both techniques were applied to the segments of the path driven by the vehicles. Various hyper-parameters are present in the analysis and they are further described here:

- the specific segment of the path driven by the vehicle. Each segment described in Figure 1 has its own characteristics and it is interesting to evaluate how the characteristics of each segment may impact the vehicle classification. The analysis is conducted on the seven segments identified in Figures 1 and 7.
- the IMU component and the direction element (X, Y, Z). An initial selection is based on the consideration that only directions which are directly related to the surface of the road can be used for vehicle authentication as the other directions are biased by the latitudinal and longitudinal movements of the vehicle (e.g., when turning a bend). Then, the optimal directions would be Accelerometer in the Z direction (i.e., the vertical direction) and the Gyroscope in the X and Y directions (the horizontal directions).
- the sample rate of the IMU recordings on which the feature extraction or the spectral domain is applied.

- another relevant choice in the features-based approach is the selection of the statistical features. Because it is a small dataset (20 paths for 12 vehicles), a limited number of features should be used for classification to avoid the curse of the dimensionality [31]. In this paper, a wrapper feature selection approach is used based on the RelieFF algorithm [33] to select the optimal subset of features.
- in the spectral (i.e., frequency domain) based approach, there is also the need to select a limited number of features (i.e., the spectral bands). In this case, the selection is on the number of bands. As shown in the subsequent sections, the RelieFF approach is also used for a large number of frequency bands.
- the hyper-parameters used in the machine learning algorithms used to perform the classification must also be optimized (e.g., K in the KNN algorithm).

The comparison of the hyper-parameters will be described for each technique—statistical features Section 5.2 and spectral domain Section 5.3.

5.2. Statistical Features Approach

Here we provide the results for the application of the statistical features approach. As mentioned before, one of the hyper-parameters (common to both to the features approach and the spectral approach) is the choice of the component and the direction.

Figure 9 shows a comparison of the different components (e.g., accelerometers and gyroscopes) and directions (e.g., Z, Y) for the application of the statistical features approach on the basis of different sample rates. The SVM machine learning algorithm was used. As discussed before, only the Accelerometer in the Z direction (Acc_z in the rest of this paper), the Gyroscope in the X direction ($Gyro_x$ in the rest of this paper) and the Gyroscope in the Y direction ($Gyro_y$ in the rest of this paper) are considered. As discussed before, the direction has been adjusted from the initial recordings to be in the direction of travel of the vehicle. Figure 9 shows that the highest classification accuracy is obtained by using $Gyro_y$, which is reasonable, because the pitch of the vehicle is more strongly stimulated by irregularities in the road surface. Figure 9 also shows that the accuracy improves with the increase of the sample rate. This result may be explained by the consideration that a higher sample rate provides more details and more discriminating power in the application of the features or the spectral transform with the machine learning classification. Evidence from studies on device identification in general supports this hypothesis [26]. As the classification is higher with $Gyro_y$, the subsequent results are based on $Gyro_y$.

In the rest of this section, it is discussed more in detail how the statistical features were selected to produce the results of Figure 9.

As discussed in the methodology Section 4, the approach based on the statistical features uses the 10 features described in Section 3, but a subset of features is selected using the RelieFF algorithm, where the best four features are selected. The number four was chosen because it was found out that other features do not contribute significantly on the basis of the weight ranking and to avoid the curse of dimensionality because of the small sampling set.

The histogram in Figure 10 shows the occurrences of the best four ranked features of RelieFF across all the best seven selected segments and the different samples rates (i.e., for each segment and each sample rate the highest four ranking features are selected) for $Gyro_y$. The histogram shows that the most relevant features are—feature 1 (Variance), feature 8 (Permutation Entropy with $M = 4$) and feature 9 (Approximate Entropy).

A visual representation on how the features identified from Figure 10 are relevant for the classification of the specific vehicles is shown in Figure 11, where a scatter plot for three selected features mentioned above (i.e., Variance, Permutation Entropy and Approximate Entropy) is shown. The scatter plot is based on Gyroscope Y with a sample rate of 500 Hz. Each of the points shown in the scatter plot of Figure 11 represents each of the EL for each vehicle.

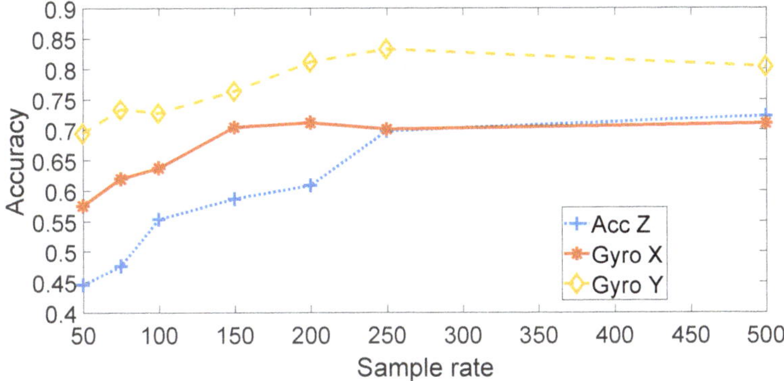

Figure 9. Comparison of different elements and directions of the IMU with feature based approach (sample rate in Hz).

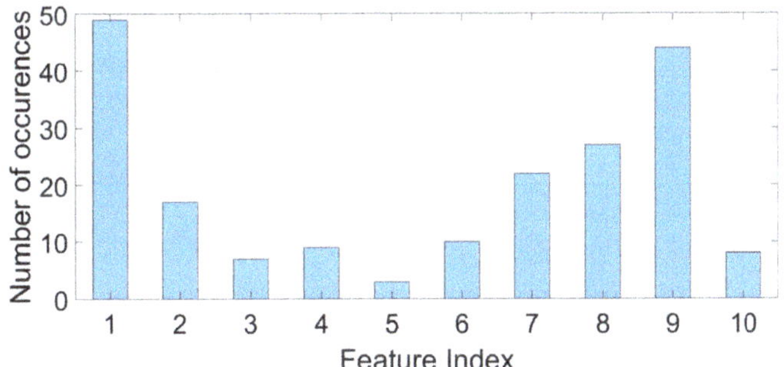

Figure 10. Histogram of the features across all the segments and sample rates.

Figure 12 shows the detail on the accuracy obtained foreach segment using $Gyro_y$ and SVM with optimized hyper-parameters ($\gamma = 2^3$ and $C = 2^7$). The average values across segments are also indicated with a transparent bar for each sample rate. The Figure 12 provides two significant results—the first is that the sample rate greatly impacts the classification accuracy. a higher sample rate provides more discriminative value than a lower sample rate as the identification accuracy increase steadily with the sample rate from 50 Hz to sample rate equal to 250 Hz. Then, an even higher sample rate does not improve the identification accuracy. This is an important result, because it shows that there is no need to use very high sample rates, which would not be practical for a deployment of continuous authentication using smartphones (i.e., current mass market smartphones have a sample rate around 100 Hz–200 Hz). The second result is that the choice of the segment can impact the identification accuracy. The Figure shows that the identification accuracy can vary greatly among the different segments and these results for the statistical features approach (a similar analysis is provided below for the spectral approach) can provide an insight into which road segments can be preferred in this authentication approach. For example, it can be seen that the segment III provides across all the sample rates a lower identification accuracy than the other segments. If we compare the results in Figure 12 with the results in Figure 7, it can be seen that segment II provides in general a higher identification accuracy (especially with sample rates at 150 Hz and 200 Hz) which can be related to a higher variance of the accelerometer in the Z direction due to the presence of irregularities of the road surface including the speed bump SB03. Then, we can conclude that the presence of irregularities on the road surface or a greater road roughness can provide a higher identification accuracy. This also means that smooth surfaces like a highway may provide a lower identification accuracy. Segment III, which does not provide a high classification accuracy, does not have specific road surface irregularities (see the map in Figure 1) and it could be considered similar to the Highway case.

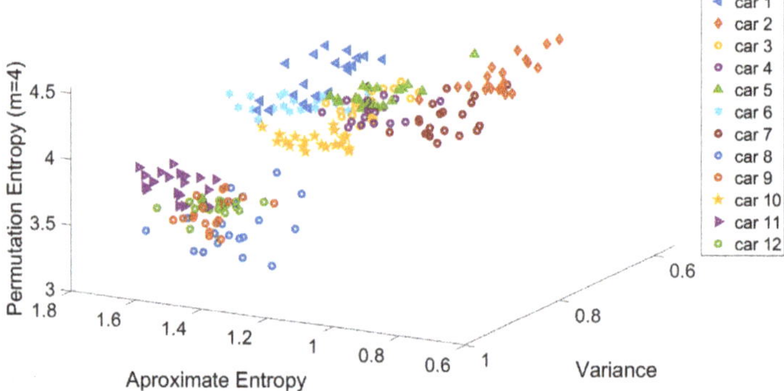

Figure 11. Scatter plot based on the three main features derived from the RelieFF algorithm. Gyroscope Y direction at 500 Hz and Segment I.

The results, shown in the previous figures, confirm that it is possible to obtain a significant classification accuracy—higher than 80% in the case of a sample rate of 200, 250 and 500 Hz for SVM and $Gyro_y$. On the other side, it will be demonstrated in the subsequent sections that it is possible to obtain an even higher classification accuracy using the spectral approach whose results are presented in Section 5.3.

As discussed in the Section 4, different machine learning algorithms are used to produce the results. A comparison of the results using the $Gyro_y$ data are provided in Figure 13. Each of the machine learning algorithms have been optimized in relation to their hyper-parameters. The optimized ML parameters are—SVM with $\gamma = 2^3$ and $C = 2^7$, KNN with Euclidean Distance and $K = 1$ and Decision Trees with maximum number of splits $= 3$. The Random Forest and Ada Boost algorithms (based on the Decision Tree) have also been optimized using the auto-optimization function of Matlab. Figure 13 shows that SVM has a better performance accuracy than the other machine learning algorithms, in particular KNN and the Decision Tree. Random Forest and Ada Boost performs better than the Decision Tree but less than SVM. The result is consistent for all the sample rates. Similar results are obtained for Acc_z and $Gyro_x$ but they are not presented to avoid an excessive length of this paper.

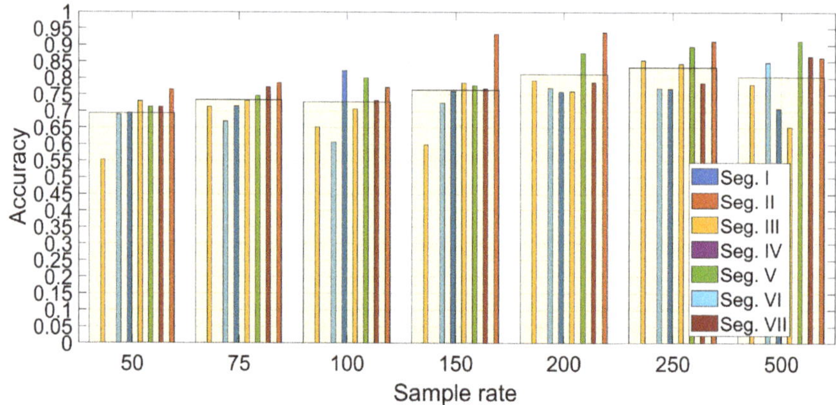

Figure 12. Comparison of the different segments and different sample rates (sample rate in Hz) for the features approach.

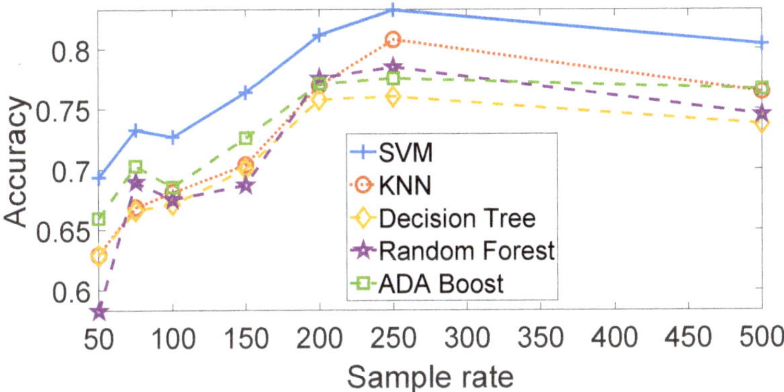

Figure 13. Comparison of machine learning algorithms with feature based approach (sample rate in Hz). Gyroscope Y direction. Results are based on the average across all segments.

As the identification accuracy may not provide a comprehensive view of the false positive or false negative, confusion matrices are provided for $Gyro_y$ at different sample rates. In all confusion matrices, the SVM algorithm was used.

The confusions matrices are shown in Figure 14a–c (respectively at 100, 200 and 500 Hz) for $Gyro_y$ and they are calculated for Segment I (similar results are obtained for the other segments) for the sample rate and IMU components indicated in the figure. They confirm the previous results on the identification accuracy where smaller sample rates decrease the identification accuracy and provide more insights on how the specific vehicles are classified. In particular, the confusion matrices show that the first 6 vehicles (e.g., of the same model Fiat Panda) are more difficult to classify with high accuracy as the number of False Positive and False Negatives is relevant. It is also noted that the 7th car (e.g., Fiat Punto) is of the same brand and it shares similar features of the Panda (e.g., age and vehicle class).

We have also calculated the time needed to process the data, perform the training and testing for a specific segment (i.e., segment II) for the statistical features approach. This evaluation is useful for a practical application of this technique. We present the result for the Ada Boost algorithm (based on the Decision Tree) as this algorithm had the longest computational time among all the algorithms. The results are presented in the bar stacked in Figure 15a for the training phase and in the bar stacked in Figure 15b for the testing phase for different sample rates. We note that the calculation of the statistical features require the larger portion of the processing time, while the selection feature with ReliefF is negligible (it can be detected with difficulty in the figures). Obviously, the computational time increases with the sample rate as a greater sample rate means a higher number of sample to be processed by the feature extraction process. The time needed to identify and authenticate a car is only few seconds on the computing platform used to perform the computation—a laptop with Intel I7-8550U with a clock at 1.8 GHz with 16 GB of RAM. A more powerful computing platform would be able to considerably reduce this processing time. Taking in consideration that the optimal sample rate for the accuracy was 250 Hz, the identification time would be approximately 2 s in the optimal case. Similar considerations are repeated for the spectral approach in the next section with a much shorter time-frame because of absence of a feature extraction process.

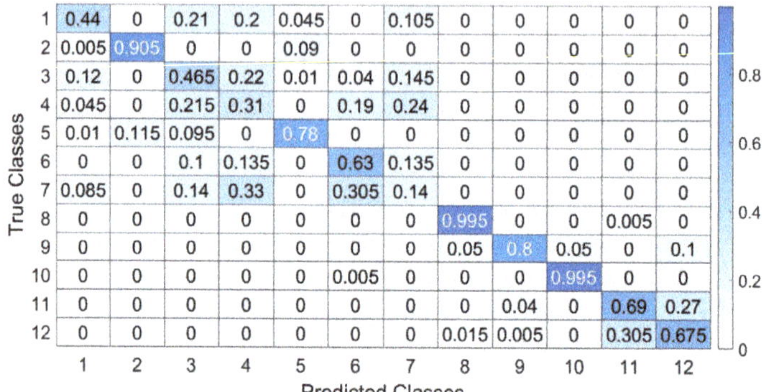

(**a**) Confusion Matrix with Gyroscope Y direction at 100 Hz.

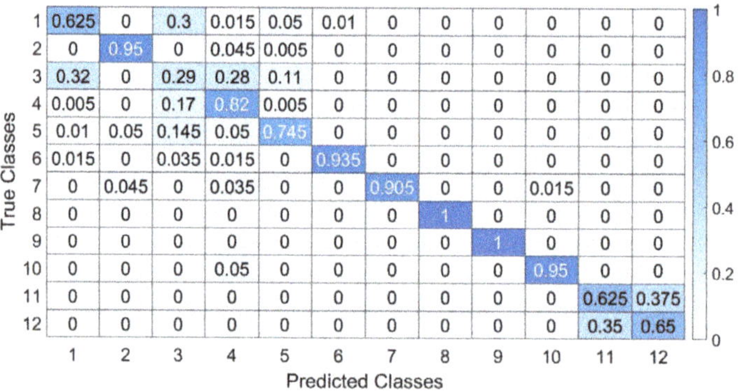

(**b**) Confusion Matrix with Gyroscope Y direction at 200 Hz.

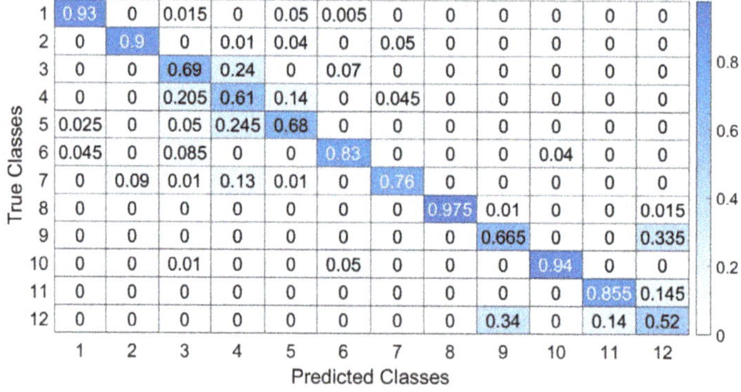

(**c**) Confusion Matrix with Gyroscope Y direction at 500 Hz.

Figure 14. Confusion Matrices with Features approach. The Y axis represents the True classes and the X axis represents the Predicted classes (**a**–**c**).

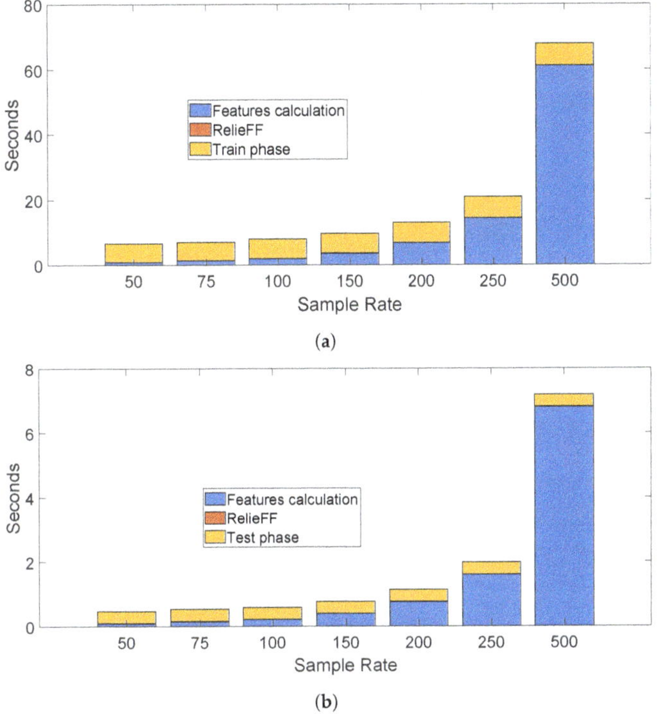

(a)

(b)

Figure 15. Time needed to execute the statistical features approach (sample rate in Hz). (**a**) Time needed to execute the training phase using the statistical features approach and the Ada Boost algorithm; (**b**) Time needed to execute the testing phase using the statistical features approach and the Ada Boost algorithm.

5.3. Spectral Approach

As described in the methodology Section 4 the spectral approach (i.e., in the frequency domain) is based on the application of the FFT to the IMU recordings. Then, the frequency representation is divided in an equal number of frequency bands and the total amplitude power is calculated for each band. The obtained values (i.e., the coefficients) are used for classification. As in the case of the features, the dimensionality reduction is significant: $S_{rate} -> N_{freq}$ (e.g., from 200 Hz to 6 coefficients).

In this approach, the hyper-parameter is the number of coefficients in the spectral domain. Then, the optimal value of N_{freq} must be identified but N_{freq} cannot be large because of the curse of the dimensionality, then it is decided that N_{freq} is limited to 6. On the other side, a feature selection process as in the feature approaches can be used with a value of N_{freq} greater than 6. Figure 16 shows a comparison of different values of N_{freq} using $Gyro_y$ for different values of the sample rates, where only a limited set is used when N_{freq} is greater than 6. It can be seen that the optimal performance curve is obtained for N_{freq}. We note that a value of $N_{freq} = 7$ was also used to make a comparison (thus contrary to the rule defined above), but it can be seen that there is no significant improvement of $N_{freq} = 7$ against $N_{freq} = 6$. The use of higher numbers than $N_{freq} = 7$ ($N_{freq} = 12$, $N_{freq} = 16$ and $N_{freq} = 20$ in combination with RelieFF shows that the identification accuracy is worst than with $N_{freq} = 6$. By comparison, the results obtained in the previous section for the feature approach, it can be concluded that a higher accuracy is obtained by using the spectral approach than the feature approach. Two potential explanations are possible. One explanation is related to the physical response of the vehicle against irregularities in the road as different vehicles types have different automotive suspension components (e.g., coil springs and control arms), which produce a different frequency response on the same road segment. This can be a more distinguishing factor for vehicle types than the designed statistical features used in the feature approach. The second aspect is that the selected features in the feature approach are derived from the research literature on continuous authentication of human beings, which may not be fully appropriate to the continuous authentication of vehicles.

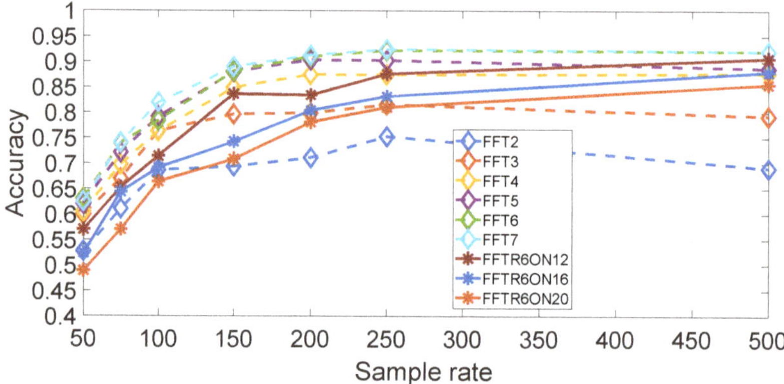

Figure 16. Comparison for different values of with spectral approach (sample rate in Hz). Gyroscope Y direction. Results are based on the average across all segments.

The results are obtained using the $Gyro_y$ direction and by averaging the results across the 7 segments selected in the study. A more detailed study for the different segments is shown is provided in Figure 17 for $N_{freq} = 6$, which confirms the previous results.

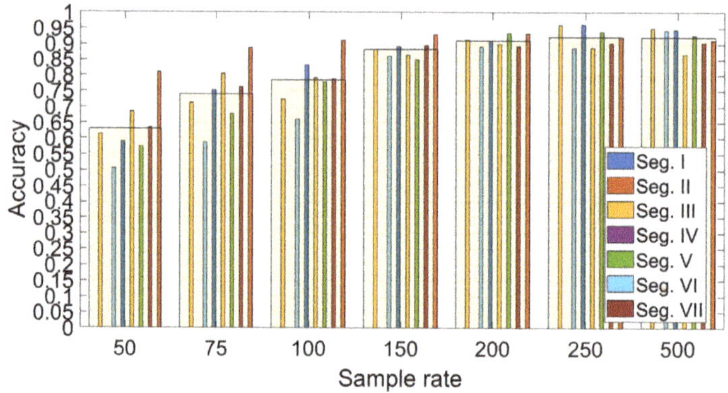

Figure 17. Comparison of the different segments and different sample rates for the spectral approach (sample rate in Hz).

In a similar way to the features approach, a comparison was made among the main IMU components and directions and the results are presented in Figure 18, which confirms the result of the features approach, as the best identification accuracy is obtained with $Gyro_y$. The results are obtained by averaging the results across all segments, using SVM withe optimized values of the machine learning hyper-parameters.

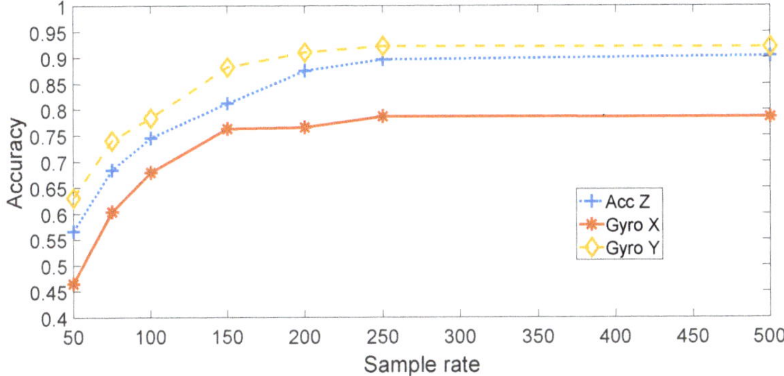

Figure 18. Comparison of different elements and directions of the IMU with spectral approach (sample rate in Hz). Results are based on the average across all segments.

In comparison to the feature based approach, another hyper-parameter to select is the amplitude or phase component as the FFT provides complex values. An evaluation using $Gyro_y$, $N_{freq} = 6$ and averaged on all segments for different sample rates is shown in Figure 19. It can be seen that the amplitude component is much more significant than the phase component. The reason is that the amplitude part is directly related to the frequency response of the mechanical components (i.e., coil springs and wheels) of the vehicle while the phase is not (e.g., the reaction of vehicle to a bumper is mostly related to the speed of the vertical acceleration).

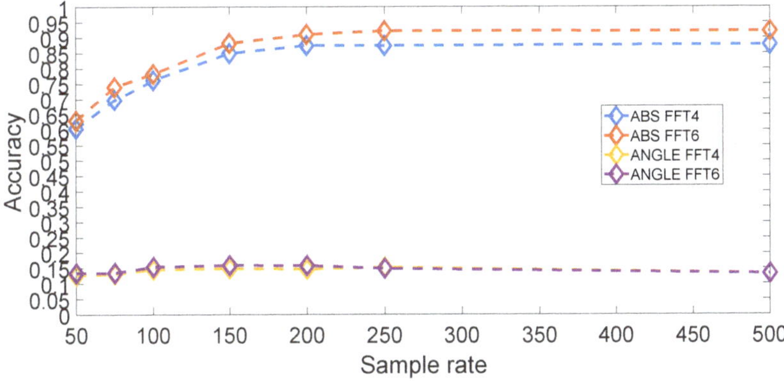

Figure 19. Comparison of magnitude and phase components in the spectral approach using $N_{freq} = 6$ for the Gyroscope Y direction (sample rate in Hz). Results are based on the average across all segments.

It could also be imagined that other transforms rather than the FFT transform could give better classification results. This hypothesis is evaluated in Figure 20 where a wavelet based transform (i.e., based on a Daubechies wavelet of order 10) is compared against the frequency domain approach for different sample rates, $Gyro_y$ and $N_{freq} = 6$ both for wavelet and frequency domains representations. In both cases, the amplitude component of both transforms have been used. The result in Figure 20 shows clearly that the FFT based approach is significantly better than the wavelet based approach. Different wavelets have also been used with similar results.

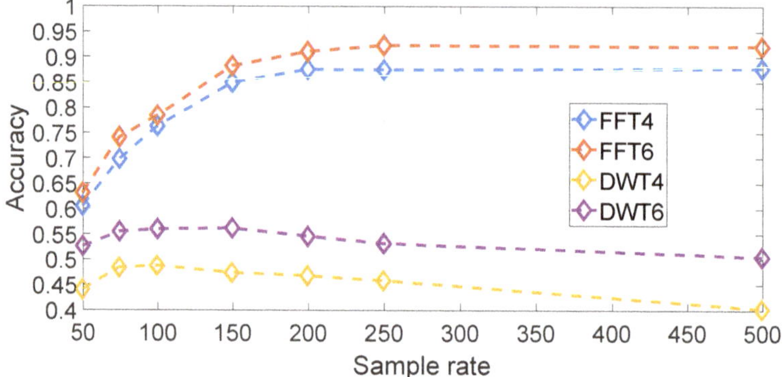

Figure 20. Comparison of spectral and wavelet approaches (sample rate in Hz). Gyroscope Y direction at 500 Hz. Results are based on the average across all segments.

All the previous results are obtained by using the SVM machine learning algorithm. As in the feature based approach, a comparison is performed among the SVM, KNN and Decision Tree algorithms and the results are shown in Figure 21. The results are created by averaging the classification accuracy results for all the segments. It can be seen that SVM outperforms KNN, Decision Tree and the related ensemble methods like Random Forest and AdaBoost. The optimized ML parameters are: SVM with $\gamma = 2^4$ and $C = 2^8$, KNN with Euclidean Distance and $K = 1$ and Decision Trees with maximum number of splits = 5. The Random Forest and Ada Boost algorithms (based on the Decision Tree) have also been optimized using the auto-optimization function of Matlab.

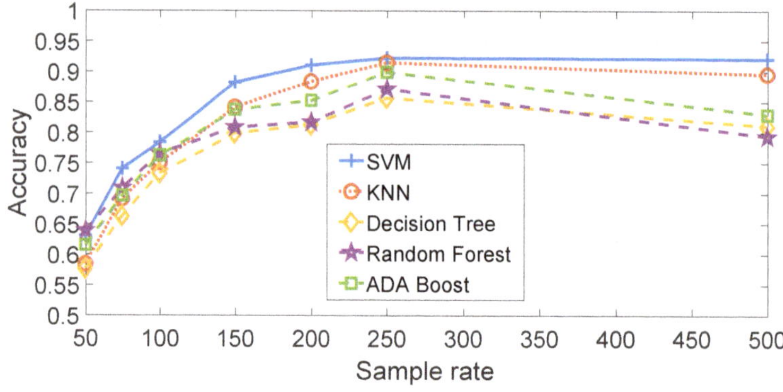

Figure 21. Comparison of machine learning algorithms with spectral based approach (sample rate in Hz). Gyroscope Y direction at 500 Hz and $N_{freq} = 6$. Results are based on the average across all segments.

As in the case of the feature based approach, confusion matrices are presented in the rest of this section. The confusion matrices are calculated by averaging the results across all the segments for the sample rate and IMU component indicated in the respective figure.

Figure 22a–c show the confusion matrices using the spectral approach with Gyroscope Y for Segment I (similar results are obtained for the other segments) and with an IMU recording sampled respectively at 100 Hz, 200 Hz and 500 Hz. The confusion matrices confirm the previous Figures obtained for the identification accuracy. As in the case of the feature based approach, the first 6 vehicles (i.e., of the same Panda model) are more difficult to distinguish than the other vehicles as expected because they have similar mechanical features.

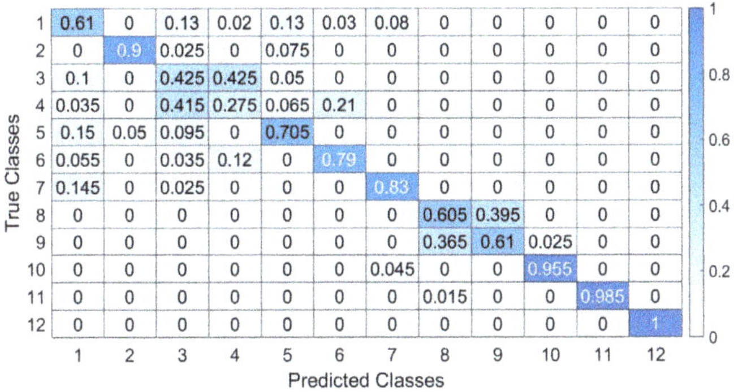

(**a**) Confusion Matrix with Gyroscope Y direction at 100 Hz.

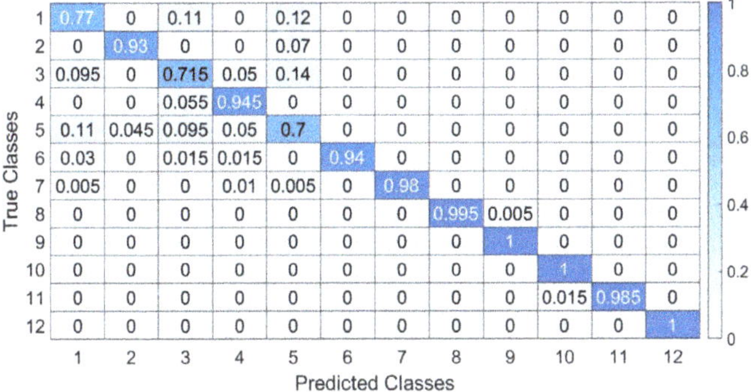

(**b**) Confusion Matrix with Gyroscope Y direction at 200 Hz.

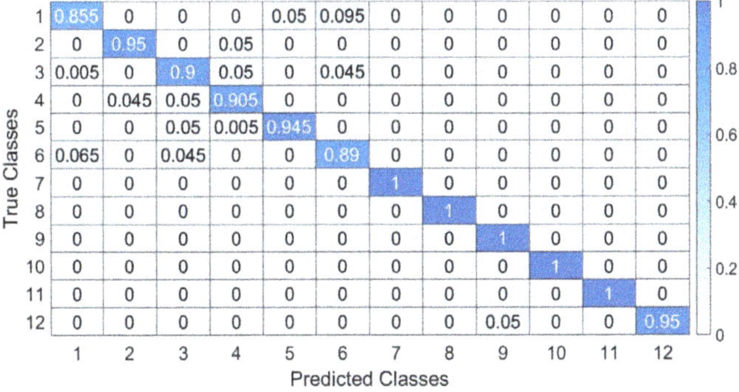

(**c**) Confusion Matrix with Gyroscope Y direction at 500 Hz.

Figure 22. Confusion Matrices with Spectral approach. The Y axis represents the True classes and the X axis represents the Predicted classes.

As in the previous case of the statistical feature approach, we have also calculated the time needed to process the data, perform the training and testing for a specific segment (i.e., segment II) for the spectral approach. We present the result for the Ada Boost algorithm (based on the Decision Tree) as this algorithm had the longest computational time among all the algorithms. The results are presented in the bar stacked Figure 23a for the training phase and in bar stacked Figure 23b for the testing phase for different sample rates. In comparison to the statistical feature approach, the processing time is much lower and it is possible to perform an identification and authentication in less than a second (0.4 s for sample rate at 250 Hz).

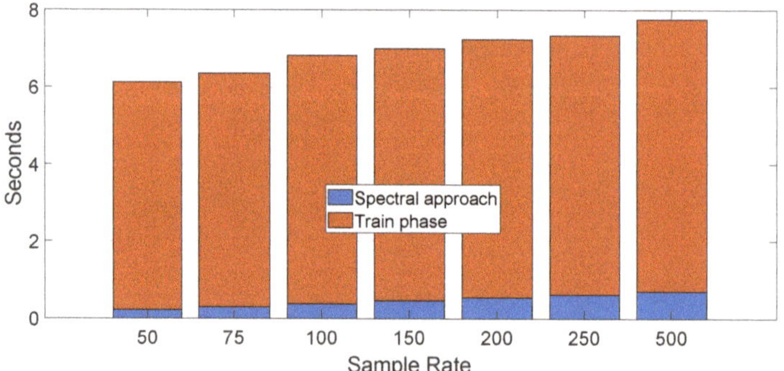

(**a**) Time needed to execute the training phase using the spectral approach and the Ada Boost algorithm

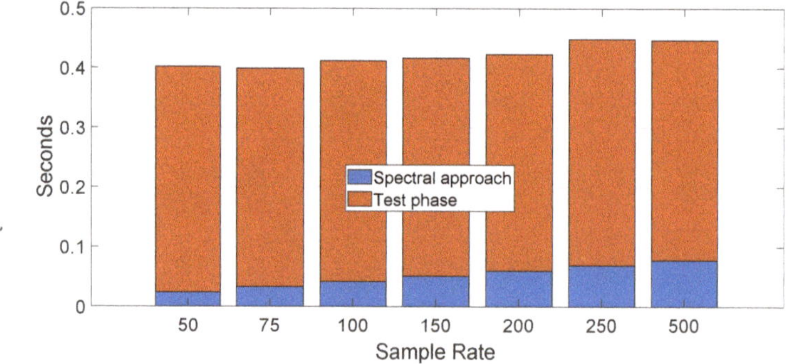

(**b**) Time needed to execute the testing phase using the spectral approach and the Ada Boost algorithm

Figure 23. Time needed to execute the spectral approach (sample rate in Hz). (**a**) Time needed to execute the training phase using the spectral approach and the Ada Boost algorithm. (**b**) Time needed to execute the testing phase using the spectral approach and the Ada Boost algorithm.

6. Discussion and Conclusions

The results presented in the previous section confirmed that the continuous authentication approach already proposed for human beings on the basis of their gait and IMU data can also be applied to vehicles as the spectral approach provides an identification accuracy higher than 90% and even higher than 95% for some specific road segments. The results are obtained using extensive measurements campaigns with 12 different vehicles on tens of kilometers of road. Even if the data were collected at a high sample rate using professional IMU, the analysis was performed with lower sample rates, which are obtainable with modern mobile phones (e.g., 200 Hz). The same driver was used to collect the recording in all the vehicles to avoid the introduction of the bias of the driver. We highlight that the proposed approach for continuous authentication of vehicles was derived from the approaches (e.g., statistical features or spectral approaches) already defined in the literature for continuous authentication of human beings. A comprehensive analysis and optimization of the hyper-parameters were performed on the dataset to select the optimal values, approaches and algorithms. Such optimization can be useful for future research activities by the research community. The results show that the spectral approach provides a higher identification accuracy than the feature approach and we recommend its use. Future developments may take in consideration more sophisticated time-frequency representation even if the amount of data to be analyzed will be higher. The presented approach is based on a strong dimensionality reduction from the initial IMU recording, which make it suitable for practical applications as only a small subset of data (e.g., 6 amplitude values in the spectral domain) for a road segment of hundreds of meters must be sent to the remote authentication system.

A complementary point of view is that the identification of vehicles can also generate a privacy threat. Then, a complementary research activity would be to investigate techniques to remove the discriminating features used to identify the vehicle, but still preserve the information needed for traffic applications (e.g., road surface maintenance, traffic analysis).

Future developments will investigate the combination of data from different components and directions to obtain a higher identification accuracy. The application of de-noising techniques to improve the quality of the recorded data could also be used to improve the classification accuracy but it must be ensured that the application of de-noising algorithms does not remove the discriminating characteristics of the vehicle. Finally, the entire dataset is made available to the research community to perform additional studies in this area.

Future extensions of this paper will also investigate the impact of different variables on the identification and authentication accuracy. For example, if the same vehicle has different numbers of passengers or if the driver is different as the driving style can impact the continuous authentication approach.

Author Contributions: Conceptualization, G.B.; methodology, G.B., F.G., R.G.; software, F.G. and G.B.; validation, G.B., F.G., R.G.; investigation, G.B., F.G., R.G.; resources, F.G., G.B.; writing—original draft preparation, G.B. and F.G.; writing—review and editing, G.B., F.G., R.G.; supervision, G.B.

Funding: This research received no external funding.

Acknowledgments: We acknowledge the support of the JRC Transport Office and the JRC units C.4, I.5 to provide the vehicles used to collect the data.

Conflicts of Interest: The authors declare no conflict of interest.

Abbreviations

The following abbreviations are used in this manuscript:

ApEn	Approximate entropy
BNW	Background Noise Window
DistEn	Distribution entropy
DTW	Dynamic Time Warping
EL	Experimental Lap
FFT	Fast Fourier Transform
GNSS	Global Navigation Satellite Systems
IMU	Inertial Measurement Unit
ITS	Intelligent Transport Systems
JRC	Joint Research Centre
KNN	K-Nearest Neighbors
OP	Orientation Point
PE	Permutation Entropy
PIN	Personal Identification Number
PKI	Public Key Infrastructure
RBF	Radial Basis Function
RF	Road Feature
RMS	Root Mean Square
SB	Speed Bump
SVM	Support Vector Machine

References

1. Patel, V.M.; Chellappa, R.; Chandra, D.; Barbello, B. Continuous user authentication on mobile devices: Recent progress and remaining challenges. *IEEE Signal Process. Mag.* **2016**, *33*, 49–61. [CrossRef]
2. Sitová, Z.; Šeděnka, J.; Yang, Q.; Peng, G.; Zhou, G.; Gasti, P.; Balagani, K.S. HMOG: New behavioral biometric features for continuous authentication of smartphone users. *IEEE Trans. Inf. Forensics Secur.* **2015**, *11*, 877–892. [CrossRef]
3. Harbach, M.; Von Zezschwitz, E.; Fichtner, A.; De Luca, A.; Smith, M. It'sa hard lock life: A field study of smartphone (un) locking behavior and risk perception. In Proceedings of the 10th Symposium On Usable Privacy and Security (SOUPS 2014), Menlo Park, CA, USA, 9–11 July 2014; pp. 213–230.
4. Frank, M.; Biedert, R.; Ma, E.; Martinovic, I.; Song, D. Touchalytics: On the applicability of touchscreen input as a behavioral biometric for continuous authentication. *IEEE Trans. Inf. Forensics Secur.* **2012**, *8*, 136–148. [CrossRef]
5. Singh, S. Keystroke Dynamics for Continuous Authentication. In Proceedings of the 2018 8th International Conference on Cloud Computing, Data Science & Engineering (Confluence), Noida, India, 11–12 January 2018; pp. 205–208.
6. Visvizi, A.; Lytras, M.D. *Smart Cities: Issues and Challenges: Mapping Political, Social and Economic Risks and Threats*; Elsevier: Hoboken, NJ, USA, 2019; ISBN 978012816639.
7. Lu, D.N.; Nguyen, D.N.; Nguyen, T.H.; Nguyen, H.N. Vehicle mode and driving activity detection based on analyzing sensor data of smartphones. *Sensors* **2018**, *18*, 1036. [CrossRef] [PubMed]
8. Baldini, G.; Sportiello, L.; Chiaramello, M.; Mahieu, V. Regulated applications for the road transportation infrastructure: The case study of the smart tachograph in the European Union. *Int. J. Crit. Infrastruct. Prot.* **2018**, *21*, 3–21. [CrossRef]
9. MacDonald, G. A review of low cost accelerometers for vehicle dynamics. *Sens. Actuators A Phys.* **1990**, *21*, 303–307. [CrossRef]
10. Wahlström, J.; Skog, I.; Händel, P. Smartphone-based vehicle telematics: A ten-year anniversary. *IEEE Trans. Intell. Transp. Syst.* **2017**, *18*, 2802–2825. [CrossRef]
11. Aljaafreh, A.; Alshabatat, N.; Al-Din, M.S.N. Driving style recognition using fuzzy logic. In Proceedings of the 2012 IEEE International Conference on Vehicular Electronics and Safety (ICVES 2012), Istanbul, Turkey, 24–27 July 2012; pp. 460–463.
12. Gonzalez-Manzano, L.; Fuentes, J.M.D.; Ribagorda, A. Leveraging User-related Internet of Things for Continuous Authentication: A Survey. *ACM Comput. Surv. (CSUR)* **2019**, *52*, 53. [CrossRef]
13. Li, Y.; Hu, H.; Zhou, G.; Deng, S. Sensor-based continuous authentication using cost-effective kernel ridge regression. *IEEE Access* **2018**, *6*, 32554–32565. [CrossRef]
14. Muaaz, M.; Mayrhofer, R. An analysis of different approaches to gait recognition using cell phone based accelerometers. In Proceedings of the International Conference on Advances in Mobile Computing & Multimedia, Vienna, Austria, 2–4 December 2013; p. 293.
15. Thang, H.M.; Viet, V.Q.; Thuc, N.D.; Choi, D. Gait identification using accelerometer on mobile phone. In Proceedings of the 2012 International Conference on Control, Automation and Information Sciences (ICCAIS), Ho Chi Minh City, Vietnam, 26–29 November 2012; pp. 344–348.
16. Juefei-Xu, F.; Bhagavatula, C.; Jaech, A.; Prasad, U.; Savvides, M. Gait-id on the move: Pace independent human identification using cell phone accelerometer dynamics. In Proceedings of the 2012 IEEE Fifth International Conference on Biometrics: Theory, Applications and Systems (BTAS), Arlington, VA, USA, 23–27 September 2012; pp. 8–15.
17. Malik, M.N.; Azam, M.A.; Ehatisham-Ul-Haq, M.; Ejaz, W.; Khalid, A. ADLAuth: Passive Authentication Based on Activity of Daily Living Using Heterogeneous Sensing in Smart Cities. *Sensors* **2019**, *19*, 2466. [CrossRef] [PubMed]
18. Dolev, S.; Krzywiecki, Ł.; Panwar, N.; Segal, M. Vehicle authentication via monolithically certified public key and attributes. *Wirel. Netw.* **2016**, *22*, 879–896. [CrossRef]
19. Guo, H.; Wu, Y.; Bao, F.; Chen, H.; Ma, M. UBAPV2G: A unique batch authentication protocol for vehicle-to-grid communications. *IEEE Trans. Smart Grid* **2011**, *2*, 707–714. [CrossRef]

20. Tan, H.; Ma, M.; Labiod, H.; Boudguiga, A.; Zhang, J.; Chong, P.H.J. A secure and authenticated key management protocol (SA-KMP) for vehicular networks. *IEEE Trans. Veh. Technol.* **2016**, *65*, 9570–9584. [CrossRef]
21. Lonc, B.; Cincilla, P. Cooperative its security framework: Standards and implementations progress in europe. In Proceedings of the 2016 IEEE 17th International Symposium on A World of Wireless, Mobile and Multimedia Networks (WoWMoM), Coimbra, Portugal, 21–24 June 2016; pp. 1–6.
22. Ďurech, J.; Franeková, M.; Holečko, P.; Bubeníková, E. Performance analysis of authentication protocols used within Cooperative-Intelligent Transportation Systems with focus on security. In *International Conference on Transport Systems Telematics*; Springer: Berlin, Germany, 2015; pp. 220–229.
23. Baldini, G.; Mahieu, V.; Fovino, I.N.; Trombetta, A.; Taddeo, M. Identity-based security systems for vehicular ad-hoc networks. In Proceedings of the 2013 International Conference on Connected Vehicles and Expo (ICCVE), Las Vegas, NV, USA, 2–6 December 2013; pp. 672–678.
24. Sysoev, M.; Kos, A.; Guna, J.; Pogačnik, M. Estimation of the driving style based on the users' activity and environment influence. *Sensors* **2017**, *17*, 2404. [CrossRef] [PubMed]
25. Wu, M.; Zhang, S.; Dong, Y. A Novel Model-Based Driving Behavior Recognition System Using Motion Sensors. *Sensors* **2016**, *16*, 1746. [CrossRef]
26. Baldini, G.; Steri, G. A survey of techniques for the identification of mobile phones using the physical fingerprints of the built-in components. *IEEE Commun. Surv. Tutor.* **2017**, *19*, 1761–1789. [CrossRef]
27. Bandt, C.; Pompe, B. Permutation Entropy: A Natural Complexity Measure for Time Series. *Phys. Rev. Lett.* **2002**, *88*, 174102. [CrossRef]
28. Lu, L.; Zhang, D. Based on multiscale permutation entropy analysis dynamic characteristics of EEG recordings. In Proceedings of the 2016 35th Chinese Control Conference (CCC), Chengdu, China, 27–29 July 2016; pp. 9337–9341. [CrossRef]
29. Li, Y.; Wang, X.; Wu, J. Fault diagnosis of rolling bearing based on permutation entropy and Extreme Learning Machine. In Proceedings of the 2016 Chinese Control and Decision Conference (CCDC), Yinchuan, China, 28–30 May 2016; pp. 2966–2971. [CrossRef]
30. Li, P.; Liu, C.; Li, K.; Zheng, D.; Liu, C.; Hou, Y. Assessing the complexity of short-term heartbeat interval series by distribution entropy. *Med. Biol. Eng. Comput.* **2015**, *53*, 77–87. [CrossRef]
31. Keogh, E.; Mueen, A. Curse of dimensionality. *Encyclopedia of Machine Learning*; Springer: Berlin, Germany, 2010; pp. 257–258.
32. Theodoridis, S.; Koutroumbas, K. *Pattern Recognition*; Academic Press: London, UK, 1998; p. 625.
33. Robnik-Šikonja, M.; Kononenko, I. Theoretical and empirical analysis of ReliefF and RReliefF. *Mach. Learn.* **2003**, *53*, 23–69. [CrossRef]

Sample Availability: The data samples of the IMU recordings from all the 12 vehicles are available from the authors.

Article

Simultaneous Optimization of Vehicle Arrival Time and Signal Timings within a Connected Vehicle Environment

Wei Wu [1], Ling Huang [2],* and Ronghua Du [3]

[1] Key Laboratory of Highway Engineering of Ministry of Education, Changsha University of Science & Technology, Changsha 410114, Hunan, China; jiaotongweiwu@csust.edu.cn
[2] School of Civil Engineering and Architecture, Changsha University of Science & Technology, Changsha 410114, Hunan, China
[3] College of Automotive and Mechanical Engineering, Changsha University of Science & Technology, Changsha 410114, Hunan, China; csdrh@csust.edu.cn
* Correspondence: huangling1010@stu.csust.edu.cn

Received: 30 November 2019; Accepted: 26 December 2019; Published: 29 December 2019

Abstract: Most existing signal timing plans are optimized given vehicles' arrival time (i.e., the time for the upcoming vehicles to arrive at the stop line) as exogenous input. In this paper, based on the connected vehicle (CV) technique, vehicles can be regarded as moving sensors, and their arrival time can be dynamically adjusted by speed guidance according to the current signal status and traffic conditions. Therefore, an integrated traffic control model is proposed in this study to optimize vehicle arrival time (or travel speed) and signal timing simultaneously. "Speed guidance model at a red light" and "speed guidance model at a green light" are presented to model the influences between travel speed and signal timing. Then, the methods to model the vehicle arrival time, vehicle delay, and number of stops are proposed. The total delay, which includes the control delay, queuing delay, and signal delay, is used as the objective of the proposed model. The decision variables consist of vehicle arrival time, starting time of green, and duration of green for each phase. The sliding time window is adopted to dynamically tackle the problems. Compared with the results optimized by the classical actuated signal control method and the fixed-time-based speed guidance model, the proposed model can significantly decrease travel delays as well as improve the flexibility and mobility of traffic control. The sensitivity analysis with the communication distance, the market penetration of connected vehicles, and the compliance rate of speed guidance further demonstrates the potential of the proposed model to be applied in various traffic conditions.

Keywords: traffic signal control; speed guidance; vehicle arrival time; connected vehicle

1. Introduction

Traffic signal control is one of the most effective methods to alleviate traffic congestion [1,2]. The existing signal control strategies can be divided into three categories, namely fixed time, actuated, and adaptive [3]. Fixed-time-based traffic control utilizes the historical traffic volume at an intersection to calculate the parameters for signal timing (e.g., cycle length, green time split, and offset) [4,5]. Fixed-time-based traffic control may not be able to adapt to demand fluctuation caused by the stochastic nature of traffic arrivals, which may degrade its benefits of signal control. In order to tackle the real-time demand fluctuation, actuated and adaptive traffic control methods have been proposed. By collecting real-time traffic arrivals from preinstalled sensors (e.g., loops, microwaves, and videos), actuated control methods can adjust their signal timing on the basis of a few simple rules, that is, green extension and minimum and maximum green duration constraints [6–9]. For adaptive control methods, they predict traffic arrivals first by utilizing the data collected from preinstalled sensors and then optimize the predefined metrics (e.g., minimize delays, maximize throughputs) [10–12]. Whether actuated control or adaptive control is involved, their signal plans are optimized based on the data collected from preinstalled sensors. However, these sensors (e.g., loops, microwaves, and videos) are only installed at a few fixed points near the intersection and can only collect traffic data at those limited points. Indeed, intersection signal controllers are not able to grasp the accurate and complete information of traffic arrivals based on these limited sensors, which may result in the inefficiency of the traffic signal control. Furthermore, the preinstalled sensors (e.g., loops, microwaves, and videos) can only monitor vehicles as they are approaching an intersection; they are not able to influence or change traffic arrivals based on the current signal timings to achieve better performance.

With the development of wireless communication technology such as radio-frequency identification (RFID), dedicated short-range communication (DSRC), and 5G (the fifth generation of cellular network technology), a connected vehicle (CV) technique has emerged and is being employed in recent years [13–19]. By equipping vehicles with an on-board unit (OBU), the connected vehicles are able to communicate with surrounding vehicles and infrastructures (e.g., signal controllers) in real time [3,20–22], and these connected vehicles can be regarded as moving sensors. Therefore, first, the trajectory data collected from the connected vehicle such as real-time speed, acceleration, and location can be transferred to the signal controller, then signal plans can be optimized with more precision [23,24]. Second, with connected vehicles, some critical metrics such as queue length can be detected or estimated [25–28], which can be incorporated to reflect the performance and influence the optimization of signal plans. This may further improve the efficiency of traffic control.

Most existing studies merely focus on how to optimize signal plans based on the trajectory data collected from the connected vehicle [3,20,29,30]. In particular, based on the trajectory data, vehicles' arrival times (i.e., the time for vehicles to arrive at the stop line) around the intersection can be estimated with more precision, then signal plans can be optimized by taking these vehicles' arrival times as exogenous input. However, based on the two-way communication between vehicles and the signal controller in a connected vehicle environment, a speed guidance or speed advisory can be deployed. For example, Wan et al. [31] proposed a speed advisory system (SAS) for reducing idling and improving energy efficiency at a red light. Ramezani and Benekohal [32] claimed that the accuracy of data and the effectiveness of speed guidance could be enhanced based on a connected vehicle technique, and they developed an optimization program to determine advisory speeds for connected vehicles in work zones. Tajalli and Hajbabaie [33] regulated the movement of vehicles to achieve a "smoother" flow of traffic by harmonizing the speed of connected vehicles.

By using a speed guidance or speed advisory, the arrival time of connected vehicles can be dynamically optimized by adjusting their travel speed according to the current signal status and traffic conditions [34–36]. This means that not only signal plans can be adjusted to adapt to traffic arrivals but also traffic arrivals can be adjusted likewise to adapt to the signal plans in a connected vehicle environment. Specifically, Tang et al. [35] for the first time introduced driver's bounded rationality into a speed guidance model. The impacts of driver's bounded rationality on fuel consumption and emissions were studied. Tang et al. [36] further introduced a speed guidance strategy into a car-following model. Wu et al. [34] proposed a speed guidance model for a fixed-time-based signal controller. However, signal timing plans were not simultaneously optimized along with the speed guidance strategy in those studies. In particular, the existing studies focused mainly on modelling speed guidance strategies without optimizing signal timing plans at the same time, especially when many connected vehicles were coming from conflicting movements (e.g., through vehicles from the east and through vehicles from the south).

This paper fills the mentioned gap and contributes to the literature in several ways. First, this study examines the dynamic interaction between signal timing and vehicle arrival time (or travel speed). This has been rarely modelled in the literature. In particular, two speed guidance strategies associated with dynamic signal timing plans and three types of vehicle delays (i.e., control delay, queuing delay, and signal delay) are modelled. Second, by using the Standard North American NEMA (National Electrical Manufacturers Association) Dual-Ring structure, the proposed model is among the first to simultaneously optimize intersection signal timings and recommended travel speed for connected vehicles coming from conflicting movements at an intersection in a united framework. Specifically, the proposed model calculates the optimal signal timing to decrease the total delay for all approaching connected vehicles as well as optimizes the travel speed for each connected vehicle at the same time. It should be noted that both signal timing and travel speed adjustments may improve travel efficiency for one vehicle; however, it remains a great challenge to simultaneously allocate green time to each phase and calculate the optimal travel speed for all approaching connected vehicles, especially when these connected vehicles are coming from conflicting movements.

The remainder of this paper is organized as follows. Section 2 starts with a simple example of a connected vehicle to clear the intersection with either signal timing or speed adjustments. Notations and descriptions are also provided in Section 2. Section 3 models the problem, and the speed guidance model, delay model, and objective function are proposed to simultaneously optimize vehicle arrival time and signal timing plans. Numerical examples are presented in Section 4 to illustrate the model and analysis. Section 5 concludes the paper.

2. Problem Description and Parameter Definition

Taking the end of the green light as an example, as shown in Figure 1, if the vehicle proceeds without speed adjustment, it reaches the intersection at T_a and stops because the signal turns red at T_r. If the vehicle needs to pass the intersection without stopping, the common practice is to extend the duration of the green light for $T_a - T_r$ seconds, which means the existing methods need to obtain the vehicle arrival time as inputs to optimize the signal timing and extend the green light. However, vehicles could also clear the intersection without stopping by other means. For example, guiding the driving speed from V_0 to V_1 can help the vehicle arrive at the intersection during green time without recalculating the signal timing. This means that either optimizing the signal timing or adjusting the arriving time (or speed) can make the vehicle clear the intersection without stopping. However, most existing studies focus mainly on signal timing optimization or speed guidance. In this paper, both signal timing and speed are assumed to be optimized simultaneously. Note that when these two variables are optimized for each vehicle, the complexity and optimization space of the model are greatly increased. The research problem in this paper can be specified as follows: how to obtain the optimal vehicle arrival time and the signal timing for all approaching vehicles around the intersection to ensure the maximum performance of the intersection.

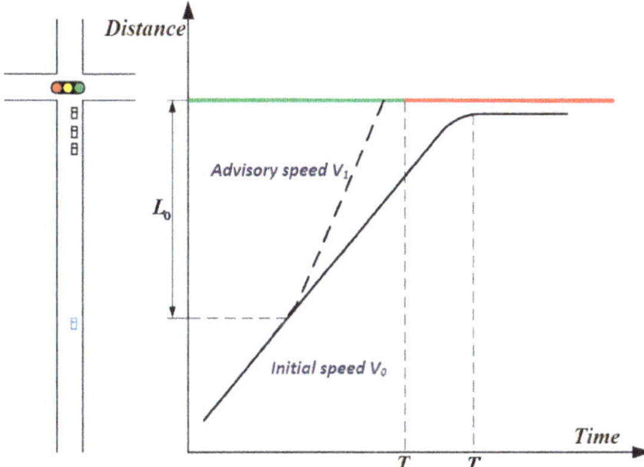

Figure 1. Signal timing and vehicle arrival.

The main notations and their definitions and descriptions are presented in Table 1.

Table 1. Notations and descriptions.

Notations	Descriptions
(i, p, j)	Vehicle j from phase p at intersection i
$V_{\max}$	The maximum speed limit
$V_{\min}$	The minimum speed limit
$g_{\max}$	The maximum green time duration for a phase
$g_{\min}$	The minimum green time duration for a phase
V_s	The speed of traffic wave spreads
$V_g(i, p, j)$	The advisory speed for vehicle j from phase p at intersection i
$V_a(i, p, j)$	The regular travel speed for vehicle j from phase p at intersection i
$V_t(i, p, j)$	The optimum travel speed for vehicle j from phase p at intersection i
$T(i, p, j)$	The time for vehicle j from phase p at intersection i to arrive at the stop line
$T_c(i, p, j)$	The time length for vehicle j from phase p at intersection i to clear the intersection
$L_d(i, p, j)$	The distance between the stop line and vehicle j from phase p at intersection i
$L_s(i, p, j)$	The vehicle length of vehicle j from phase p at intersection i
$d(i, p, j)$	The total delay of vehicle j from phase p at intersection i
$d_c(i, p, j)$	The control delay of vehicle j from phase p at intersection i
$d_q(i, p, j)$	The queuing delay of vehicle j from phase p at intersection i
$d_s(i, p, j)$	The signal delay of vehicle j from phase p at intersection i
$\phi(i, p, j, k)$	The binary variable, $\phi(i, p, j, k) = 1$, denotes vehicle j from phase p at intersection i will traverse the intersection in cycle k, zero otherwise
t_n	The current time
$t(i, p, k)$	The starting time of green light for phase p at intersection i in cycle k
$g(i, p, k)$	The time duration of green light for phase p at intersection i in cycle k
$v(i, p, k)$	The time duration of phase p at intersection i in cycle k, which is equal to the summation of the duration of green light and green interval of phase p

3. Optimization Model

3.1. Speed Guidance Associated with Dynamic Signal Timing Plans

When modelling speed guidance, it is assumed that the travel paths for each connected vehicle are preplanned. Therefore, through two-way communication between the signal controller and connected vehicles, the signal controller can acquire the vehicle's turning decision (i.e., left turn, through, or right turn) in advance. Furthermore, vehicles shift to their chosen lane in advance when approaching an intersection according to their turning decisions, thus the negative impacts of lane changing were not incorporated into this research. In this context, in order to guide the vehicles to reach the intersection at the best time, this paper proposes two speed guidance models and establishes the mathematical relationship between the optimal driving speed and the starting time of the green light. Note that the starting times of the green light for each of the traffic movements are also variables and must be optimized (discussed in Section 3.5). According to different signal timings and traffic conditions, different speed guidance models can be established. There are mainly two models as follows:

Speed guidance model at a red light. When the vehicle's requesting phase is yellow or red, or it is green but the queue has not dissipated, the proposed speed guidance model is called the "red-light guidance model".

Speed guidance model at a green light. When the vehicle's requesting phase is green and the queue has dissipated, the proposed speed guidance model is called the "green-light guidance model".

The biggest difference between the red-light guidance and green-light guidance models is that the queue is not dissipated when the red-light speed guidance is implemented. In order to determine the vehicle's optimal speed, we defined a "reference point" for each vehicle and assumed that the travel speed for this vehicle travelling from the current location to its "reference point" was constant. Under the red-light speed guidance, the reference point should take the queue length into account, whereas under the green-light guidance, because the queue has dissipated, the reference point for speed guidance is therefore at the stop line.

3.1.1. Speed Guidance Model at a Red Light

The strategy of the red-light speed guidance is to ensure that when a vehicle reaches the end of the queue (caused by the red light), the shock wave is just transmitted to the end of the queue, that is, the vehicle at the end of the queue starts to move. In this context, this vehicle can smoothly clear the intersection following the shock wave. Therefore, the model for the red-light speed guidance can be written as follows:

$$V_t(i,p,j) = \frac{L_d(i,p,j) - \sum\limits_{j_1=1}^{j-1} L_s(i,p,j_1)}{t(i,p,k) + \sum\limits_{j_1=1}^{j-1} L_s(i,p,j_1)/V_s - t_n} \tag{1}$$

In the above equation, the numerator indicates the distance between the vehicle and the stop line minus the road space occupied by all the preceding vehicles (queue length), and the denominator indicates the time duration between the time that the last vehicle in the queue begins to move and the current time. Note that the suggested speed (i.e., $V_t(i,p,j)$) and the vehicle's arrival time (i.e., $t(i,p,k)$) are two variables to be optimized simultaneously.

3.1.2. Speed Guidance Model at a Green Light

The effectiveness of the speed guidance model at a green light is presented in Figure 2. Note that the red line in Figure 2 is the trajectory of the guided vehicle, which means it can pass through the intersection without stopping by increasing the vehicle speed in the current cycle or by decreasing the vehicle speed and then passing through in the next cycle. Therefore, the speed guidance model at a green light can be specified as follows:

$$V_t = L_d(i,p,j)/\left\{ \sum\limits_{j_1=1}^{j-1} L_s(i,p,j)/V_a(i,p,j) \cdot \phi(i,p,j,k) + [t(i,p,k+1) - t_n + \sum\limits_{j_1=1}^{j-1} L_s(i,p,j)/V_a(i,p,j)] \cdot \phi(i,p,j,k+1) \right\}. \tag{2}$$

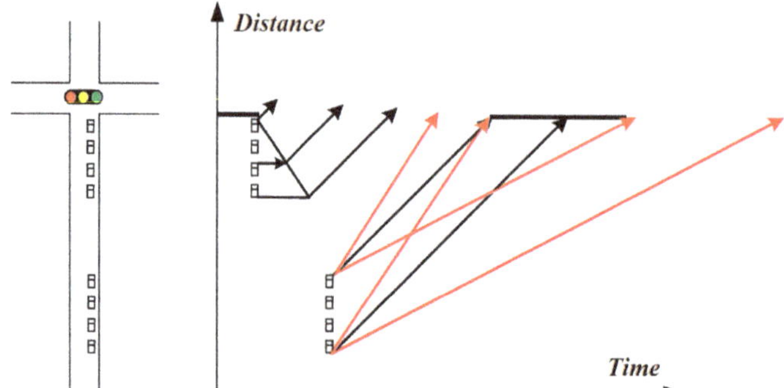

Figure 2. Speed guidance at a green light.

In the above equation, ϕ is a binary variable, which indicates whether the vehicle passes through the intersection in the current cycle or not. The optimization model in this paper only considers two cycles at a time, that is, vehicles in the entrance lane should pass through the intersection in the current cycle or in the next cycle, in other words as follows:

$$\phi(i,p,j,k) + \phi(i,p,j,k+1) = 1 \tag{3}$$

Whether in the model of the red-light speed guidance or the green-light speed guidance, the suggested driving speed for a vehicle after guidance must have the following:

$$V_g(i,p,j) = \begin{cases} V_t(i,p,j), & V_{min} \leq V_t \leq V_{max} \\ V_{max}, & V_t(i,p,j) > V_{max} \\ V_{min}, & V_t(i,p,j) < V_{min} \end{cases} . \tag{4}$$

Equation (4) illustrates that the suggested speed must be constrained within the upper and lower bounds, namely V_{max} and V_{min}.

3.2. Vehicle Arrival Time

3.2.1. Vehicle Arrival Time Model under the Red-Light Guidance

If $V_g(i,p,j) = V_t(i,p,j)$, it means that the vehicle can approach the intersection at an ideal speed. When it travels to the end of the queue, the last queued vehicle starts to move. Then, this vehicle follows the vehicle at the end of the queue to clear the intersection without stopping. In this case, the arrival time of the vehicle can be estimated as follows:

$$T(i,p,j) = t(i,p,k) + \frac{\sum_{j_1=1}^{j-1} L_s(i,p,j_1)}{V_s} + \frac{\sum_{j_1=1}^{j-1} L_s(i,p,j_1)}{V_a(i,p,j_1)}. \tag{5}$$

Equation (5) demonstrates that the arrival time of a vehicle is equal to the time of its preceding vehicle passing the stop line.

If $V_g(i,p,j) = V_{max}$, it means that even if the vehicle travels at the maximum speed, it cannot catch up with the vehicle at the maximum queue point, that is, the queue has completely dissipated when the vehicle travels near the intersection, and its arrival time can be estimated by the following:

$$T(i,p,j) = t_n + L_d(i,p,j)/V_{max}. \tag{6}$$

Equation (6) demonstrates that the arrival time of a vehicle is equal to the current time plus the duration of traveling to the stop line at the highest speed.

If $V_g(i, p, j) = V_{min}$, it means that even if the vehicle travels at the lowest speed, it will also move to the maximum queuing point and then stop. In this case, the arrival time of the vehicle can also be calculated by Equation (5), which means that the arrival time of the vehicle is equal to the time of its preceding vehicle passing through the stop line. However, in this case, the vehicle will stop because it arrives at the maximum queuing point before the queue is dissipated.

3.2.2. Vehicle Arrival Time Model under the Green-Light Guidance

In the case of the green-light guidance, if $V_g(i, p, j) = V_t(i, p, j)$, it means that whether the vehicle passes through the intersection in the current cycle or in the next cycle, it can approach the intersection at an ideal speed without stopping. In this scenario, if the vehicle passes through the intersection within the rest of the green time in the current cycle, the arrival time of the vehicle can also be calculated by the Equation (5), that is, following the preceding vehicle to pass through the intersection. If the vehicle passes the green light in the next cycle (please refer to Figure 2), the arrival time of the vehicle can be estimated as follows:

$$T(i, p, j) = t(i, p, k+1) + \frac{\sum_{j_1=1}^{j-1} L_s(i, p, j_1)}{V_s} + \frac{\sum_{j_1=1}^{j-1} L_s(i, p, j_1)}{V_a(i, p, j_1)}. \tag{7}$$

Equation (7) demonstrates that the arrival time is equal to the time of the vehicle smoothly passing the stop line following the preceding vehicle in the next cycle.

If $V_g(i, p, j) = V_{max}$, it means that even if the vehicle travels at the maximum speed, it cannot catch up with the preceding vehicle when passing through the intersection in the current cycle. In this context, its arrival time can be estimated by Equation (6). If $V_g(i, p, j) = V_{min}$, it means that even if the vehicle runs at the lowest speed, it will pass through the intersection in the next cycle; then, the arrival time of the vehicle can be estimated by Equation (7).

3.3. Prediction of Vehicle Delay

Vehicle delay includes control delay, queuing delay, and signal delay, where control delay is caused by the change in speed due to speed guidance, queuing delay is caused by the queue, and signal delay is caused by the red light [1].

3.3.1. Control Delay

In the case of the red-light guidance, where the point of the maximum queue length is taken as the reference point for the speed guidance, the control delay caused by the speed change can be calculated by the following:

$$d_c(i, p, j) = \frac{L_d(i, p, j) - \sum_{j_1=1}^{j-1} L_s(i, p, j_1)}{V_g(i, p, j_1)} - \frac{L_d(i, p, j) - \sum_{j_1=1}^{j-1} L_s(i, p, j_1)}{V_a(i, p, j_1)}. \tag{8}$$

In the case of the green-light guidance, if the vehicle speed is guided with the stop line as the reference point, the control delay caused by the speed change can be specified as follows:

$$d_c(i, p, j) = L_d(i, p, j)/V_g(i, p, j) - L_d(i, p, j)/V_a(i, p, j). \tag{9}$$

3.3.2. Queuing Delay and Signal Delay

If $V_g(i,p,j) = V_t(i,p,j)$ or $V_g(i,p,j) = V_{max}$, vehicles can smoothly pass the intersection without stopping, and the queuing delay and signal delay are zero. If $V_g(i,p,j) = V_{min}$, under the guidance of a red light, queuing delay and signal delay are shown in Figure 3.

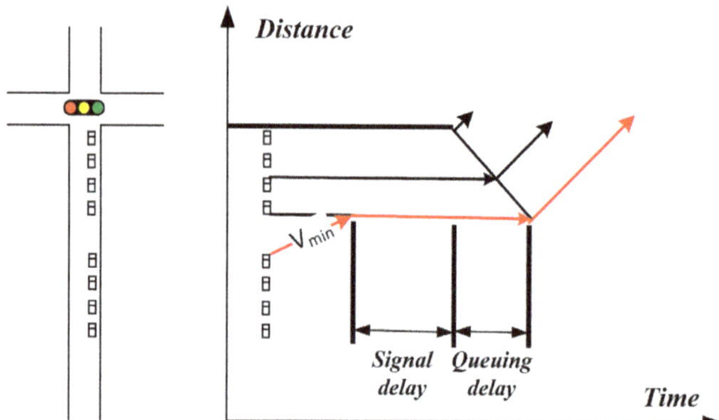

Figure 3. Queuing delay and signal delay caused by traffic control.

In this case, the signal delay and queuing delay can be written as follows:

$$d_s(i,p,j) = t(i,p,k) - t_n - \frac{L_d(i,p,j) - \sum\limits_{j_1=1}^{j-1} L_s(i,p,j_1)}{V_{min}}, \tag{10}$$

$$d_q(i,p,j) = \sum\limits_{j_1=1}^{j-1} L_s(i,p,j_1) / V_s. \tag{11}$$

In the case of the green-light guidance, the queuing delay and signal delay are similar, except that the starting point of the green light in Equation(10) (i.e., $t(i,p,k)$) should be replaced by $t(i,p,k+1)$, which means that when $V_g(i,p,j) = V_{min}$ under the guidance of a green light, this vehicle will pass through the intersection in the next cycle.

3.4. Prediction of Number of Stops

The number of stops is another important metrics to measure the effectiveness of traffic control. In this paper, the number of stops was reduced by adopting a speed guidance to adjust the arrival of traffic flow. Indeed, if $V_g(i,p,j) \geq V_{min}$ in the proposed model, vehicles will pass through the intersection without stopping; if the guided speed is equal to the minimum speed limit, the vehicle will stop, in other words as follows:

$$s(i,p,j) = \begin{cases} 0, & V_g(i,p,j) \geq V_{min} \\ 1, & V_g(i,p,j) = V_{min} \end{cases}. \tag{12}$$

3.5. Intersection Signal Control Model

The intersection signal control model in this paper adopted the Standard North American NEMA Dual-Ring structure, as shown in Figure 4. The left diagram shows the basic shape of the intersection and the distribution of phases in each entrance lane. The right diagram shows the phases contained in each ring, where the first ring contains phases 1, 2, 3, and 4, and the second ring contains phases 5, 6, 7, and 8. All phases are divided into two groups: the first group includes phases 1, 2, 5, and 6, and the second group includes phases 3, 4, 7, and 8. The phases of the second group can only be operated until all phases in the first group finish. Note that based on the NEMA structure, phase skipping and phase insertion are not allowed in this study.

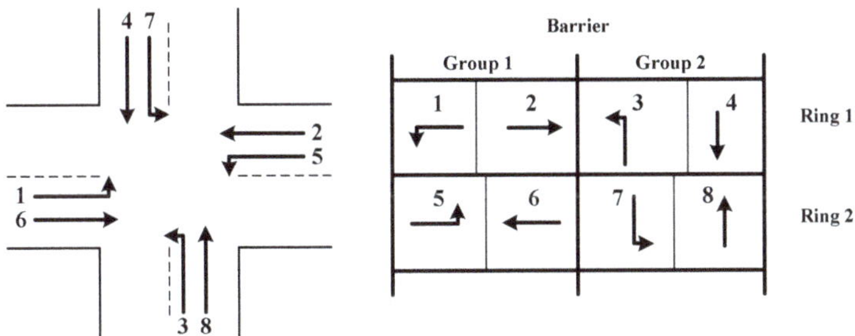

Figure 4. NEMA dual-ring structure for traffic signal control.

Similar to Head et al. [37] and He et al. [30], the dual-ring structure can be modelled according to the time sequence of phase execution, as shown in the following formula.

$$t(i,1,1) = 0 \tag{13}$$

$$t(i,5,1) = 0 \tag{14}$$

$$t(i,2,k) = t(i,1,k) + v(i,1,k) \tag{15}$$

$$t(i,6,k) = t(i,5,k) + v(i,5,k) \tag{16}$$

$$t(i,3,k) = t(i,2,k) + v(i,2,k) \tag{17}$$

$$t(i,3,k) = t(i,6,k) + v(i,6,k) \tag{18}$$

$$t(i,7,k) = t(i,2,k) + v(i,2,k) \tag{19}$$

$$t(i,7,k) = t(i,6,k) + v(i,6,k) \tag{20}$$

$$t(i,4,k) = t(i,3,k) + v(i,3,k) \tag{21}$$

$$t(i,8,k) = t(i,7,k) + v(i,7,k) \tag{22}$$

$$t(i,1,k+1) = t(i,4,k) + v(i,4,k) \tag{23}$$

$$t(i,1,k+1) = t(i,8,k) + v(i,8,k) \tag{24}$$

$$t(i,5,k+1) = t(i,4,k) + v(i,4,k) \tag{25}$$

$$t(i,5,k+1) = t(i,8,k) + v(i,8,k) \tag{26}$$

In the above, phases 1 and 5 are taken as starting points of optimization. The starting time of the green light of phase 2 (i.e., $t(i,2,k)$), is equal to the starting time of the green light of phase 1 (i.e., $t(i,1,k)$), plus the time of duration of phase 1 (i.e., $v(i,1,k)$). Note that the phase duration is equal to the sum of the duration of the green light and the green interval. Furthermore, the duration of the green light must be constrained within the upper and lower bounds, in other words as follows:

$$g_{min} \le g \le g_{max}. \tag{27}$$

3.6. Objective Function

The objective function used in the proposed model was to minimize the total delay of all approaching vehicles around the intersection, which can be specified as follows:

$$\min \sum_{(i,p,j)} d(i,p,j),$$

where $d(i,p,j)$ is the total delay of vehicle j from phase p at intersection i, which consisted of its control delay, queuing delay, and signal delay, in other words as follows:

$$d(i,p,j) = d_c(i,p,j) + d_q(i,p,j) + d_s(i,p,j). \tag{28}$$

The control variables are the following:

Starting time of green light: $t(i,p,k)$
Green light's duration: $g(i,p,k)$
Cycle selection parameters: $\phi(i,p,j,k)$
Vehicle's arrival time: $T(i,p,j)$

In the above control variables, the starting time of the green light (i.e., $t(i,p,k)$) is optimized for all the signal phases of an intersection. $t(i,p,k)$ is mainly restricted by the constraints in Section 3.5 (i.e., intersection signal control model). The vehicle's arrival time (i.e., $T(i,p,j)$) is decided by the vehicle's speed. $T(i,p,j)$ is mainly restricted by the constraints in Section 3.1 (i.e., speed guidance associated with dynamic signal timing plans). Note that the starting time of the green light and the vehicle's arrival time are both control variables that are required to be optimized simultaneously to decrease the intersection delays.

3.7. Optimization Methods

A sliding time window was used to dynamically optimize the arrival time of vehicles and the starting and duration of the green light of each phase. The implementation of the sliding time window is shown in Figure 5.

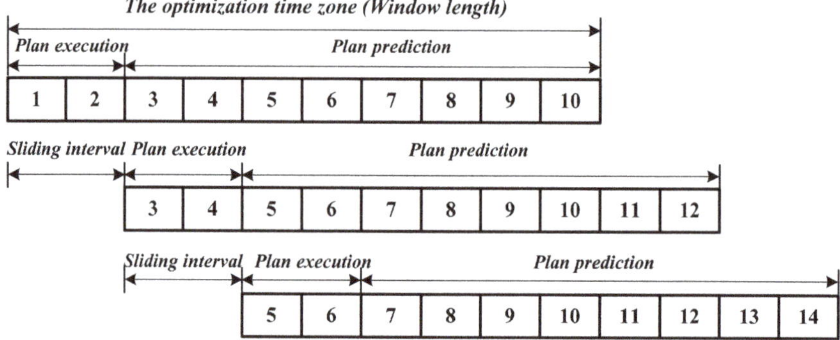

Figure 5. Illustration of the sliding time window method.

As shown in Figure 5, by using the sliding time window, signal timing and the vehicle's speed are optimized in the optimization time zone (i.e., window length). The sliding interval is always identical with the plan execution time. The time duration of the window length is always longer than the sliding interval. This means that the outcomes after each optimization are not implemented for the whole window length. After a sliding interval (or plan execution time), a new optimization for a next window length is activated.

3.8. Optimization Procedures

In this paper, the optimization was based on VISSIM 5.3 simulation software (Planung Transport Verkehr AG, Karlsruhe, Germany) and the C++ programming environment in the Microsoft Visual Studio 2010 platform (Redmond, WA, USA). The COM interface was used to connect VISSIM and the C++ programming environment. The NEMA signal control module was used as the signal control execution module in the VISSIM simulation. With the sliding time window, the dynamic results of the signal control program optimized in the C++ program were transmitted to the NEMA control module for execution. In order to guide the vehicle speed, it was necessary to use the C++ program to modify the "desired speed" of the vehicle in VISSIM. The optimization mainly includes the following steps.

Step 1. Simulation initialization. Build the network model for the intersection in VISSIM software, input traffic flow and initial basic parameters into the NEMA control module.

Step 2. Simulation runs. Using C++ and the COM interface to connect VISSIM software and run the simulation. If the simulation runs to the starting point of a sliding time window, step 3 is executed.

Step 3. Recording the position and the speed of each vehicle in the entrance lane of the intersection, and recording the signal status of the current phase, including the light color (default 1 in VISSIM software is red light, 3 is green light, 4 is yellow light) and the time duration that the current signal has been executed.

Step 4. The model is optimized and solved to generate vehicle speed and signal timing scheme. In C++, all possible signal control schemes that meet all constraints in the current cycle and the next cycle are circulated, mainly including the parameters of the green light's starting time and of the green light's duration. The optimal guiding speed of the vehicle under all possible signal control schemes is calculated, and then the total delay of the vehicle under this guiding speed is calculated. Eventually, the signal control scheme and the speed guidance scheme with the minimum total delay of the vehicle are selected as the optimal scheme to be executed for the next sliding optimization step.

Step 5. Implementing the speed guidance scheme and the signal timing scheme. Note that the implementation of the signal control scheme is achieved by modifying the "Detection" attribute of the NEMA signal control module in the VISSIM–COM interface through the C++ program. The implementation of the speed guidance scheme is achieved by modifying the "Desired Speed" attribute in the VISSIM–COM interface through the C++ program. When step 5 is completed, go to step 2.

4. Model Verification

4.1. Experimental Design

In order to verify the effectiveness of the model proposed in this paper, based on VISSIM microsimulation software, a basic intersection consisting of two approaches (with two lanes at each approach) was used for the simulation. The saturation flow for each lane was 1900 pcu/h. For each entrance lane, the travel demand was 830 vehicles pcu/h for the benchmark case (i.e., the degree of saturation equals 1.0). Furthermore, we varied the travel demand for the performance analysis. For example, if the degree of saturation was λ, $\lambda \geq 0$, then the travel demand was 830λ. The method proposed in this paper was compared with the classical actuated signal control method and the speed guidance with fixed-time-based traffic signal control. The parameters used in each method are as follows.

Case 1. Actuated signal control method. In this method, a loop detector with a length of 18 m was placed at each entrance lane. The duration of the yellow light of each phase was set to 3 s and the all-red time was set to 2 s. Gap-out time of each phase was set to 3 s. The minimum duration of the green light for each phase was set to 5 s and the maximum duration of the green light was set to 35 s.

Case 2. Speed guidance with fixed-time-based traffic signal control proposed by Wu et al. [34]. In Case 2, the cycle time was 80 s. The green duration for each phase was 35 s. The duration of yellow and the all-red times were all identical with those in Case 1.

The proposed method. The length of the time window was set to 10 s. The duration of the yellow light, the duration of the all-red time, and the minimum and maximum duration of the green light were all identical with those in Case 1. The minimum speed limit was 20 km/h and the maximum speed limit was 60 km/h. The communication distance was set to 400 m, that is, when the distance between a connected vehicle and the stop line was less than 400 m, the signal controller was aware of this vehicle's approach and optimized the suggested speed for this vehicle. Later, the communication distance was varied for the sensitivity analysis.

In order to eliminate the impacts of the length of the entrance lane on the simulation results, the length of the entrance lane was set to 2000 m in the simulation. Each simulation ran for 2700 s (45 min), and the first 900 s were the pre-heating period of the simulation. The metrics including average delay and average number of stops were collected in the half-hour period between 900 and 2700 s. The simulation of each set of parameters used 20 different random seeds, and the average performance (i.e., average delay and average number of stops) under these different random seeds was used for comparison and analysis. The simulation results are shown in Figure 6.

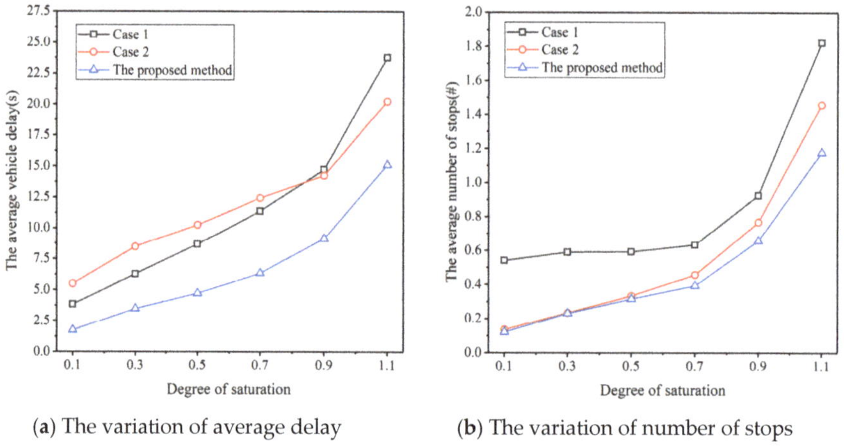

(**a**) The variation of average delay   (**b**) The variation of number of stops

Figure 6. Comparison of average delay and average number of stops at different traffic saturation degrees.

A few observations can be made from Figure 6. First, with the increase in the degree of saturation, the average vehicle delay and the average number of stops of Case 1, Case 2, and the proposed method increased accordingly. However, whether in low saturation or high saturation, if we compare the proposed method with Case 1 (i.e., actuated signal control method), the average delay and the average number of stops are greatly reduced. The reduction range of the average vehicle delay is between 37.8% and 54.0%, and that of the number of stops is between 29.0% and 77.5%, which proves that the proposed method can effectively reduce the delays and number of stops and thus improve the efficiency of the intersection. This is exactly the advantage of the simultaneous optimization of vehicle arrival and signal timing proposed in this paper, which can allocate the green light and arrange vehicle arrivals in a more reasonable way, thereby improving the operation efficiency.

Second, if we compare the proposed method with Case 2 (i.e., speed guidance with fixed-time-based traffic signal control), the results clearly show that the proposed method outperforms Case 2 in terms of average vehicle delay. With respect to the number of stops, when the degree of saturation is relatively low (e.g., no greater than 0.5), the two methods provide almost the same performance. This means that speed guidance is extremely effective at low saturation, even when the signal timing is fixed. This result is similar to that in Wu et al. [34]. As the degree of saturation increases, and by simultaneously optimizing vehicle arrival and signal timing, the proposed method dominates Case 2 in terms of the number of stops.

Third, Case 2 always provides a smaller number of stops when compared with Case 1. This further verifies the effectiveness of speed guidance in decreasing the number of stops. With respect to the average vehicle delay, when the degree of saturation is less than 0.9, Case 1 with the actuated signal control provided a smaller vehicle delay. This occurs because actuated signal control allocates green time in a more reasonable way when the degree of saturation is low. However, at a high degree of saturation (e.g., when the degree of saturation is 1.1), the average vehicle delay in Case 2 with speed guidance is even smaller. This is the case because, when traffic volume is extremely heavy, actuated signal control only produces limited benefits, whereas speed guidance reduces the number of delays for vehicles.

4.2. Sensitivity Analysis

4.2.1. Sensitivity Analysis of the Communication Distance

Communication distance is critical to the success of the proposed model. The signal controller can only acquire the vehicles' arrival information within the communication distance. In addition, the speed guidance can only be conducted in the range of the communication distance. Therefore, the performance of the proposed model under different communication distances had to be investigated. By varying the communication distance from 100 to 500 m and varying the degree of saturation from 0.1 to 1.1, we obtained the simulation results that are shown in Figure 7.

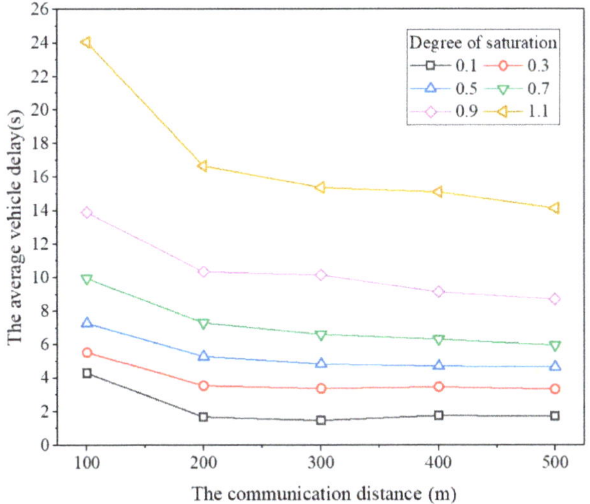

Figure 7. Comparison of average delay at different communication distances.

In Figure 7, it is clearly shown that at a given degree of saturation, the average delay decreases with respect to the communication distance, that is, a larger communication distance provides a smaller average delay. This result is as expected. Furthermore, at a given degree of saturation, the average delay decreases obviously when the communication distance varies from 100 to 200 m. When the communication distance varies from 200 to 500 m, the decrements of average delay are not significant.

4.2.2. Sensitivity Analysis of the Market Penetration of Connected Vehicles

When promoting two-way communication between vehicles and vehicles/infrastructures, it is inevitable that the number of connected vehicles will vary from fewer to more. The market penetration of connected vehicles will reach 100% over a long period of time. In the United States, it may take around 25 years for the connected vehicle's occupancy rate to reach 100% after the new car is installed in the factory [3]. Therefore, it is necessary to discuss the effectiveness of the traffic control method proposed in this paper when connected vehicles and regular vehicles are mixed on the roads. Therefore, the impacts of the variation in the CV penetration on the proposed method were studied.

We fixed the degree of saturation to 0.7 and varied the market penetration of connected vehicles from 10% to 100%. Figure 8a shows the variation of average delays of all vehicles including connected vehicles and regular vehicles at the intersection for the three cases.

Figure 8a shows that as the market penetration of connected vehicles increases, the average vehicle delay for Case 1 (i.e., actuated signal control) stays the same because the actuated signal timing plan is loop-detector-based and does not depend on connected vehicles. For Case 2 and the proposed method, the average vehicle delay decreases with the increase of the market penetration of connected vehicles. In particular, by using the proposed method, as the market penetration of connected vehicles increases from 10% to 100%, the reduction of average vehicle delay is about 43.5%. Furthermore, when the CV penetration is very small, for instance 10%, in terms of vehicle delays, the performance of the proposed method can be better or even worse according to different traffic arrival patterns (or different random seeds; note that different random seeds will result in different traffic arrival patterns). With respect to the average vehicle delay, when CV penetration is 10%, the performance of the proposed method is almost identical to that of the actuated signal control method. This is the case because, when the CV penetration is small, the signal controller can only obtain information from fewer vehicles on the road, and speed guidance is difficult due to the obstruction between vehicles, which leads to poor efficiency. However, when the CV penetration reaches 30%, the efficiency of the traffic control method proposed in this paper is better than that of the actuated signal control method.

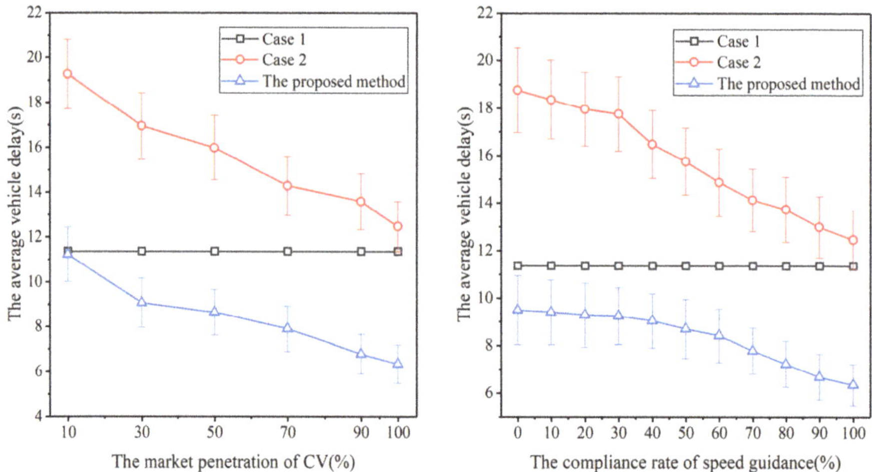

(**a**) The impacts of connected vehicle (CV) penetration (**b**) The impacts of compliance rates

Figure 8. The variation of average delay under different connected vehicle (CV) penetration and compliance rates.

4.2.3. Sensitivity Analysis of the Compliance Rate of Speed Guidance

One may consider that, in a real road traffic environment, not all drivers follow the speed advice from the traffic control system. This means that some drivers may not drive at the recommended speed. Therefore, the effectiveness of the traffic control model proposed in this paper had to be analyzed in this scenario. We fixed the degree of saturation to 0.7 and varied the compliance rate of speed guidance from 0% to 100%. The variation of the average vehicle delay for the three cases is shown in Figure 8b.

Figure 8b shows that as the driver's compliance rate of speed guidance increases, the average vehicle delay decreases for Case 1 and the proposed method (i.e., cases with speed guidance). For

Case 2 without speed guidance, the average delay stays the same. For the proposed method, from a compliance rate of 0% to 100%, the reduction of average vehicle delay is 33.2%, which can be considered as the effectiveness of vehicle speed guidance for the proposed method. In particular, when the compliance rate is 0%, which means all drivers do not follow the recommended speed, all the benefits are generated by signal control. In this extreme case, the proposed method can still reduce the average vehicle delay by about 16.4% compared with the vehicle delay by actuated signal control method. This further verifies that by using the trajectory data from connected vehicles, the proposed traffic control method can better allocate space–time resources and improve traffic efficiency compared with the actuated signal control method.

5. Conclusions

In this paper, an integrated traffic control model was established to optimize the vehicle arrival time and signal timing simultaneously, which overcomes the disadvantages of most existing traffic signal control models, which only passively adapt to vehicle arrivals. Based on the two-way communication between vehicle and signal controller, a speed guidance model, a vehicle arrival model, and a prediction model of delays and number of stops were proposed. The objective function was to minimize the average vehicle delay for all approaching vehicles. A sliding time window technique was used for dynamic optimization. VISSIM simulation results show that the proposed method can outperform both the classical actuated signal control method and the fixed-time-based speed guidance model. In particular, compared with the actuated signal control model, the proposed method can significantly reduce vehicle delays by about 37.8%–54.0% and the number of stops by about 29.0%–77.5%. A sensitivity analysis of the communication distance, the market penetration of connected vehicles, and the compliance rate of speed guidance further shows the effectiveness of the proposed model. Specifically, the proposed model can obtain similar or even better benefits compared with the classical actuated signal control when the CV penetration or the compliance rate of speed guidance is low. This verifies the advantages of the proposed traffic control model that simultaneously optimizes the vehicle arrival time and signal timing.

In order to highlight the effectiveness of the proposed traffic control model, this paper only used the trajectory data that could be obtained based on two-way communication. In fact, other sources of data such as a loop detector or video can also be collected. Further studies should be conducted to make use of each data source and improve the benefits of traffic control by data fusion methods, especially when the market penetration of connected vehicles is low. The case study used in this paper only analyzes a basic signalized intersection; more intensive performance testing of the proposed model in large-scale road networks while considering multiple road users such as pedestrians and cyclists may also be investigated in a future study.

Author Contributions: In this paper, W.W. participated in the development of optimization model and carried out the simulation. L.H. committed to the data and result validation of the numerical and sensitivity analysis, she also completed the writing work of the whole manuscript. R.D. participated in the development of the research ideas and designed the research programs. All authors have read and agreed to the published version of the manuscript.

Funding: This research was partially supported by the National Natural Science Foundation of China under Grant Nos. 61773077 and 61973047, the Research Foundation of Education Bureau of Hunan Province under Grant No. 19A023, and the Open Fund of the Key Laboratory of Highway Engineering of the Ministry of Education (Changsha University of Science and Technology) under Grant No. kfj170201.

Acknowledgments: We wish to thank the three anonymous reviewers for their valuable and constructive comments, which helped to substantially improve both the technical quality and the presentation of this paper.

Conflicts of Interest: The authors declare no conflict of interest.

References

1. Mirchandani, P.; Head, L. A real-time traffic signal control system: Architecture, algorithms, and analysis. *Transp. Res. Part C Emerg. Technol.* **2001**, *9*, 415–432. [CrossRef]

2. Lo, H.K. A novel traffic signal control formulation. *Transp. Res. Part A Policy Pract.* **1999**, *33*, 433–448. [CrossRef]

3. Feng, Y.; Head, K.L.; Khoshmagham, S.; Zamanipour, M. A real-time adaptive signal control in a connected vehicle environment. *Transp. Res. Part C Emerg. Technol.* **2015**, *55*, 460–473. [CrossRef]

4. Serafini, P.; Ukovich, W. A mathematical model for the fixed-time traffic control problem. *Eur. J. Oper. Res.* **1989**, *42*, 152–165. [CrossRef]

5. Muralidharan, A.; Pedarsani, R.; Varaiya, P. Analysis of fixed-time control. *Transp. Res. Part B Methodol.* **2015**, *73*, 81–90. [CrossRef]

6. Hao, P.; Wu, G.; Boriboonsomsin, K.; Barth, M.J. Eco-approach and departure (EAD) application for actuated signals in real-world traffic. *IEEE Trans. Intell. Transp. Syst.* **2018**, *20*, 30–40. [CrossRef]

7. Shiri, M.S.; Maleki, H.R. Maximum Green Time Settings for Traffic-Actuated Signal Control at Isolated Intersections Using Fuzzy Logic. *Int. J. Fuzzy Syst.* **2017**, *19*, 247–256. [CrossRef]

8. Liu, Z.; Bie, Y. Comparison of hook-turn scheme with U-turn scheme based on actuated traffic control algorithm. *Transp. A Transp. Sci.* **2015**, *11*, 484–501. [CrossRef]

9. Wu, W.; Liu, Y.; Liu, W.; Zhang, F.; Rey, D.; Dixit, V. An integrated approach for optimizing left-turn forbiddance decisions at multiple intersections. *Transp. B Transp. Dyn.* **2019**, *7*, 1481–1504. [CrossRef]

10. Mannion, P.; Duggan, J.; Howley, E. An experimental review of reinforcement learning algorithms for adaptive traffic signal control. In *Autonomic Road Transport Support Systems*; Springer: Berlin/Heidelberg, Germany, 2016; pp. 47–66.

11. Lee, S.; Wong, S.; Varaiya, P. Group-based hierarchical adaptive traffic-signal control part I: Formulation. *Transp. Res. Part B Methodol.* **2017**, *105*, 1–18. [CrossRef]

12. Zeng, J.; Hu, J.; Zhang, Y. Adaptive traffic signal control with deep recurrent Q-learning. In Proceedings of the 2018 IEEE Intelligent Vehicles Symposium (IV), Changshu, China, 26–30 June 2018; pp. 1215–1220.

13. Liu, J.; Cai, B.; Wang, J. Cooperative localization of connected vehicles: Integrating GNSS with DSRC using a robust cubature Kalman filter. *IEEE Trans. Intell. Transp. Syst.* **2016**, *18*, 2111–2125. [CrossRef]

14. Abboud, K.; Omar, H.A.; Zhuang, W. Interworking of DSRC and cellular network technologies for V2X communications: A survey. *IEEE Trans. Veh. Technol.* **2016**, *65*, 9457–9470. [CrossRef]

15. Wu, X.; Subramanian, S.; Guha, R.; White, R.G.; Li, J.; Lu, K.W.; Bucceri, A.; Zhang, T. Vehicular communications using DSRC: Challenges, enhancements, and evolution. *IEEE J. Sel. Areas Commun.* **2013**, *31*, 399–408.

16. Camacho, F.; Cárdenas, C.; Muñoz, D. Emerging technologies and research challenges for intelligent transportation systems: 5G, HetNets, and SDN. *Int. J. Interact. Des. Manuf.* **2018**, *12*, 327–335. [CrossRef]

17. Wang, J.; Ni, D.; Li, K. RFID-based vehicle positioning and its applications in connected vehicles. *Sensors* **2014**, *14*, 4225–4238. [CrossRef]

18. Tang, T.; Shi, W.; Shang, H.; Wang, Y. A new car-following model with consideration of inter-vehicle communication. *Nonlinear Dyn.* **2014**, *76*, 2017–2023. [CrossRef]

19. Wu, W.; Head, L.; Yan, S.; Ma, W. Development and evaluation of bus lanes with intermittent and dynamic priority in connected vehicle environment. *J. Intell. Transp. Syst.* **2018**, *22*, 301–310. [CrossRef]

20. Guo, Q.; Li, L.; Ban, X.J. Urban traffic signal control with connected and automated vehicles: A survey. *Transp. Res. Part C Emerg. Technol.* **2019**, *101*, 313–334. [CrossRef]

21. Harding, J.; Powell, G.; Yoon, R.; Fikentscher, J.; Doyle, C.; Sade, D.; Lukuc, M.; Simons, J.; Wang, J. *Vehicle-to-vehicle Communications: Readiness of V2V Technology for Application*; National Highway Traffic Safety Administration: Washington, DC, USA, 2014.

22. Lozano Domínguez, J.M.; Mateo Sanguino, T.J. Review on V2X, I2X, and P2X Communications and Their Applications: A Comprehensive Analysis over Time. *Sensors* **2019**, *19*, 2756. [CrossRef]

23. Omidvar, A.; Pourmehrab, M.; Emami, P.; Kiriazes, R.; Esposito, J.C.; Letter, C.; Elefteriadou, L.; Crane, C.D., III; Ranka, S. Deployment and testing of optimized autonomous and connected vehicle trajectories at a closed-course signalized intersection. *Transp. Res. Rec.* **2018**, *2672*, 45–54. [CrossRef]

24. Goli, S.A.; Far, B.H.; Fapojuwo, A.O. Vehicle Trajectory Prediction with Gaussian Process Regression in Connected Vehicle Environment. In Proceedings of the 2018 IEEE Intelligent Vehicles Symposium (IV), Changshu, China, 26–30 June 2018; pp. 550–555.

25. Tiaprasert, K.; Zhang, Y.; Wang, X.B.; Zeng, X. Queue length estimation using connected vehicle technology for adaptive signal control. *IEEE Trans. Intell. Transp. Syst.* **2015**, *16*, 2129–2140. [CrossRef]

26. Li, J.-Q.; Zhou, K.; Shladover, S.E.; Skabardonis, A. Estimating queue length under connected vehicle technology: Using probe vehicle, loop detector, and fused data. *Transp. Res. Rec.* **2013**, *2356*, 17–22. [CrossRef]

27. Christofa, E.; Argote, J.; Skabardonis, A. Arterial queue spillback detection and signal control based on connected vehicle technology. *Transp. Res. Rec.* **2013**, *2366*, 61–70. [CrossRef]

28. Gao, K.; Han, F.; Dong, P.; Xiong, N.; Du, R. Connected Vehicle as a Mobile Sensor for Real Time Queue Length at Signalized Intersections. *Sensors* **2019**, *19*, 2059. [CrossRef]

29. Goodall, N.J.; Smith, B.L.; Park, B. Traffic signal control with connected vehicles. *Transp. Res. Rec.* **2013**, *2381*, 65–72. [CrossRef]

30. He, Q.; Head, K.L.; Ding, J. PAMSCOD: Platoon-based arterial multi-modal signal control with online data. *Transp. Res. Part C Emerg. Technol.* **2012**, *20*, 164–184. [CrossRef]

31. Wan, N.; Vahidi, A.; Luckow, A. Optimal speed advisory for connected vehicles in arterial roads and the impact on mixed traffic. *Transp. Res. Part C Emerg. Technol.* **2016**, *69*, 548–563. [CrossRef]

32. Ramezani, H.; Benekohal, R. Optimized speed harmonization with connected vehicles for work zones. In Proceedings of the 2015 IEEE 18th International Conference on Intelligent Transportation Systems, Washington, DC, USA, 15–18 September 2015; pp. 1081–1086.

33. Tajalli, M.; Hajbabaie, A. Dynamic speed harmonization in connected urban street networks. *Comput.-Aided Civ. Infrastruct. Eng.* **2018**, *33*, 510–523. [CrossRef]

34. Wu, W.; Li, P.; Zhang, Y. Modelling and simulation of vehicle speed guidance in connected vehicle environment. *Int. J. Simul. Model.* **2015**, *14*, 145–157. [CrossRef]

35. Tang, T.; Zhang, J.; Liu, K. A speed guidance model accounting for the driver's bounded rationality at a signalized intersection. *Phys. A Stat. Mech. Appl.* **2017**, *473*, 45–52. [CrossRef]

36. Tang, T.; Yi, Z.; Zhang, J.; Wang, T.; Leng, J. A speed guidance strategy for multiple signalized intersections based on car-following model. *Phys. A Stat. Mech. Appl.* **2018**, *496*, 399–409. [CrossRef]

37. Head, L.; Gettman, D.; Wei, Z. Decision model for priority control of traffic signals. *J. Transp. Res. Rec.* **2006**, *1978*, 169–177. [CrossRef]

Article

An Autonomous Path Planning Model for Unmanned Ships Based on Deep Reinforcement Learning

Siyu Guo, Xiuguo Zhang *, Yisong Zheng and Yiquan Du

School of Information Science and Technology, Dalian Maritime University, Dalian 116026, China;
guosy@dlmu.edu.cn (S.G.); zhengyisong@dlmu.edu.cn (Y.Z.); duyiquan@dlmu.edu.cn (Y.D.)
* Correspondence: zhangxg@dlmu.edu.cn; Tel.: +86-185-5305-9562

Received: 27 November 2019; Accepted: 2 January 2020; Published: 11 January 2020

Abstract: Deep reinforcement learning (DRL) has excellent performance in continuous control problems and it is widely used in path planning and other fields. An autonomous path planning model based on DRL is proposed to realize the intelligent path planning of unmanned ships in the unknown environment. The model utilizes the deep deterministic policy gradient (DDPG) algorithm, through the continuous interaction with the environment and the use of historical experience data; the agent learns the optimal action strategy in a simulation environment. The navigation rules and the ship's encounter situation are transformed into a navigation restricted area, so as to achieve the purpose of planned path safety in order to ensure the validity and accuracy of the model. Ship data provided by ship automatic identification system (AIS) are used to train this path planning model. Subsequently, the improved DRL is obtained by combining DDPG with the artificial potential field. Finally, the path planning model is integrated into the electronic chart platform for experiments. Through the establishment of comparative experiments, the results show that the improved model can achieve autonomous path planning, and it has good convergence speed and stability.

Keywords: unmanned ships; deep reinforcement learning; DDPG; autonomous path planning; end-to-end; collision avoidance

---

1. Introduction

With the density of maritime traffic increasing, various types of marine accidents frequently occur. 80% of marine accidents are caused by human factors, according to the world maritime disaster records that were filed by international maritime organization (IMO) from 1978 to 2008 [1]. Therefore, improving the autonomous driving level of ships has become an urgent problem to be solved. On the other hand, the environment that is faced by ships is more and more complex. In some cases, it is not suitable for manned ships to go to the workplace to carry out tasks, while unmanned ships are more suitable for dealing with the complex and changeable harsh environment at sea. This requires unmanned ships to have the ability of autonomous path planning and obstacle avoidance, so as to efficiently complete tasks and enhance the comprehensive operation ability [2].

With their strong autonomy and adaptability, unmanned ships have gradually become the new research direction pursued by the current industry [3]. Unmanned ships can not only perform tasks independently in dangerous sea areas, but also cooperate with manned ships to improve work efficiency. They are widely used in marine exploration, military tasks, material transportation, and other fields. Autonomous path planning is an essential key technology in order to improve the self-determination ability of unmanned ships. The autonomous path planning of unmanned ships needs to optimize the path in the safe navigation area according to certain navigation rules and crew experience. It can safely avoid obstacles and independently plan an optimal trajectory from the known starting point to the target point. Unmanned ships often face complex and variable navigational environments. Therefore, it is necessary to adopt a continuous and effective method for controlling the trajectory of the ship during the voyage, thus ensuring the safety of the ship [4]. The research directions of unmanned ships are mainly divided into autonomous path planning, navigation control, autonomous collision avoidance, and semi-autonomous task execution. Autonomous path planning plays a key role in ship automation and practical application, as the basis and premise of autonomous navigation [5]. In the actual navigation process, ships often meet with other ships, which requires reasonable methods to guide ships to avoid other ships and achieve the target point. Therefore, it is valuable to consider how to avoid dynamic obstacles in the process of path planning. The unmanned ship path planning method can guide the ship to take the optimal action and avoid other obstacles. At the same time, it can also be divided into the navigable area and the obstacle area by the method according to the obstacle information to realize the function of avoiding the obstacle in the local area. Many scholars have completed relevant research and experiments to solve the problem of autonomous path planning for unmanned ships. However, traditional path planning methods usually require relatively complete environmental information as prior knowledge, and it is quite difficult to obtain surrounding environment information in an unknown sea environment. Moreover, the traditional algorithm has a large amount of computation, which makes it difficult to realize the real-time behavior decision-making of ships, resulting in large mistake information of path planning.

In recent years, artificial intelligence technology has been rapidly developed and applied. Deep Learning (DL) [6] and Reinforcement Learning (RL) [7] have achieved great success in many fields. DL has a strong perceptual ability and RL has decision-making ability. DRL [8] is obtained by combining the advantages of DL and RL, which provides a solution to the perceptual decision-making problem of complex systems. DRL can effectively solve the problem of continuous state space and action space. It directly takes original data as input and output results as execution action, realizes an end-to-end learning mode, and greatly improves the efficiency and convergence of the algorithm. At present, DRL has been widely used in robotic control [9], automatic driving [10–12], financial prediction [13], traffic control [14], and other fields. Artificial intelligence technology is gradually infiltrating into various fields and the concept of unmanned autonomy is getting closer. As an important part of the transportation field, unmanned ships are moving forward in the direction of intelligence and autonomy.

RL has attracted widespread attention in recent years, emphasizing the learning of agents from the environment to behavioral mapping and seeking the most accurate or optimal action decisions by maximizing the value function. Mnih, V et al. [15] proposed a Deep Q-Network (DQN) algorithm, which opened a new era of DRL. The DQN algorithm utilizes the powerful function fitting ability of the deep neural network to avoid the huge storage space of the Q table, and the experience replay memory and target network are used to enhance the stability of the training process. At the same time, DQN implements an end-to-end learning method, and only original data are used as input, and the output result is the Q value of each action. The DQN algorithm has achieved great success in discrete action, but it is difficult to achieve high-dimensional continuous action. If continuously changing action is infinitely split, the number of actions exponentially increases with increasing degrees of freedom, which leads to the problem of latitude catastrophe and can cause great training difficulties. In addition, simply discretizing the action removes important information regarding the structure of the action domain. The Actor-Critic (AC) algorithm [16] is capable of handling continuous action problems and it is widely used in continuous action spaces. The AC algorithm network structure includes Actor network and Critic network. The Actor network is responsible for outputting the probability value of the action. The Critic network evaluates the output action. In this way, the network parameters are continuously optimized and the optimal action strategy is obtained, but the random strategy of the AC algorithm makes the network difficult to converge. Lillicrap, T.P et al. [17] offered the Deep Deterministic Policy Gradient (DDPG) algorithm to solve the problem of DRL in continuous state action space. DDPG is a model-free algorithm that combines the advantages of the DQN algorithm with an experience replay memory and a target network. At the same time, the AC algorithm that is based on the Deterministic Policy Gradient (DPG) is used to make the network output result a certain action value, which ensures that DDPG can be applied to the continuous action space field.

The DDPG can be easily applied to complex problems and larger network structures with simplicity and convergence. Zhu, M et al. [11] proposed a framework for human-like autonomous car-following planning based on DDPG. In this framework, unmanned cars learn from the environment through trial and error. Finally, the path planning model of unmanned cars was obtained and it has good experimental results. This research shows that DDPG can gain insight into driver behavior and can help to develop human-like autopilot algorithms and traffic flow models.

This paper proposes an autonomous path planning model for unmanned ships that are based on the DRL method. The essence of the model is that the agent independently finds the most efficient path through the enumerated method, which might be closer to human manipulation. At the same time, the ship motion is transformed into a continuous motion control problem, so that it conforms to the actual ship's motion characteristics. The above ideas have been implemented in this paper. Firstly, the interaction mode between ships and environment is analyzed, and a virtual computing environment for unmanned ships autonomous path planning close to the real world is established. Secondly, the model is defined and the network structure and parameters of the DDPG algorithm are set, and the action exploration strategy and reward function are designed. At the same time, the international maritime collision avoidance rules (COLREGS) and crew's experience are quantified, and the attraction strategy to the target point is added, to ensure the standardization of navigation and avoid the algorithm falling into local optimum. Finally, historical experience data is stored in the memory pool and the neural network parameters are updated by random extraction, thereby reducing the relevance of the data and improving the learning efficiency of the algorithm. Besides, this paper combines the artificial potential field (APF) with the DDPG to obtain the APF-DDPG based autonomous ship autonomous path planning model. The APF-DDPG model has higher decision making power and faster convergence. The model is integrated with the electronic chart platform in order to evaluate the validity and accuracy of the model, and experiments are carried out under the conditions of a single ship encounter and multi-ship encounter, respectively. Moreover, five sets of verification experiments of the DQN, AC, DDPG, APF-DDPG, and Q-learning [18] algorithms are designed as comparative cases. The results show that the APF-DDPG has high convergence speed and planning efficiency, the planned path is more in line with navigation rules, and the autonomous path planning of unmanned ships is realized.

The rest of this paper is structured, as follows. Section 2 reviews the related works. Section 3 presents the autonomous path planning model for unmanned ships that are based on DRL. Section 4 presents a comparative analysis of the simulation experiment process and experimental results. Finally, Section 5 concludes the paper.

2. Related Research

At present, research on autonomous path planning for unmanned ships has been carried out at domestic and foreign. The methods include traditional algorithms, such as APF, velocity obstacle method, and A* algorithm, as well as some intelligent algorithms, such as ant colony optimization algorithm, genetic algorithm, neural network algorithm, and other related algorithms of DRL.

In terms of traditional algorithms, Petres, C et al. [19] used APF to construct a virtual gravitational field to guide the autonomous surface vehicles (ASV) to the target point. The ASV avoids obstacles and performs path planning in a complex navigation environment by transforming the navigation restricted area into a virtual obstacle area. The APF has the advantages of high computational efficiency and simple algorithm, but it is necessary to set reasonable potential field parameters to avoid falling into local minimum. The practicality and effectiveness of the algorithm can be guaranteed in the process of unmanned ships autonomous path planning, combined with the COLREGS [20]. Kuwata, Y et al. [21] adopted the velocity obstacle method with COLREGS, and presented a path planning method for unmanned surface vehicles (USV) to navigate safely in dynamic, cluttered environments. The experimental results show that USV can achieve better obstacle avoidance and path planning. Campbell, S et al. [22] showed a USV real-time path planning method that was based on improved

A* algorithm, which is integrated with decision-making framework and combined with COLREGS. The results show that the method realizes real-time path planning of the USV in complex navigation environment. However, the A* algorithm relies on the design of the grid map, and the size and number of grids will directly affect the calculation speed and accuracy of the algorithm. Xue, Y et al. [23] introduced a path planning method of unmanned ships that are based on APF, and combined with COLREGS. The experimental showed that the method could effectively realize unmanned ships path search and collision avoidance in complex environments. However, this method is difficult to deal with ships autonomous path planning and collision avoidance problems in an unknown restricted navigation environment.

In addition, many intelligent algorithms, such as genetic algorithms, ant colony optimization algorithms, and neural network algorithms, have also been used in the autonomous path planning problem of unmanned ships. For example, Vettor, R et al. [24] used the optimization genetic algorithm to calculate the environmental information as the initial population to obtain the navigation path that satisfies the requirements. Lazarowska, A et al. [25] proposed the ant colony optimization algorithm to transform ships path planning and collision avoidance problems into dynamic optimization problems, with collision risk and range loss as the objective function. The optimal planning path and collision avoidance strategy were obtained based on the motion prediction of dynamic obstacles. Xin, J et al. [26] adopted the improved genetic algorithm to increase the number of offspring by using the multi-domain inversion. The result shows that the algorithm is superior with a desirable balance between the path length and time-cost, and it has a shorter optimal path, a faster convergence speed, and better robustness. Xie, S et al. [27] proposed a predictive collision avoidance method for under-actuated surface ships based on the improved beetle antena search (BAS) method. A predictive optimization strategy for real-time collision avoidance is established and COLREGS is used as a constraint condition while considering the minimization of safety and economic cost. The simulation experiments verify the effectiveness of the improved BAS method. However, such intelligent algorithms are usually computationally intensive, and they are mainly used in offline global path planning or auxiliary decision making, and they are difficult to be used for real-time ship action decision problems.

In the field of intelligent ships, the application of DRL to the control of unmanned ships has gradually become a new research field. For example, Chen, C et al. [18] introduced a path planning and maneuvering method for unmanned cargo ships that were based on Q-learning. This method can learn the action reward model and it obtains the best action strategy. After enough rounds of training, the ship can find the right path or navigation strategy by itself. Fu, K et al. [28] proposed a ship rotation detection model based on a feature fusion pyramid network and DRL (FFPN-RL). The ship's independent guidance and docking function is realized by applying the Dueling Q network to the inclined ship detection task. Yang, J et al. [29] showed autonomous navigation control that is based on relative value iterative gradient (RVIG) algorithm for unmanned ships, and designed the navigation environment of the ship while using Unity3D game engine software. The simulation results showed that the unmanned ships could successfully avoid obstacles and reached the destination in a complex environment. Shen, H.Q et al. [30] offered a method that was based on the Dueling DQN algorithm for automatic collision avoidance of multiple ships, and combined ship manoeuvrability, crew's experience, and COLREGS to verify the path planning and collision avoidance capability of unmanned ships. Zhang, R.B et al. [31], based on the Sarsa on-policy algorithm, proposed a behavior-based USV local path planning and obstacle avoidance method, and tested in real marine environment. WANG, Y et al. [32] introduced a USV course tracking control plan by combining DDPG algorithm and achieved good experimental results. Zhang, X et al. [33] proposed an adaptive navigation method that was based on maritime autonomous surface ships for hierarchical DRL. The method is combined with the ship maneuverability and navigation rules COLREGS for training and learning. The results show that the method can effectively improve the navigation safety and avoid collision. Zhao, L et al. [34] used the proximal policy optimization (PPO) algorithm, combined with the ship motion model and navigation rules, the unmanned ship autonomous collision avoidance model in the multi-ship environment is proposed. The experimental results show that the model could obtain the time efficiency and collision-free path of multiple ships, and it has good adaptability to unknown complex environments. In addition, as the human factor is composed of navigators' subjectivity and indeterminacy of the navigation situation in the actual navigation process, it is usually has a game character in the actual ship control process [35]. The DRL overcomes the shortcomings of usual intelligent algorithm, which requires a certain number of samples. At the same time, it has less error and response time.

Many critical autonomous path planning methods have been proposed in the field of unmanned ships. However, these methods are mainly focused on the research of small and medium-sized USV, while the research on unmanned ships is relatively rare, and few experts currently apply the DDPG to unmanned ships path planning. This paper chooses DDPG for unmanned ships path planning, because it has powerful deep neural network function fitting ability and better generalized learning ability. The fitting ability can achieve higher precision when approaching the motion of the ship, and the learning ability can enable the ship to obtain the action output close to the human characteristic from the decision behavior, which is helpful in further developing the human-like traffic models. In addition, the outputs of algorithms, such as Q-learning and DQN, are discrete behaviors, which will lead to the discontinuity and incompleteness of the actions. The DDPG has the advantages of fast convergence speed and continuous action space. When compared with the traditional path planning, the method proposed in this paper has more continuous action output and less decision error when the unmanned ship is sailing.

3. Construction of Autonomous Path Planning Model for Unmanned Ships Based on DRL

3.1. Deep Reinforcement Learning

As an important field of machine learning, DL can extract high-precision feature samples from the original input data. This method has greatly promoted the development of visual object recognition, object detection, and voice recognition. DL uses deep neural networks (DNN) to automatically learn the features of high-dimensional data. The core idea of DL is to update the network parameters through the back propagation method and then discover the distributed characteristics of the data from the training process.

RL, as another field of machine learning, is widely used in robot control, game simulation, and scheduling optimization. The purpose of RL is to maximize the cumulative reward that was obtained by agents in the training process and learn the optimal strategy to achieve the goal. At the same time, the agent directly interacts with the environment information to replace the large amount of training data by evaluating the action value. The principle is shown in Figure 1. The agent observes the state s_t from the environment and makes action a_t that is based on a certain policy. The environment then makes feedback reward r_t for the executed action and it moves to the new state s_{t+1}. The agent uses the new state and reward r_{t+1} to select the action again and keep repeating to maximize the long-term accumulation index of reward.

The DRL is an end-to-end learning method that combines DL and RL, while using the perception ability of DL and the decision-making ability of RL [36]. It has made great progress in continuous motion control and it can effectively solve the shortcomings of traditional unmanned ships in path planning. DDPG is a type of algorithm in DRL that can be used to solve continuous action spaces problems. Among them, deep refers to the deep network structure and policy gradient is the strategy gradient algorithm, which can randomly select actions in the continuous action space according to the learned strategies (action distribution). The purpose of deterministic is to help the policy gradient avoid random selection and to output a specific action value. The DDPG is being gradually applied to the transportation sector, especially in the areas of driverless cars and unmanned ships. These vehicles (buses, small cars, boats, and cargo ships, etc.) often require continuous motion control as well as traffic rules and human operating experience to ensure driving safety. The DDPG can solve the above problems well with its powerful self-learning ability and function fitting ability. Therefore, DDPG has broad prospects and expansion space in the maritime domain.

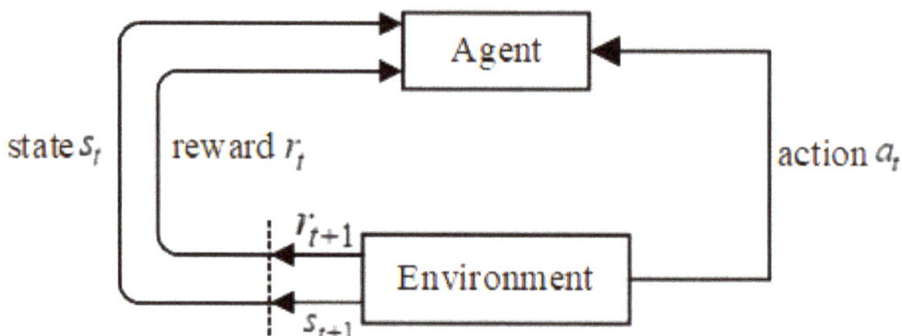

Figure 1. The Principle of Reinforcement Learning.

3.2. The Principle of DDPG Algorithm

3.2.1. AC Algorithm

DDPG is based on AC algorithm, whose structure is shown in Figure 2.

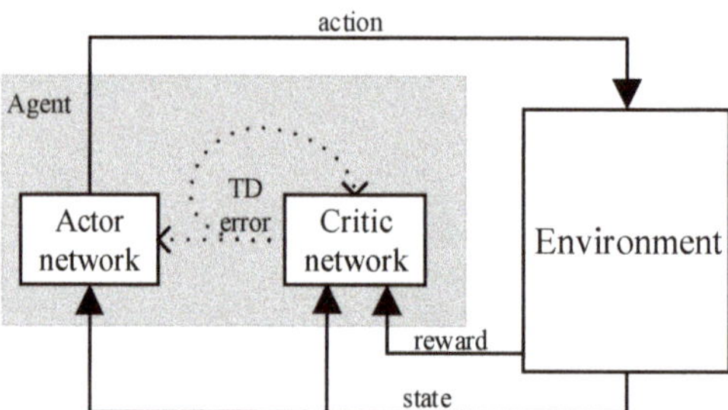

Figure 2. Actor-Critic (AC) algorithm structure.

The network structure of the AC framework includes a policy network and an evaluation network. The policy network is called an Actor network, and the evaluation network is called a Critic network. The Actor network is used to select actions corresponding to the DDPG, and the Critic network evaluates the merits and demerits of the selected actions by calculating the value function.

The Actor network and the Critic network are two separate networks that share state information. The Actor network uses state information to generate actions, while the environment feeds back the resulting actions and outputs reward. The Critic network uses state and reward to estimate the value of the current action and constantly adjust its own value function. Meanwhile, the Actor network updates its action strategy in the direction of improving the value of the action. In this cycle, the Critic network evaluates the action strategy by means of a value function, giving the Actor network a better gradient estimate, and finally obtaining the optimal action strategy. It is very important to evaluate the action strategy in the Critic network, which is more conducive to the convergence and stability of the current Actor network. The above features ensure that the AC algorithm can obtain the optimal action strategy with the gradient estimation at a lower variance.

3.2.2. DDPG Algorithm Structure

Figure 3 shows the structure of DDPG algorithm.

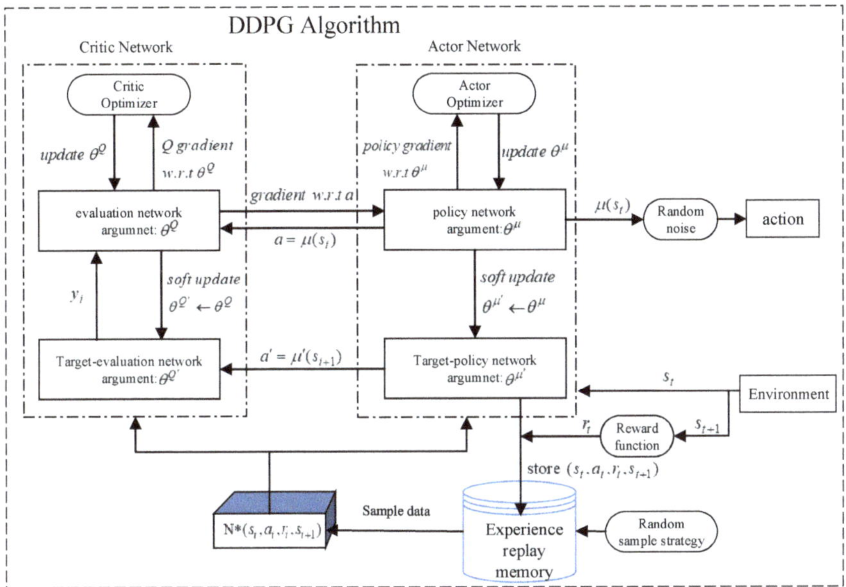

Figure 3. Deep Deterministic Policy Gradient (DDPG) algorithm structure.

The DDPG algorithm takes the information of the initial state as the input, and the output result is the action strategy $\mu(s_t)$ calculated by the algorithm. Afterwards, the random noise is added to the action strategy to obtain the final output action, and this process is a typical end-to-end learning mode. When starting the task, the agent outputs an action according to the current state s_t. The reward function is designed and the action is evaluated in order to verify the validity of the output action, thereby obtaining a feedback reward r_t of the environment. The action that is beneficial to the agent to achieve the goal gives a positive reward and, on the contrary, gives a negative reward. Afterwards, the current state information, the action, the reward, and the state information of the next time (s_t, a_t, r_t, s_{t+1}) are stored in the experience buffer pool. At the same time, the neural network trains experience and continuously adjusts action strategy by randomly extracting sample data from the experience buffer pool, and uses the gradient descent approach to update and iterate network parameters, so as to further enhance the stability and accuracy of the algorithm.

DDPG combines with DQN on the premise of AC algorithm in order to further enhance the stability and effectiveness of network training, which makes it more conducive in the filed of solving the problem of continuous state and action space. Additionally, DDPG uses DQN's the experience replay memory and the target network to solve the problem of non-convergence when using neural network to approximate the function value. Meanwhile, DDPG subdivides the network structure into online network and target network. The online network is used to output the actions in real time, evaluate actions, and update network parameters through online training, which includes online Actor network and online Critic network, respectively. The target network includes target Actor network and target Critic network, which are used to update the value network system and the Actor network system, but not carry out online training and updating of network parameters. The target network and the online network have the same neural network structure and initialization parameters. In the training process, the parameters of the target Actor network and target Critic network are updated in a way of slow change (Soft Replace), instead of directly copying the parameters of online Actor network and online Critic network, so as to further enhance the stability of the training process. The following formulas describe how to update the parameters.

$$\begin{cases} \theta^{Q'} = \tau\theta^Q + (1-\tau)\theta^{Q'} \\ \theta^{\mu'} = \tau\theta^\mu + (1-\tau)\theta^{\mu'} \end{cases} \tag{1}$$

where, θ^μ and θ^Q are the parameters of the online Actor network and the online Critic network, $\theta^{\mu'}$ and $\theta^{Q'}$ are the parameters of the target Actor network and the target Critic network, $\tau \ll 1$ is the approximation coefficient. Figure 4 is the flow chart of the DDPG algorithm.

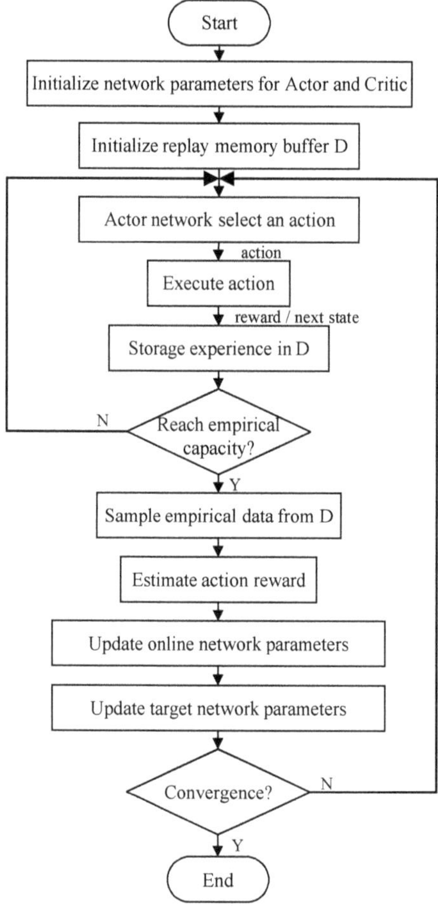

Figure 4. DDPG algorithm flow chart.

3.3. Structure Design of Autonomous Path Planning Model for Unmanned Ships Based on DDPG Algorithm

The DDPG algorithm is a combination of deep learning and reinforcement learning. Based on this algorithm, this paper designs an autonomous path planning model for unmanned ships. The algorithm structure mainly includes AC algorithm, experience replay mechanism, and neural network. The output action of the model is gradually accurate by using the AC algorithm to output and judge the ship's action strategy. By using the experience replay mechanism, the historical information of the DDPG algorithm executed in the experimental environment is remembered, and the data are randomly extracted for training and learning. In the course of the experiment, the number of environmental states and behavioral states of the ship is large, so it is necessary to use neural networks for fitting and generalization. The current state of the unmanned ship obtained in the environment is used as the input of the neural network, and the output is the Q value of actions that the ship can perform in the current state. The model can learn the optimal action strategy in the current state by continuously training and adjusting neural network parameters.

3.3.1. Model Structure

Figure 5 shows the model structure of unmanned ships path planning based on the DDPG algorithm. The model mainly includes AC algorithm, Environment (Ship Action Controller and Ship Navigation Information Fusion Module), and Experience Replay Memory. Among them, the model obtains the environmental information and the ship's state data through the Ship Navigation Information Fusion Module, and it serves as the input state of the AC algorithm. The optimal ship action strategy is output, which satisfies the ship's maximum cumulative return during the learning process, by randomly extracting data from the experience buffer pool for repeated training and learning. Finally, unmanned ships can avoid obstacles and reach the destinations with the help of the Ship Action Controller.

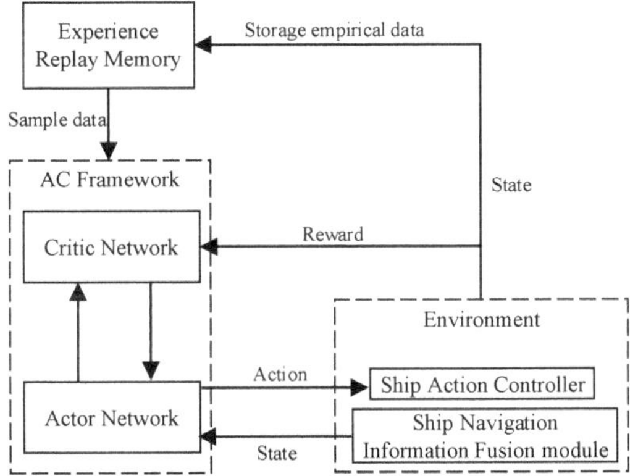

Figure 5. Structure of DDPG algorithm based unmanned ships path planning model.

- AC Algorithm

AC Algorithm includes Actor network and Critic network, which respectively include online network and target network. The two networks have the same network structure and they adopt the experience replay technology to randomly extract ship sample data from the experience buffer pool for network training.

- Environment

The training environment of autonomous path playing model of unmanned ships mainly includes Ship Action Controller and Ship Navigation Information Fusion Module. When the ship performs the action the environment feedbacks the reward and the state of the ship.

- Ship Action Controller

The Ship Action Controller converts the model output action to deflection heading and speed increment. These actions indicate that the unmanned ship should perform in the current state.

- Ship navigation information fusion module

The module can receive and process global positioning system (GPS), automatic identification system (AIS), depth sounder, and anemometer in real time (the information can be integrated and displayed on the electronic chart platform), and provide information of the unmanned ships state, including the ship's position, heading, speed, the distance between the ship and the obstacle, and the angle of the ship and the target point.

- Experience replay memory

The function of this module is to store the state of the ship, the action, the reward, and the state of the next moment. When the maximum capacity of the experience buffer pool is reached, the old data will be replaced by the new data, and the buffer pool is updated in this way.

The DDPG algorithm pseudo code is as follows.

Algorithm 1: Pseudo code of the DDPG algorithm.

1: Randomly initialize Critic network $Q(s, a|\theta^Q)$ and Actor network $\mu(s|\theta^\mu)$ with weight parameters θ^Q and θ^μ.

2: Initialize target network Actor $Q'(s, a|\theta^{Q'})$ and Critic $\mu'(s|\theta^{\mu'})$ with weight parameters $\theta^{Q'} \leftarrow \theta^Q$ and $\theta^{\mu'} \leftarrow \theta^\mu$.

3: Initialize experience replay memory D.

4: for t = 1 to M do

5:    Initialize the random process $\mathbb{N}$ in the action exploration strategy.

6:    Input initial unmanned ships and environment observation state s_1: ship latitude and Longitude, ship heading, ship speed, angle with target point, distances from obstacles.

7:    for t = 1 to T do

8:       Choose ship heading and ship speed $a_t = \mu(s_t|\theta^\mu) + \mathbb{N}_t$ based on current strategy $\mu(s_t)$ and exploring noise $\mathbb{N}$.

9:       Implement output action a_t to get the reward r_t and the new state s_{t+1}.

10:      Save transition (s_t, a_t, r_t, s_{t+1}) into D.

11:      Sample random batch of N transitions (s_i, a_i, r_i, s_{i+1}) from D.

12:      Set $y_i = r_i + \gamma Q'(s_{i+1}, \mu'(s_{i+1}|\theta^{\mu'})|\theta^{Q'})$

13:      Update online Critic network by minimizing loss: $L = \frac{1}{N}\Sigma_i \left(y_i - Q(s_i, a_i|\theta^Q)\right)^2$

14:      Update online Actor network using sampled policy gradient:

$$\nabla_{\theta^\mu} J \approx \frac{1}{N}\Sigma_i \nabla_a Q(s, a|\theta^Q)|_{s=s_i, a=\mu(s_i)} \nabla_{\theta^\mu} \mu(s|\theta^\mu)|_{s_i}$$

15:      Update target Actor network and target Critic network:
$$\theta^{\mu'} \leftarrow \tau\theta^\mu + (1-\tau)\theta^{\mu'}$$
$$\theta^{Q'} \leftarrow \tau\theta^Q + (1-\tau)\theta^{Q'}$$

16:   end for

17: end for

3.3.2. Structural Design of AC Algorithms

The input parameters of the Actor network and the Critic network are all from the ship state data that were provided by the Ship Navigation Information Fusion Module. The Actor network and the Critic network continuously calculate and update their network parameters by continuously inputting ship state information, and finally get better network output results, which are respectively the action of unmanned ships and the value of action. Figure 6 shows the specific network structure of the AC algorithm.

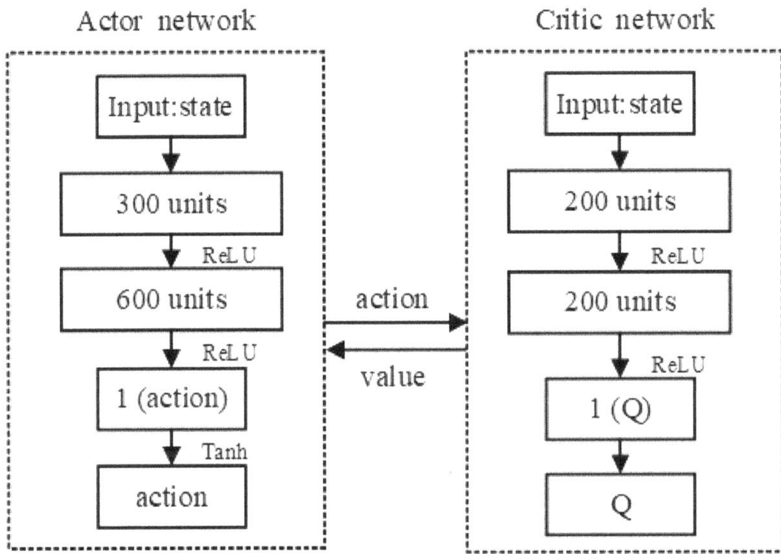

Figure 6. Network structure of AC algorithms.

The Actor network structure contains two fully connected hidden layers. The number of neurons is 300 and 600, respectively. The output nodes of each hidden layer are nonlinearly processed by the activation function to limit the number of output actions to one. First, the Actor network enters the initial ship state information s_t, calculates it through two hidden layers of neurons, and then uses the *ReLU* activation function to limit the output of each layer. At the same time, in the last layer of the network, the *Tanh* activation function is used to limit the network single output action value a_t between $[-1, 1]$, and the Ship Action Controller is then used to convert the network output action into the actual operation action.

The Critic network and The Actor network have the same network structure. The number of neurons of the two hidden layers is 200, and the output node of each hidden layer also adopts the activation function for non-linear processing. The evaluation value of the output action of Actor network is the final network result, which is known as the Q value. Different from the input parameters of the Actor network, the Critic network takes the initial state of the ships s_t, and the output action of the Actor network a_t as the input parameters. After two hidden layers' calculation, the output is processed by *ReLU* activation function, but when the final action value Q is output, no activation function is used to ensure that the output result of network is the definite action value. Additionally, the action value is used to evaluate the output action of Actor network.

3.3.3. Action Strategy Design

In the process of DRL, the relationship between exploitation and exploration needs to be correctly handled. The appropriate action exploration strategy can enable the agent to try more new actions, so as to avoid falling into local optimization. In this paper, the action strategy of unmanned ships mainly includes action control strategies and action exploration strategies. When designing a neural network structure, the random noise is added into the output action of the Actor network as an action exploration strategy, and the output action is converted into specific execution as the control strategy according to the actual physical meaning.

- Action control strategy

In the aspect of action control strategy, one of the output actions of Actor network is used to indicate the deflection heading $a_{steering}$ and the *Tanh* activation function is used to control the range of the output value is $[-1, 1]$. $D_{\max}$ indicates the maximum deflection angle, the value range is $[0°, 35°]$, and *steering* is used to represent the actual deviation sailing value of the ships. The calculation formula is as follows:

$$steering = Tanh(a_{steering})D_{\max} \tag{2}$$

Another output action of the Actor network is ship speed increment $a_{shifting}$, which is processed by activation function *Tanh*. The output range is $[-1, 1]$. $V_{\max}$ is the maximum value of the ship's speed change, the range of value is $[0, 15]$ kn, and *shifting* is the actual change of ship speed. The calculation formula is as follows:

$$shifting = Tanh(a_{shifting})V_{\max} \tag{3}$$

- Action exploration strategy

In the aspect of action exploration strategy, the effective implementation of action exploration in the continuous control space can enable the unmanned ships to find a better action since DDPG is an off-strategy algorithm. The method of adding random noise into the output action of the neural network is used to realize the exploratory process of the action. Random noise is defined, as follows:

$$\mu'(s_t) = \mu(s_t) + \mathbb{N}_t \tag{4}$$

where, μ' is the exploration strategy and $\mathbb{N}_t$ is the added random noise.

The Ornstein–Uhlenbeck (OU) process is a sequential correlation process. It is often used as the random noise of DDPG algorithm and it has good effect in continuous action space. In this paper, this method is used as random noise of unmanned ships action strategy and it is specifically defined as:

$$dx_t = \theta(\mu - x_t)dt + \sigma dW_t \tag{5}$$

In the above formula, θ is the speed of the variable approaching the average value, μ is the action mean value, σ is the fluctuation degree of the random process, and W_t is the Wiener process.

3.3.4. Data Preparation

Based on the ship AIS system, this paper collects the real AIS data of 50 ships from August 1st to 15th, 2018 in the Dalian-Yantai route. Through the analysis of data, complete ship navigation data can be obtained, for example: Longitude, Latitude, speed over ground (SOG), course over ground (COG), heading (HDG), and other information. Table 1 shows the main information of AIS.

Table 1. Automatic identification system (AIS) data attribute table.

ID	Data Type	Data Value	Data Sources
1	Longitude	121°30.43′ E	AIS
2	Latitude	37°47.30′ N	AIS
3	SOG	17.5	AIS
4	COG	10.70°	AIS
5	HDG	10.00°	AIS
6	DRIFT	0.7°	AIS

This paper mainly selects the longitude, latitude, heading, and speed information of the ship as the input parameters of the model. The trajectory of each ship can be clearly displayed since the data is collected in real time. Their total mileage is 133,550 nautical miles and the data volume is around 3G. This paper randomly selects the data of 20 ships from the first 50 ships as the input data of the model training since these data are collected in real time. These data were normalized prior to model training to speed up the training of the model and improve the accuracy of the model.

3.3.5. Design of Reward Function

On the one hand, the unmanned ships needs to maintain the exploration ability to obtain better action, on the other hand, it also needs to use the learned action to obtain more reward from the environment, in order to fully explore the environment space to obtain the optimal action strategy.

The reward function is also known as the immediate reward or enhancement signal R. When an unmanned ship performs an action, the environment will make feedback information based on the action to evaluate the performance of the action. The environment and decision makers design the reward function. It is usually a scalar, with positive value as reward and negative value as punishment. It is very important to obey the navigation rules for the practicability of the algorithm in the autonomous path planning process of unmanned ships.

According to the COLREGS, the ship encounter situation can generally be divided into three types: head-on situation, crossing situation, and overtaking situation. Figure 7 shows the situation of the ship encounter situation.

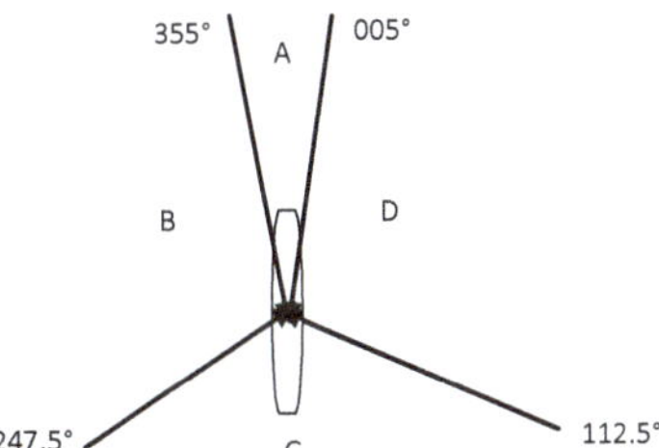

Figure 7. Ship encounter situation chart.

Different ship encounters will be divided according to COLREGS when the sea environment has good visibility, as shown in Figure 7. The range of the A region is (0°, 005°) and (355°, 360°); it is called head-on situation. The ship should take a steering action that is greater than 15° to the right. The B region is a crossing situation with a range of (247.5°, 355°). The C region is the overtaking situation, and the range is (112.5°, 247.5°). The ship is overtaken and it does not usually take action. The D region is a crossing situation with a range of (005°, 112.5°). The ship should take right-turning action. In addition, in situation where the visibility of the sea environment is restricted, there will be no responsibility separation between the stand-on vessel and the give-way vessel. The COLREGS made the following rules for the situation of the ship at this time: For coming ships in the range of (0°, 90 °) and (270°, 360°), the self-ship will turn to the right. Self-ship takes a turn towards other ships for coming ships in the range of (90°, 180°) and (180°, 270°). We mainly study the situation of good visibility at sea in this article.

The rules need to be mathematically quantified and the rules are converted into navigational restrictions in order to integrate the COLREGS and crew's experience into the model. Set the navigation limit area when the distance between the ship and the other ship is less than six nautical miles, otherwise it will not be set. The following describes the rule conversion in the three scenarios:

1. Head-on situation. Under this circumstance, both ships should take the action of turning to the right to avoid the boat. The method of rule conversion is as follows: a navigational limit line with a length of three times the ship length and a direction of 10° from the bow is drawn clockwise in

the bow direction of the two ships. The environment will punish the ship if the ship crosses the navigation limit line. The specific performance is that the ship receives a negative reward.

2. Crossing situation. If there is a ship on the right side of the other ship, the ship should pass through the tail of the other ship. The method of rule conversion is as follows: draw a navigation limit line that is equal to four times the length of the ship in the direction of the bow of the other ship, to avoid the self-ship passing through the bow of the other ship. The ship will be punished and receive a negative reward if the ship crosses the navigation limit of the other ship.

3. Overtaking situation. The ship should pass from both sides of the other ship when the ship is behind the other ship and needs to overtake the other ship. The method of rule conversion is as follows: a navigation limit line of 1.5 times the length of the ship is drawn at the tail of the other ship in the vertical direction of the course. If the ship crosses the navigation limit line, it will be punished and get a negative return value.

For the conversion of crew's experience, this paper adds a virtual area for obstacles based on crew's experience in order to enable the ship to take actions to avoid obstacles in advance. When the distance between the ship and the obstacle is more than 1.5 times the length of the ship, the obstacle area is set, otherwise it is not set. The same as embedding COLREGS, we also convert the crew's experience into a restricted area. We set the obstacle as an approximate regular pattern in order to simplify the calculation and unify the criteria (a circular). The method of setting the virtual obstacle area is as follows: the center of the obstacle is the circle and the longer side of the obstacle is used as the radius. Once the ship enters the obstacle area, the environment will punish it. In this case, the reward obtained by the ship is the inverse of the distance between the ship and the obstacle. The punishment is stronger when the ship is closer to the obstacle and, on the contrary, it will be smaller. The reward dynamically changes until the ship is outside the obstacle area.

By setting the navigation restriction area, the ship's requirements for driving according to COLREGS and crew's experiences are realized, thereby constraining and guiding the ship to select correct and reasonable behavior. The navigation rules are converted into navigational restrictions through the quantitative treatment of COLREGS and crew's experiences. When the ship crosses a restricted navigation area or collides with an obstacle, it will be punished with a negative value. When the ship reaches the target point, the return value is positive. At the same time, the goal-based attraction strategy is adopted, and the reward is positively correlated with the distance between the ship and the target at the adjacent time. The reward function is designed, as follows:

$$
R_t \begin{cases} 2, & d_{t-goal} < D_{g-\min} \\ -1, & d_{t-obs} < D_{o-\min} \\ (d_{t'-goal} - d_{t-goal}) - 0.1, & other \end{cases} \tag{6}
$$

In the above formula, d_{t-goal} is the distance between the ship and the target point at the current time t, the $d_{t'-goal}$ is the distance between the ship and the target point at the previous time t', the d_{t-obs} is the dangerous distance between the ship and the obstacle at the current time t, the $D_{g-\min}$ is the minimum threshold for the ship to reach the target point, and the $D_{o-\min}$ is the minimum dangerous distance threshold between the ship and the obstacle.

During the training process, when the ship reaches the target range, that is $d_{t-goal} < D_{g-\min}$, the reward is set to 2. When reaching the obstacle range, that is $d_{t-obs} < D_{o-\min}$, the reward is set to −1. In other cases, the attraction strategy is to attract the ship to the target point as soon as possible, in order to maximize the return value of the action. The current ship's distance from the target point is subtracted from the distance between the ship and the target point at the previous time to determine whether the ship is approaching the target point. As the simulation environment is based on a 800×600 pixels two-dimensional map, 1 pixel represents 10m in the actual environment. Here, $(d_{t'-goal} - d_{t-goal})$ is normalized, that is, the result of dividing the distance between the current position of the ship and the target point by the distance of the diagonal of the two-dimensional map is used as the reward value, and the range of the reward value is between [0, 1]. At the same time, the reduction of 0.1 in each step is to reduce the number of steps to reach the target point and avoid the redundancy in the planned path. The reward is calculated as $(d_{t'-goal} - d_{t-goal}) - 0.1$. There are many training rounds in the experiment. The current round ends when the ship reaches the target point. If it collides with obstacles, the ship is stepped back and re-selected action. At the same time, the maximum number of steps in each round is set to 400 steps. The current round is ended and the next round is re-entered when the maximum number of steps is reached.

3.4. Model Execution Process

The path planning model of unmanned ships based on the DDPG algorithm is used to abstract the real complex environment and then transform it into a simple virtual environment through the model. At the same time, the action strategy of this model is applied to the environment of electronic chart platform, so as to obtain the optimal planning track of unmanned ships and realize the learning process of end-to-end algorithm in the real environment. Figure 8 shows the execution flow of the autonomous path planning model for unmanned ships that are based on DDPG algorithm.

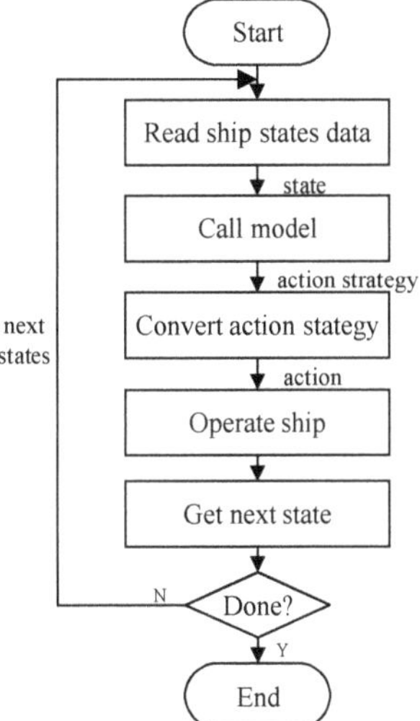

Figure 8. Execution process of unmanned ships path planning model.

The execution process of the model for unmanned ships is described, as follows:

1. When the unmanned ships path planning program starts, the system reads the ship data through the ship navigation information fusion module.
2. The system invokes the model and takes the ship data as the input state and obtains the ship's action strategy under the current state after the model processing and calculation.
3. According to the actual motion of the ship, the model further converts the action strategy into the actual action that the unmanned ships should take.
4. The ship action controller analyzes the acquired execution action and performs the action in the current unmanned ships state.
5. The model obtains the states information of the unmanned ships at the next moment after the execution action, and determines whether the ship state after the execution action is the end state.
6. The model will continue to use the state information of the ship at the next moment if it is not the end state, and then calculate and judge what action the ship should take at the next moment and

cycle through it. If it is the end state, it indicates that the unmanned ships have completed the path planning task, and the model ends the calculation and the invocation.

In this section, first, the unmanned ship uses the obtained state information as the input to the algorithm. Second, the action strategy is obtained through training based on the navigation restricted and the reward function. Subsequently, the parameters of the model are updated with historical experience data until the cumulative return is at the maximum. Finally, the ship executes the action and changes its state, and it determines whether to end the execution process by judging the current state.

4. Simulation Results and Experimental Comparison

4.1. Definition of Model Input and Output Parameters

The input parameters of the model are ship data obtained from ship navigation information fusion module, which mainly includes ship's own state, target point information, and surrounding information. The output of the model is the execution action of unmanned ship, including the deflection heading and the speed increment. Table 2 shows the input parameters names and definitions used in the model.

Table 2. Input state definition.

Name	Parameters	Description
Longitude	$[0°, 180°]$ E, $[0°, 180°]$ W	Ship longitude
Latitude	$[0°, 90°]$ N, $[0°, 90°]$ S	Ship latitude
Direction	$[0°, 360°]$	Ship heading
Speed	$[0, 30]$ kn	Ship speed
Angel	$[0°, 360°]$	Angle with target point
Tracks	$(0, 1000)$/m	Array of distances from obstacles

Here, the Angle between the current heading of the ship and the target point is designed as the input parameter, so that the unmanned ships can be quickly driven toward the target point, and the model training period is prevented from being too long. At the same time, the obstacle risk in the range of 1000 m is calculated and then stored in the set Tracks as the decision-making basis for ship obstacle avoidance. Among them, the set of distances from obstacles is differentiated according to the training situation. In the case where the self-ship heading extension line intersects with the other ship heading extension line, that is, on the premise that there is a danger of collision. If it is a single ship encounter, Tracks has only one distance information of other ships. Tracks contains the distance information of the most dangerous other ship if it is a situation where multiple ships meet. Here, the method of selecting the most dangerous other ship is: When the heading extension line of multiple other ships intersects the heading extension line of self-ship, we choose to store the smallest distance between the self-ship and other ship in Tracks. Finally, the above input states are respectively normalized to limit their range to $[0, 1]$, thereby improving the calculation efficiency of the model. The deflection heading Steering and speed increment Shifting are used as the output action of the model, which realizes the direct conversion from the input state to the output action. Table 3 shows the output parameters definition.

Table 3. Output state definition.

Name	Parameters	Description
Steering	$[-1, 1]$	Deflection angle (-1 for right full rudder, 1 for left full rudder)
Shifting	$[-1, 1]$	Speed increment (-1 for full deceleration, 1 for full acceleration)

According to the actual navigation situation, the maximum deflection heading of the unmanned ship is set to 35°. There is a situation that the ship might not be able to complete the currently required deflection angle but receive a new steering action due to the slower steering of the ship. Therefore, the same action accumulation method is used to calculate the heading deflection angle d_ψ. The heading deflection angle of the unmanned ship is obtained by Formula (7).

$$d_\psi = \sum_{i=t}^{t+k-1} a_i \tag{7}$$

where, a_i represents the unmanned ship deflection angle generated by the model at time i, and k represents the same action performed $k(k > 0)$ times in succession.

In addition, the current ship's heading ψ_d can be obtained by summing the ship's heading and deflection angle at the previous moment since the ship's heading at the previous moment ψ_p is known. The current heading of the unmanned ship is obtained by Formula (8).

$$\psi_d = \psi_p + d_\psi \tag{8}$$

4.2. Model Training Process

Python language and electronic chart platform are used for model training and simulation experiments in order to verify the validity and feasibility of the unmanned ship autonomous path planning model based on DDPG algorithm. The deep learning framework TensorFlow trains the model. The model is designed using the python language, and the convergence and accuracy of the model training can be directly observed. The simulation environment is an electronic chart platform designed and developed based on Visual Studio 2013. Additionally, the state size of the simulation environment is a two-dimensional map of 800×600 pixels, and the range of motion of the ship is set to the size of the two-dimensional map. The ship is considered to have collided if the ship crosses the boundary of the two-dimensional map. This paper first uses the DDPG algorithm to train a global route from the start point to the end point, and then trains on the situation of encountering dynamic obstacles (other ships) during the voyage. We will mainly study the part of the ship that avoids obstacles and reaches the end point since the planning of the global static route is not the focus of this paper.

After many experiments, the better structure and parameters of the neural network are designed. The network structure of the Actor and the Critic both use a fully connected neural network with two layers of hidden layers. The Actor network hyper parameters are set, as follows: the numbers of neurons in the hidden layer are 300 and 600, respectively, the learning rate of the network is 10^{-4}, the action of the output layer is the heading deflection and the speed increment respectively, and different activations are selected according to the difference of the output action range. The heading deflection uses the *tanh* activation function and its output range is $[-1, 1]$; the speed increment uses the *sigmoid* activation function, and its output range is $[0, 1]$. The Critic network hyper parameters are set, as follows: the number of neurons in the hidden layer is 200, the learning rate is 10^{-3}, and the discount factor γ of the reward function is set to 0.9. In addition, the experience buffer pool size is set to 5000 and the batch learning sample size is 64. The random noise uses the method in Equation (5); the value of μ is set to 0, θ is set to 0.6, and σ is set to 0.3. The rate of the soft update method is $\tau = 0.01$. The Actor network and the Critic network both use the Adam network optimizer. To prevent infinite training, set the maximum number of steps per round to 400, for a total of 300 rounds. The parameters of the Actor network and the Critic network are updated every 1000 steps in order to further improve the accuracy of the model in the training process. Based on the above training conditions and parameters, this paper studies and trains the path planning model of the unmanned ship. The training process and results are described, as follows.

Figure 9 shows the number of training steps per round during the unmanned ships autonomous path planning process. It can be seen that, in about the first 40 rounds, the training steps of the model reach the maximum, which indicates that the model triggers the termination condition of training and it does not realize the path planning or falls into the local obstacle area. Between 45th and 110th round, the training steps of each round began to decrease, and the average training steps are maintained at 50 steps, which indicated that unmanned ships learn more and more action strategies and plan a complete path independently. After about the 115th round, the training steps of the model are kept below 40 steps in each round, which indicates that the unmanned ship has fully learned the optimal action strategy and it has no collision risk. In the vicinity of the 275th round, the number of training steps increased due to the exploratory strategy of the algorithm, which makes unmanned ships attempt random action.

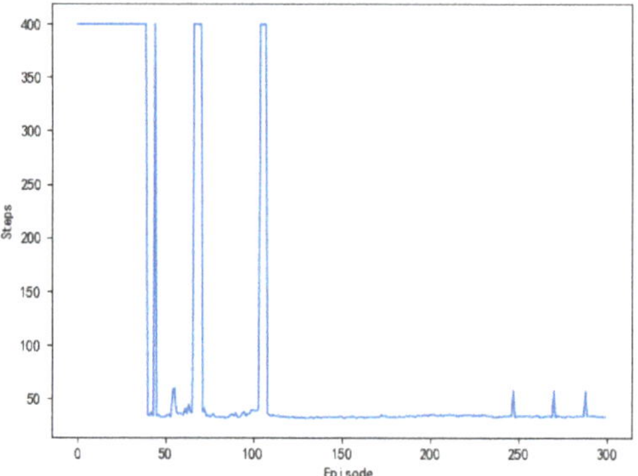

Figure 9. Number of steps per turn.

The purpose of the model is to improve the reward on the action strategy through continuous interaction with the environment. The greater the cumulative reward per round, the better the learning effect. Figure 10 shows the cumulative reward of each round of the model. In the first 40 rounds, the reward of the model is lower per round, and the fluctuation state is processed, which indicated that the unmanned ships has not found the correct path and is constantly trying new action strategy. Around the 45th round, the reward of the model began to increase, which indicated that a path to the target point is found. After the 115th round, the cumulative reward of the model per round is basically maintained at the maximum, indicating that the unmanned ships have found the optimal action strategy. The trend of cumulative reward per round is consistent with the change of steps per round when compared with Figure 9.

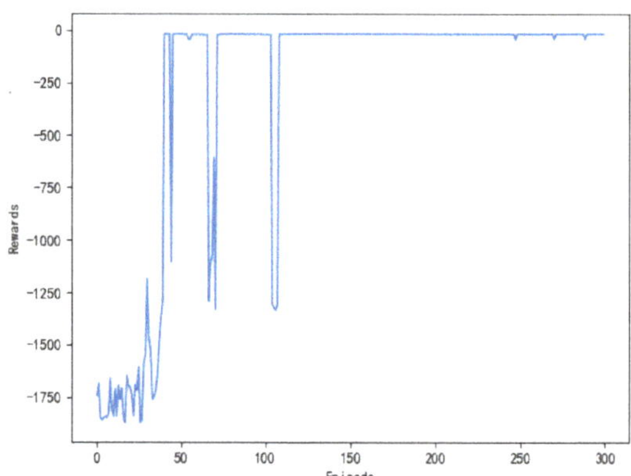

Figure 10. Cumulative reward per turn.

The average reward reflects the effect of the learning process and it also more directly observes the degree of change in the reward. Figure 11 shows the average reward of the model every 50 rounds.

As can be seen from the figure, the general trend of average compensation is upward. About the 45th round, the growth rate of average rewards began to slow down and then gradually leveled off. After the 110th round, the average reward stabilized and then remained at a large value, which indicated that the model has found the optimal action strategy at this time.

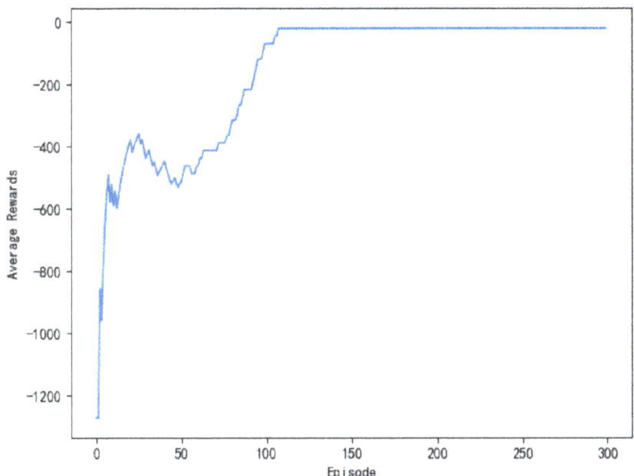

Figure 11. Average reward per rounds.

4.3. Model Integration and Simulation Experiment

The model is tested and observed in the simulation environment in order to verify the validity and correctness of the autonomous path planning model for unmanned ships in Section 4.2. Firstly, the electronic chart platform and ship navigation rules are briefly introduced. Secondly, the model is validated in three different situations to observe whether the unmanned ship is running correctly according to the navigation rules. Finally, the other two unmanned ship path planning methods are selected as comparative experiments, and the training process and simulation results of the three methods are compared and analyzed.

4.3.1. Verification Environment

It is difficult for traditional numerical simulation methods to accurately describe the environmental information because the unmanned ships have a wide range of work and its environment is complex. The electronic chart [37] platform is an important navigation tool for ships and other marine vehicles. It can provide real and complete environmental information needed in navigation, including land, ocean, water depth, obstacles, and islands, etc.

In this paper, the electronic chart platform that was developed in C++ language is used as the verification environment of the model, as shown in Figure 12. The platform has the following main features: (1) Displaying standard chart information, which can be used in any scale chart interface. (2) Use Microsoft Foundation Classes (MFC) as a dynamic link library, it provides a flexible and convenient interface. (3) Set the motion information of the ship and other ships, including longitude, latitude, heading, and speed. (4) Dynamic display of the track of all ships.

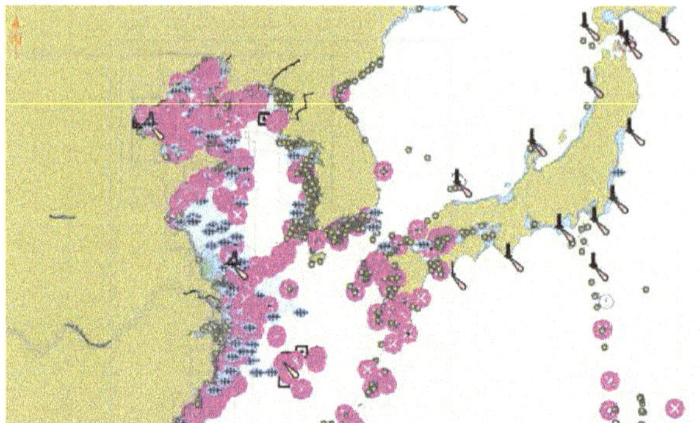

Figure 12. Electronic chart platform.

4.3.2. Validation Results

The autonomous path planning model of unmanned ships is obtained through training, and the model is invoked by the electronic chart platform to further verify the effectiveness of the algorithm. In the electronic chart platform, the self-ship is indicated by black concentric circles and Ship-1, other ships are represented by triangles and Ship-i (i = 2, 3, 4, ...), the yellow circle indicates the target point of the self-ship, and the purple circle indicates the target point of the other ship, static obstacles are indicated by irregular figures, and the trajectory of the ships are drawn during the movement. The motion parameters of self-ship are defined, as follows: the ship's captain is set $L = 12.5$ m, the ship's width is set $W = 2.1$ m, the heading change amount is $\psi_m = [-35°, 35°]$, and the shipping speed change amount is $v_m = [-15, 15]$ kn. All other ships adopt uniform motion parameters. The length of the ship is $L_o = 10.6$ m, and the ship's width is set $W_o = 1.8$ m. After setting the ship's motion parameters, they travel to the set point according to the uniform linear motion.

The head-on situation, crossing situation, overtaking situation, and multi-ship encounter situation in the ship navigation process are tested, respectively, in order to verify whether the action strategy made by the model in this paper complies with the COLREGS. Here, the parameter information in the experiment process is uniformly introduced. Ship-1 represents the self-ship and Ship-i (i = 1, 2, 3, ...) represents the other ship. Heading angle represents the initial angle of the ship and Ship Speed indicates the initial speed. The starting point and target point points are expressed in latitude and longitude.

(1) Head-on case

Table 4 sets the motion parameters of the ships in detail. Figure 13 shows the experimental results. It can be seen from the waypoints information that the Ship-1 turned to the right according to the rules, successfully avoided the other ship, and then headed for the target position. In this case, the planned path length is 6.592 nautical miles.

Table 4. Setting of ship in head-on case.

	Starting Point	**Target Point**	**Heading Angle**	**Ship Speed**
Ship-1	(N 38.104, E 121.105)	(N 38.104, E 121.134)	90°	28 kn
Ship-2	(N 38.104, E 121.113)	(N 38.104, E 121.106)	270°	18 kn

Figure 13. Verification result of ship trajectories in the head-on case.

(2) Crossing case

Tested by the crossing of two ships, Tables 5 and 6 show the specific ship motion parameters. Figures 14 and 15 show the experimental results, respectively. In Figure 14, the other ship approached from the right side, where the Ship-1 acts as a give-way vessel with respect to the Ship-2, and then deflects the heading to the right according to rule and passes the stern of the Ship-2. In this case, the planned path length is 6.663 nautical miles. In Figure 15, the other ship approached from the left side, In this case, the planned path length is 6.462 nautical miles.

Table 5. Setting of two ship in crossing case 1.

	Starting Point	Target Point	Heading Angle	Ship Speed
Ship-1	(N 38.104, E 121.105)	(N 38.104, E 121.134)	90°	28 kn
Ship-2	(N 38.193, E 121.112)	(N 38.221, E 121.112)	0°	18 kn

Table 6. Setting of two ship in crossing case 2.

	Starting Point	Target Point	Heading Angle	Ship Speed
Ship-1	(N 38.104, E121.105)	(N 38.104, E 121.134)	90°	28 kn
Ship-2	(N 38.221, E121.112)	(N 38.193, E 121.112)	180°	18 kn

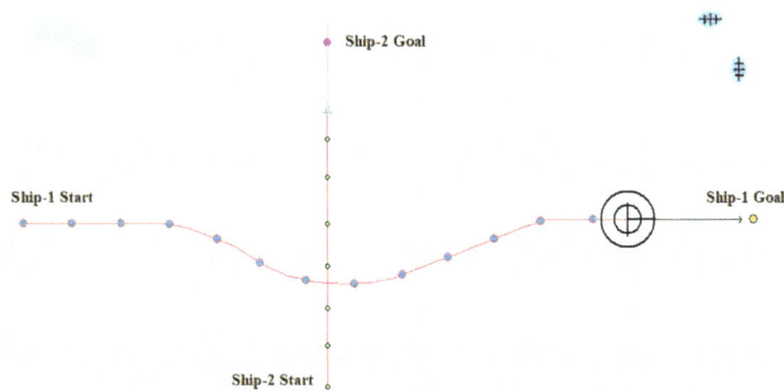

Figure 14. Verification result of ship trajectories in the crossing case 1.

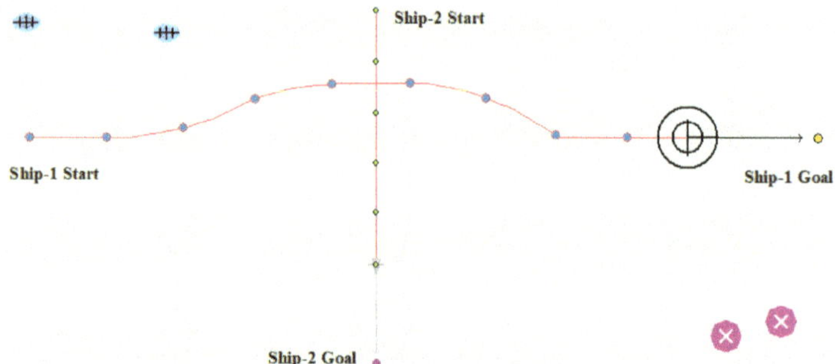

Figure 15. Verification result of ship trajectories in the crossing case 2.

(3) Overtaking case

Table 7 shows the motion parameters of the two ships. From the test results and the waypoints information, Figure 16 shows that the Ship-1 is on the port side of the Ship-2, and the Ship-1 passes as the give-way vessel from the stern of the Ship-2. Ship-2 maintains heading as a stand-on vessel. In this case, the planned path length is 6.416 nautical miles.

Table 7. Setting of ship in overtaking case.

	Starting Point	Target Point	Heading Angle	Ship Speed
Ship-1	(N 38.203, E 121.134)	(N 38.203, E 121.246)	90°	28 kn
Ship-2	(N 38.203, E 121.137)	(N 38.203, E 121.236)	90°	18 kn

Figure 16. Verification result of ship trajectories in the overtaking case.

(4) Multi-ship encounter case 1

Table 8 shows the motion parameters of the three ships. First, Ship-1 and Ship-2 form a head-on encounter, and then a crossing encounter with Ship-3. Ship-1 first takes the action of turning right, passing through the right side of Ship-2, then takes the action of turning right again, passing through the tail of Ship-3, and finally reaching the target point, according to the test results and waypoint information. Figure 17 shows the scene construction of multi ship encounter case 1, and Figure 18 shows the verification results of multi ship encounter case 1. In this case, the planned path length is 14.436 nautical miles.

Table 8. Setting of ship in multi-ship encounter case 1.

	Starting Point	**Target Point**	**Heading Angle**	**Ship Speed**
Ship-1	(N 38.203, E 121.134)	(N 38.245, E 121.257)	60°	28 kn
Ship-2	(N 38.220, E 121.174)	(N 38.221, E 121.152)	240°	18 kn
Ship-3	(N 38.232, E 121.236)	(N 38.254, E 121.232)	330°	18 kn

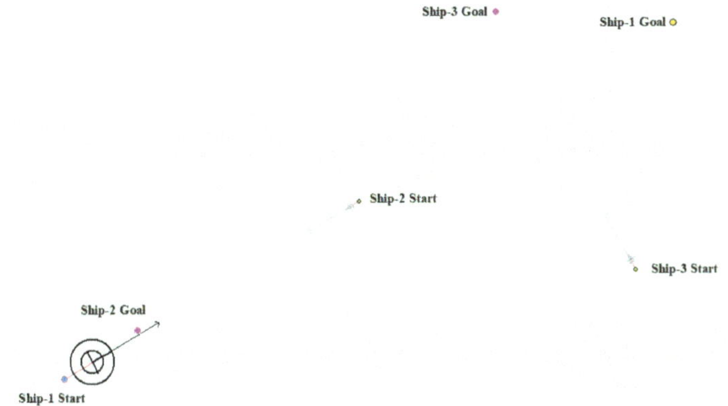

Figure 17. Multi-ship encounter case 1 scenario construction.

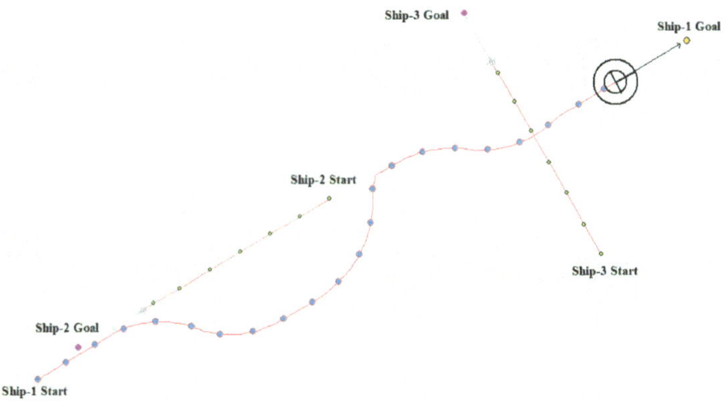

Figure 18. Verification result of ship trajectories in multi-ship encounter case 1.

(5) Multi-ship encounter case 2

Table 9 shows the motion parameters of the three ships. First, Ship-1 and Ship-2 form a crossing encounter situation and then form an overtaking situation with Ship-3. Ship-1 first takes the action of turning right, passing through the stern of Ship-2, then takes the action of turning left, passing through the right side of Ship-3, and finally reaches the target point, according to the test results and waypoint information. Figure 19 shows the scene construction of multi ship encounter case 2 and Figure 20 shows the verification results of multi ship encounter case 2. In this case, the planned path length is 13.651 nautical miles.

Table 9. Setting of ship in multi ship encounter case 2.

	Starting Point	Target Point	Heading Angle	Ship Speed
Ship-1	(N 38.203, E 121.134)	(N 38.264, E 121.264)	60°	28 kn
Ship-2	(N 38.204, E 121.185)	(N 38.235, E 121.164)	330°	18 kn
Ship-3	(N 38.231, E 121.228)	(N 38.258, E 121.248)	60°	18 kn

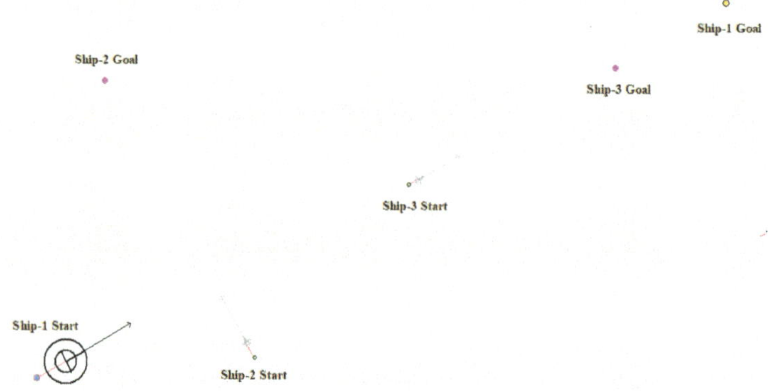

Figure 19. Multi-ship encounter case 2 scenario construction.

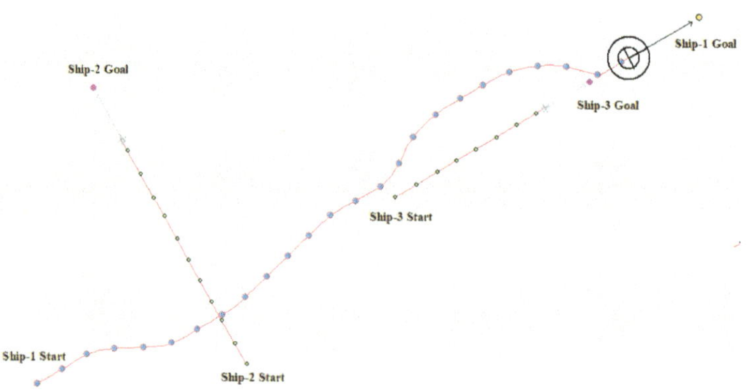

Figure 20. Verification result of ship trajectories in multi-ship encounter case 2.

(6) Multi-ship encounter case 3

Table 10 shows the motion parameters of the three ships. First, Ship-1 and Ship-2 form a crossing encounter situation and then form an overtaking situation with Ship-3. Ship-1 first takes the action of turning left, passing through the stern of Ship-2, then takes the action of turning left, passing through the right side of Ship-3, and finally reaches the target point, according to the test results and waypoint information. Figure 21 shows the scene construction of multi ship encounter case 2 and Figure 22 shows the verification results of multi ship encounter case 3. In this case, the planned path length is 15.264 nautical miles.

Table 10. Setting of ship in multi ship encounter case 3.

	Starting Point	Target Point	Heading Angle	Ship Speed
Ship-1	(N 38.203, E 121.134)	(N 38.264, E 121.264)	60°	28 kn
Ship-2	(N 38.235, E 121.164)	(N 38.204, E 121.184)	150°	18 kn
Ship-3	(N 38.231, E 121.228)	(N 38.258, E 121.248)	60°	18 kn

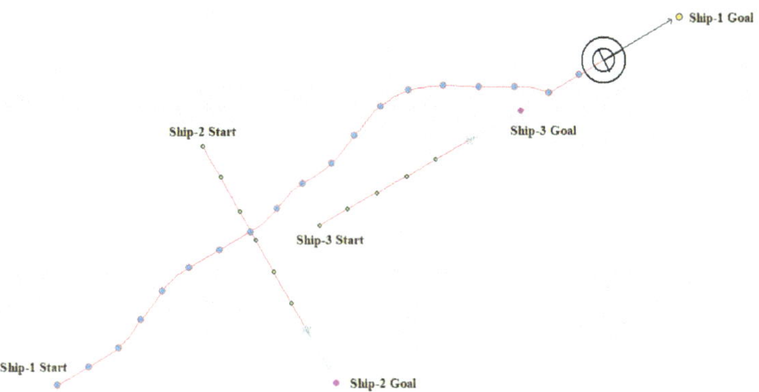

Figure 21. Multi-ship encounter case 3 scenario construction.

Figure 22. Verification result of ship trajectories in multi-ship encounter case 3.

(7) Multi-ship encounter case 4

Table 11 shows the motion parameters of the three ships. First, Ship-1 and Ship-2 form a crossing encounter situation, form a head-on encounter situation with Ship-3, and then finally form an overtaking encounter situation with Ship-4. According to the test results and waypoint information, Ship-1 first takes the action of turning right, passing from the stern of Ship-2 and the right side of Ship-3, then takes the action of turning left, passing from the left side of Ship-4, and then finally reaches the target point. Figure 23 shows the scene structure of case 3 of multi ship encounter and Figure 24 shows the verification results of case 4 of a multi ship encounter. In this case, the planned path length is 18.713 nautical miles.

Table 11. Setting of ship in multi ship encounter case 4.

	Starting Point	Target Point	Heading Angle	Ship Speed
Ship-1	(N 38.203, E 121.134)	(N 38.203, E 121.278)	90°	28 kn
Ship-2	(N 38.184, E 121. 138)	(N 38.246, E 121.141)	10°	18 kn
Ship-3	(N 38.203, E 121.164)	(N 38.203, E 121.131)	270°	18 kn
Ship-4	(N 38.202, E 121.245)	(N 38.201, E 121.257)	90°	18 kn

Figure 23. Multi-ship encounter case 4 scenario construction.

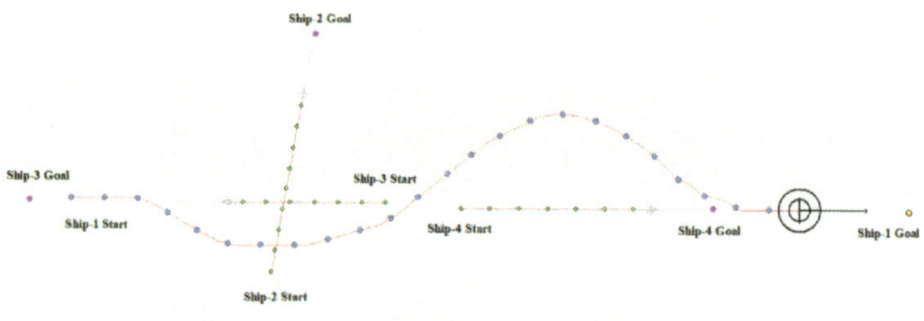

Figure 24. Verification result of ship trajectories in multi-ship encounter case 4.

Through the observation of the above verification test, it is shown that the unmanned ship based on the model that is proposed in this paper has better path planning effect in the case of single ship and multi-ship, and it can successfully avoid the obstacle according to the navigation rules. Finally, reach the target point. At the same time, by observing the time and path length used in path planning, it is further indicated that the path distance of the model planning is shorter and more in line with the actual navigation experience.

4.4. Improved Model of Autonomous Path Planning

The planned path is redundant and not sufficiently flat, although the unmanned ship autonomous path planning based on the DDPG algorithm is implemented in Section 4.3.2. By observing the training process in Section 4.2., it is found that the DDPG training time is longer, the convergence speed is slower, and the algorithm is easy to fall into the problem of local iteration. The above phenomenon is because the DDPG has no prior knowledge of the environment and the initial stage of learning can only randomly select actions. Therefore, the convergence speed of the algorithm is slow in complex environments. This section improves the DDPG and adds the APF method to obtain the unmanned ships path planning model based on APF-DDPG in order to improve the learning efficiency in the initial stage and speed up the convergence of the algorithm.

The APF is a method of virtual gravitational field and repulsive field, which has been widely used in real-time path planning of robots. The basic principle is to virtualize the simulation environment, and each state point in the environment has a corresponding potential energy value. The target point generates a gravitational potential field in the virtual environment, the obstacle generates a repulsive potential field in the virtual environment, and the total field strength is obtained by superimposing the gravitational field and the repulsive field. The object approaches the target point by using the gravitational field, and the repulsion field is used to avoid the obstacle. Generally, it is calculated by the following formula:

$$U(\mathrm{s}) = U_a(\mathrm{s}) + U_r(\mathrm{s}) \tag{9}$$

In the above formula, $U_a(s)$ is the potential energy value of the gravitational field at the point s, $U_r(s)$ is the potential energy value of the repulsive field at the point s, and $U(s)$ is the potential energy of the point s. $U_a(s)$ and $U_r(s)$ are obtained by Formulas (10) and (11), respectively.

$$U_a(s) = \frac{1}{2}k_a\rho_g^2(s) \tag{10}$$

In the Formula (10): k_a is the scale factor of gravitational field and $\rho_g(s)$ is the minimum distance between the s and the target.

$$U_r(s) = \begin{cases} \frac{1}{2}k_r\left(\frac{1}{\rho_{ob}(s)} - \frac{1}{\rho_0}\right)^2, & \rho(s) < \rho_0 \\ 0, & \rho(s) \geq \rho_0 \end{cases} \tag{11}$$

In the Formula (11): k_r is the scale factor of repulsion field and $\rho_{ob}(s)$ is the minimum distance between the s and the obstacle, where $\rho_{ob}(s)$ is the obstacle influence coefficient.

For the path planning problem of unmanned ships, the immediate reward r can only be obtained when the destination is reached or an obstacle is encountered. The sparsity of the reward function leads to low initial efficiency and multiple iterations. There are a large number of invalid iterative search spaces, especially for large-scale unknown training environments. Therefore, the APF is constructed according to the position information of the target point and the obstacle point. At this time, the potential field value of each state in the potential field represents the maximum cumulative return $V(s_i)$ of the state s_i, and the relation expression is expressed as:

$$V(s_i) = |U(s_i)| \tag{12}$$

In the Formula (12): $U(s_i)$ is the potential field value of state s_i in the virtual potential field environment. $V(s_i)$ represents the maximum cumulative return when the optimal action is taken in state s_i.

The steps of the APF-DDPG based autonomous path planning method are as follows:

1. The potential field is constructed according to the target point and the obstacle in the virtual environment, and the gravity potential field with the target point as the center of the potential field is established.
2. Define the potential energy value $U(s_i)$ in the potential energy field as the maximum cumulative report $V(s_i)$ under state s_i, according to Equation (12).
3. The ship explores the environment from the starting point and selects the action in the current state s_i. The environment status is updated to state s' and an immediate return value is received r.
4. Update the Q value according to the state value function: $Q(s_i, a) = r + \gamma V(s_i')$. Subsequently, update the online Critic network.
5. Observe whether the ship reaches the target point or reaches the set maximum number of learning. If the two meets one of them, the round of learning ends and the next iteration is started. Otherwise, return to step (3).

When the gravitational field is added to the target point position, the unmanned ship reaches the target point faster and the selection of the action strategy becomes more stable. The experimental procedure that is based on the APF-DDPG method is described below.

The path planning experimental parameters that are based on the APF-DDPG are the same as the DDPG parameters in Section 4.3.2. The APF parameters are set, as follows: $k_a = 1.6, k_r = 1.2, \rho_0 = 3.0$. We choose the environment of case (4), (5), and (7) in Section 4.3.2 as the comparison environment in order to better compare APF-DDPG with DDPG.

Figure 25 shows the training process of APF-DDPG. From Figure 25a, it can be seen that the number of training steps in each round begins to decline and converge in the 43th round, and it fluctuates in the subsequent training process, which is caused by the action exploration strategy of the algorithm. Figure 25b shows the reward value of each round. The reward value of each round starts to increase and then reaches the maximum value from the training to the 43th round, indicating that a better action strategy is found at this time. Figure 25c shows the average reward value of each round, which rapidly increases and then stabilizes at the maximum value at the beginning of the 43th round.

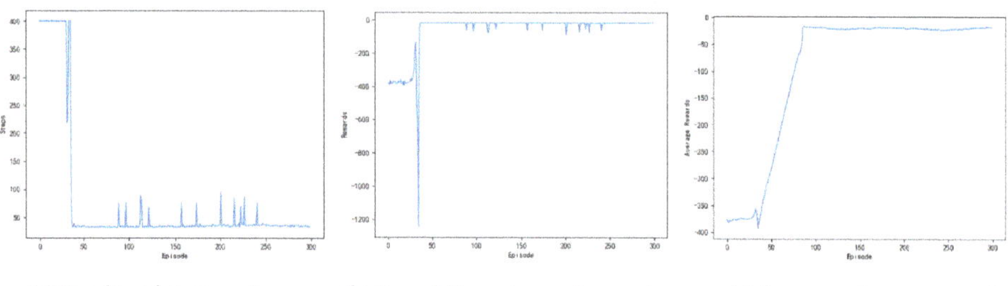

(**a**) Number of steps per turn.   (**b**) Cumulative return value per turn.   (**c**) Average return per rounds.

Figure 25. Artificial potential field-DDPG (APF-DDPG) training process.

The result is shown in Figures 18 and 26 as compared with the (4) experimental case in Section 4.3.2. Figure 18 has more redundant paths and it takes longer. The path in Figure 26 is smoother and the length of the planned path is shorter and takes less time. Similarly, when compared with the experimental case (5) (7) in Section 4.3.2, the results are shown in Figures 20, 24, 27 and 28, respectively. It can be seen that APF-DDPG is better than DDPG in the distance and time of path planning. By comparing the DDPG and the APF-DDPG, it is shown that the APF-DDPG has better decision-making level and faster convergence speed.

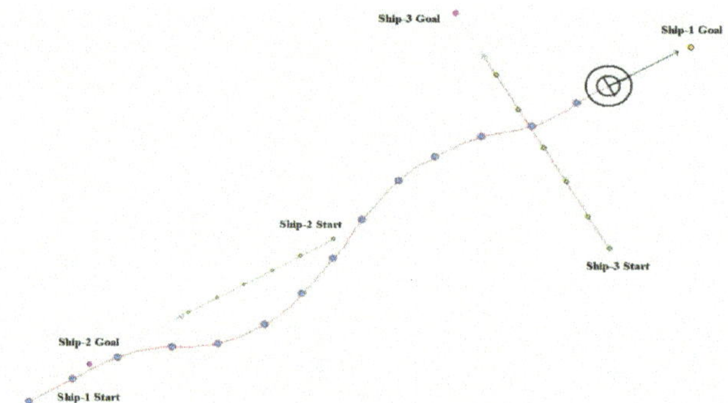

Figure 26. Experimental results based on APF-DDPG (a).

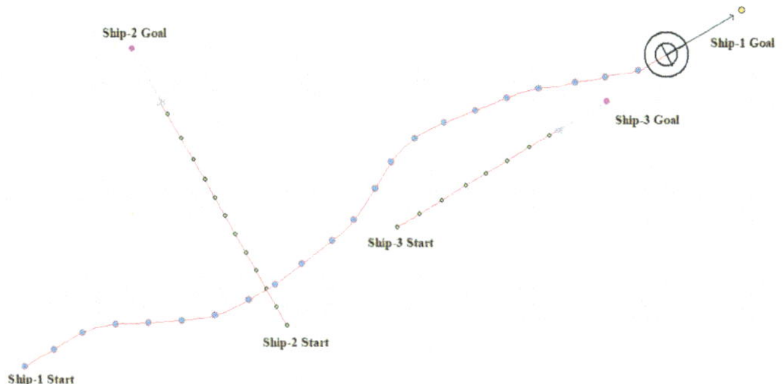

Figure 27. Experimental results based on APF-DDPG (b).

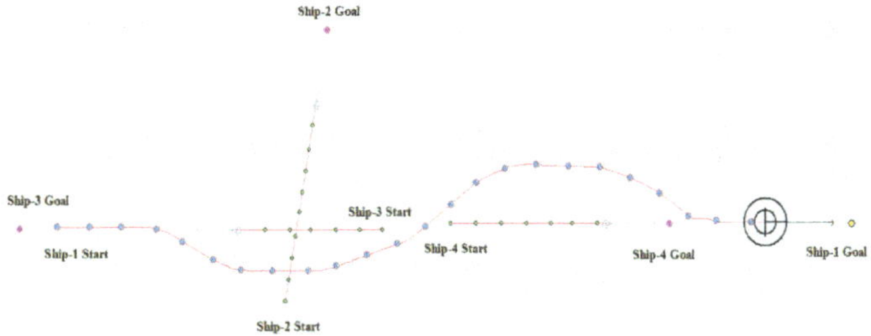

Figure 28. Experimental results based on APF-DDPG (c).

4.5. Experimental Comparison and Analysis

Currently, there are five path planning methods comparisons and experimental analyses, as described in this section. These methods are called the DQN method, AC method, DDPG method, Q-learning method [18], and the path planning method based on APF-DDPG. The five methods are trained to obtain training steps, the reward, and the average reward in the case that the ship's motion parameters and the surrounding environment are the same. Analyze and compare the experimental process and performance data of unmanned ships autonomous path planning during the training phase.

The five comparison methods set the same neural network structure and network parameters, the discount factor of the reward function is set to 0.9, and the experience buffer pool size is set to 5000. Among them, the Q-learning algorithm and DQN algorithm discretize the deflection heading action into 70 deflection angle values in the range of [−35°, 35°], and other algorithms directly output the specific action value.

Figure 29 shows the contrast experiment of autonomous path planning for unmanned ships. Figure 29a–c are the results of an experimental of unmanned ships path planning based on DQN. Figure 29d–f are the results of an experimental based on AC. Figure 29g–i are based on the DDPG. Figure 29j–l are the results of an experiment based on the Q-learning. Figure 29m–o are based on the experimental results of the APF-DDPG. Figure 30 is a comparison of the average reward of the five methods.

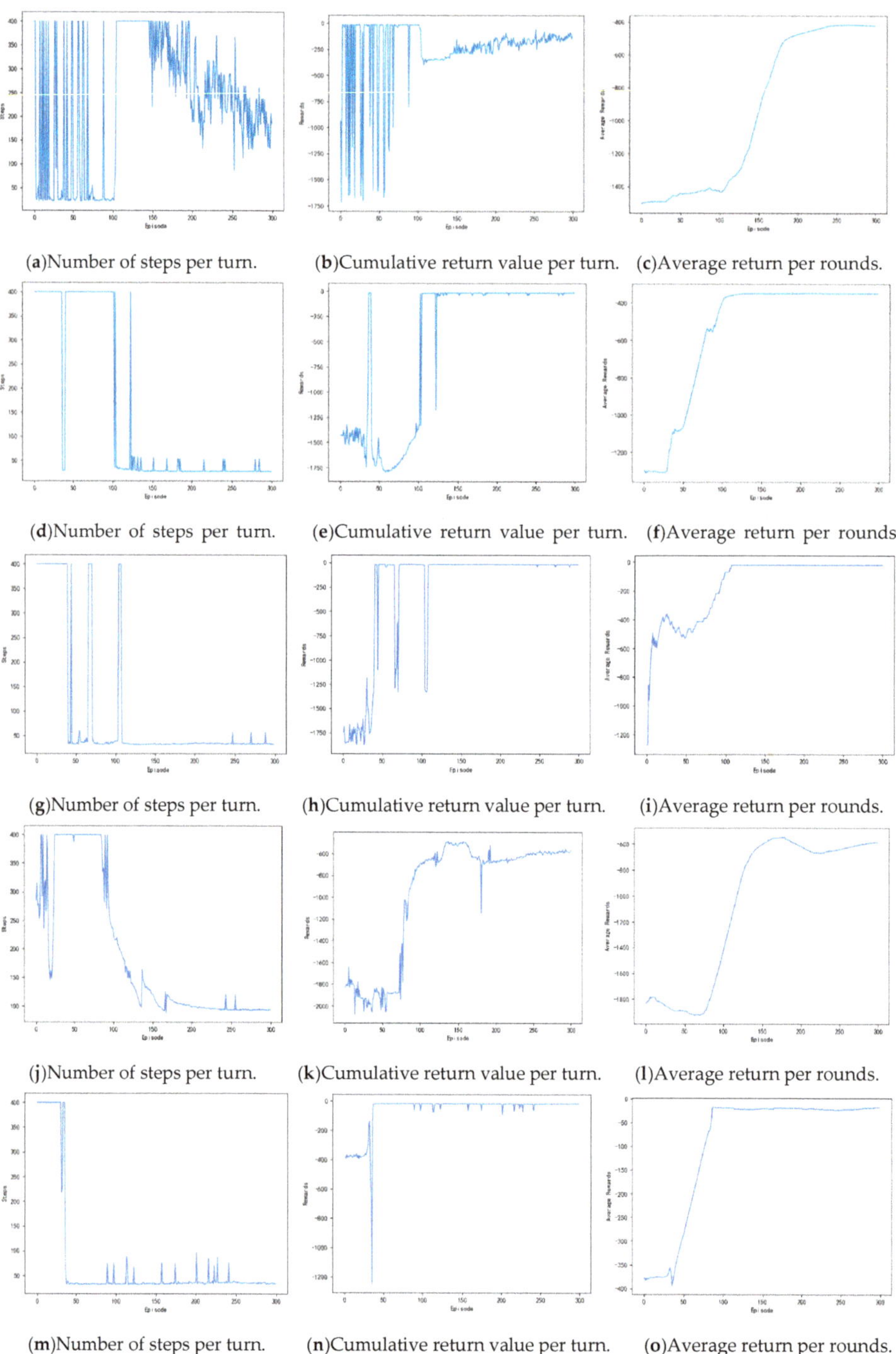

(**a**)Number of steps per turn.   (**b**)Cumulative return value per turn.   (**c**)Average return per rounds.

(**d**)Number of steps per turn.   (**e**)Cumulative return value per turn.   (**f**)Average return per rounds.

(**g**)Number of steps per turn.   (**h**)Cumulative return value per turn.   (**i**)Average return per rounds.

(**j**)Number of steps per turn.   (**k**)Cumulative return value per turn.   (**l**)Average return per rounds.

(**m**)Number of steps per turn.   (**n**)Cumulative return value per turn.   (**o**)Average return per rounds.

Figure 29. Comparative experiment of different path planning methods.

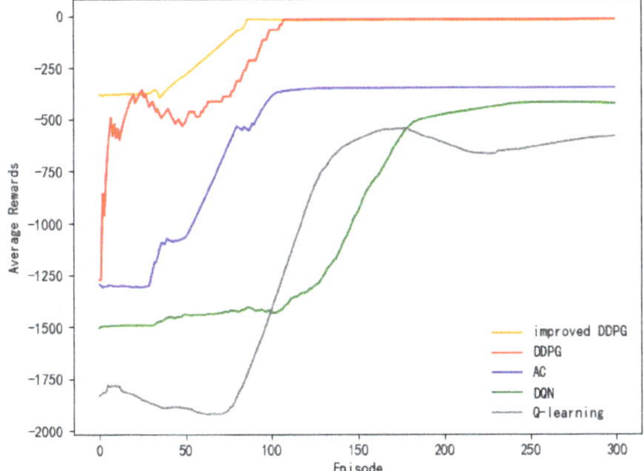

Figure 30. Comparison of average returns of five algorithms.

(a), (d), (g), (j), and (m) in Figure 29 are the experimental results of the number of steps per round obtained while using the DQN, the AC, the DDPG, the Q-learning, and the APF-DDPG, respectively. By longitudinally comparing the number of execution steps per round, it can be found that the number of steps per turn based on the DQN (a) starts to decrease at about the 150th round, while the number of steps per turn based on the AC (d) and the DDPG (g) begins to decrease and then gradually converges around the 100th round. The number of steps per turn based on the DQN does not converge to the minimum number of steps, but the AC and the DDPG have both converged to the minimum number of steps, and the DDPG produces fewer fluctuations in the latter process. In addition, the number of steps per round based on the Q-learning (j) begins to decrease around the 150th round, and the number of steps per round based on the APF-DDPG (m) begins to decrease around the 43th round. By vertically comparing the five methods, it can be found that the path planning method that is based on the APF-DDPG reaches the target point with fewer rounds, and the convergence speed is faster, which is better than the other algorithms in terms of stability.

(b), (e), (h), (k), and (n) in Figure 29 are the experimental result of the reward per round obtained by the DQN, the AC, the DDPG, the Q-learning, and the APF-DDPG, respectively. The round reward based on the DQN (b) and the AC (e) starts to increase at about the 100th round, and then gradually reaches the maximum reward, and the round reward based on the DDPG (h) starts to increase at about the 50th round and then gradually stabilizes at the maximum reward, which indicated that the unmanned ships while using DDPG found the optimal action strategy faster. The round reward based on the DQN continuously fluctuates and does not reach the maximum reward, which indicates that the unmanned ships has not learned the optimal behavior strategy, and the AC has found the maximum reward value, indicating that the optimal behavior strategy has been learned. In addition, the round reward based on Q learning (k) has been low in the initial round, indicating that the correct action has not yet been learned and is in the exploratory stage. The 100th round of the award began to gradually increase, but ultimately did not reach the maximum reward. The round reward that was based on the APF-DDPG (n) began to increase at the 43th round and reached the maximum reward value faster. Longitudinal comparison of the five experimental processes shows that the APF-DDPG achieves the maximum round bonus value with the least amount of time and it is consistently maintained. At the same time, the APF-DDPG is more stable than the other algorithms in the later stage, which indicates that the unmanned ships has learned the optimal behavior strategy.

(c), (f), (i), (l), and (o) in Figure 29 is the experimental result of the average reward per round obtained by the DQN, the AC, DDPG, the Q-learning, and the APF-DDPG, respectively. By observing the average reward, it is more intuitive to judge the efficiency and accuracy of the convergence of the three algorithms. By observing (c), it can be known that the average reward that is based on the DQN starts to increase at about the 110th round and tends to be stable around the 180th round. By observing (f), the average reward that is based on the AC is the 40th round began to increase and then stabilized around 130th round. By observing (i), the average reward that is based on the DDPG of this paper started to increase at about the 10th round, and the downward trend appeared around the 50th round. This is because the random exploration strategy of the algorithm causes the reward of each round to fluctuate, but around the 115th round, the average reward reaches the maximum. In addition, the average reward based on Q learning (l) showed a downward trend at the beginning, which indicated that it was in the exploration stage. Subsequently, there was an increase in the 86th round, but did not reach the maximum average reward finally. By observing (o), the average reward that is based on the APF-DDPG increases at the 43th round and reaches the maximum average reward value faster, which converges faster than other algorithms. Longitudinal comparison of five experimental processes shows that the APF-DDPG achieves the maximum average reward with the least amount of time and it is consistently maintained.

Figure 30 is a comparison of the average reward for the five path planning methods. It can be seen from the figure that the Q-learning and DQN methods that are based on discrete motion space converge slowly, starting to gradually increase in the 80th to 100th rounds, and fail to reach the maximum draw value. Based on the AC and DDPG methods, convergence begins on the 10th to 50th rounds. The maximum average reward value is greater than the DQN and Q-learning methods, and the DDPG algorithm has reached the maximum average reward value. The average reward value of the APF-DDPG method is higher than the other methods at the beginning, which indicated that the initial convergence rate is improved after the artificial potential field method is added. At the same time, the APF-DDPG method reached the maximum average reward value in the 70th round. In summary, the APF-DDPG method is superior to other methods in terms of convergence speed and stability.

Experimental data from five methods were extracted, and the comparison is made from the total iteration time, the optimal decision time, convergence steps, and the number of obstacle collisions to further illustrate the accuracy and effectiveness of the improved model. As shown in Table 12, DQN and Q-learning based on discrete motion space have greater overhead in terms of training time and convergence speed. In contrast, AC and DDPG have better training time and convergence effects. When compared to the above method, the decision method using APF-DDPG has less iteration time and optimal decision time. In addition, the convergence speed of the method is improved, and the trial and error rate is reduced. The simulation results show that the APF-DDPG has higher convergence speed and stability.

Table 12. Comparison of experimental data.

Comparative Experiment	Total Iteration Time (s)	Optimal Decision Time(s)	Convergence Steps	Number of Collisions
DQN	452.512	410.856	235	139
AC	361.219	332.463	136	84
DDPG	338.713	281.651	122	72
Q-learning	486.321	437.245	260	175
APF-DDPG	294.960	236.152	68	63

5. Conclusions

For the traditional path planning algorithm, the historical experience data cannot be recycled and used for online training and learning, which results in low accuracy of the algorithm, and the actual path of the plan is not smooth and redundant. This paper proposes an unmanned ships autonomous path planning method based on the DDPG algorithm. First, the ship data are acquired based on the electronic chart platform. Secondly, the model is designed and trained in combination with the

COLREGS and crew experience, and validated in three classical encounter situations on the electronic chart platform. The experiments show that unmanned ships take the best and reasonable action in unfamiliar environment, successfully complete the task of autonomous path planning, and realize the end-to-end learning method of unmanned ships. Finally, by combining APF and DDPG, an unmanned ship autonomous path planning method that is based on improved DRL is proposed. The improved DRL is compared with other classical DRL methods. The results show that the improved DRL has faster convergence speed and accuracy, can achieve continuous operation output, and has less navigation error, which further validates the effectiveness of the proposed method. However, the ship's motion model and the actual verification environment are not considered in this paper. How to consider the motion model of a ship in a complex sea area and then verify it in a real environment is the focus of the next research in this paper.

Author Contributions: Conceptualization, X.Z. and S.G. Methodology, S.G. Software, S.G. and Y.Z. Writing—original draft preparation, X.Z., S.G., Y.D. Writing—review and editing, X.Z. and S.G. All authors have read and agreed to the published version of the manuscript.

Funding: This work is supported by the National Key R&D Program of China (Grant No. 2018YFB1601500), the National Natural Science Foundation of China (Grant No. 51679025) and the Fundamental Research Funds for the Central Universities (Grant No. 3132019313).

Conflicts of Interest: The authors declare no conflict of interest.

References

1. Liu, D.G.; Zheng, Z.Y.; Wu, Z.L. Risk analysis system of underway ships in heavy sea. *J. Traffic Transp. Eng.* **2004**, *4*, 100–102.
2. Piao, Z.; Guo, C.; Sun, S. Research into the Automatic Berthing of under actuated Unmanned Ships under Wind Loads Based on Experiment and Numerical Analysis. *J. Mar. Sci. Eng.* **2019**, *7*, 300. [CrossRef]
3. Roberts, G.N.; Sutton, R.; Zirilli, A.; Tiano, A. Intelligent ship autopilots—a historical perspective. *Mechatronics* **2003**, *13*, 1091–1103. [CrossRef]
4. Perera, L.P.; Carvalho, J.P.; Soares, C.G. Autonomous guidance and navigation based on the COLREGs rules and regulations of collision avoidance. In Proceedings of the International Workshop Advanced Ship Design for Pollution Prevention, Split, Croatia, 23–24 November 2009; pp. 205–216.
5. Liu, Z.; Zhang, Y.; Yu, X.; Yuan, C. Unmanned surface vehicles: An overview of developments and challenges. *Annu. Rev. Control* **2016**, *41*, 71–93. [CrossRef]
6. Lecun, Y.; Bengio, Y.; Hinton, G. Deep learning. *Nature* **2015**, *521*, 436–444. [CrossRef]
7. Sutton, R.S.; Barto, A.G. *Introduction to Reinforcement Learning*; MIT Press Cambridge: Cambridge, MA, USA, 1998; Volume 135.
8. Mnih, V.; Kavukcuoglu, K.; Silver, D.; Rusu, A.A.; Veness, J.; Bellemare, M.G.; Graves, A.; Riedmiller, M.; Fidjeland, A.K.; Ostrovski, G.; et al. Human-level control through deep reinforcement learning. *Nature* **2015**, *518*, 529. [CrossRef]
9. Tai, L.; Paolo, G.; Liu, M. Virtual-to-real deep reinforcement learning: Continuous control of mobile robots for mapless navigation. In Proceedings of the 2017 IEEE/RSJ International Conference on Intelligent Robots and Systems (IROS), Vancouver, BC, Canada, 24–28 September 2017; pp. 31–36.
10. Al-Nima, R.R.O.; Han, T.; Chen, T. Road tracking using deep reinforcement learning for self-driving car applications. *Int. Conf. Comput. Recognit. Syst.* **2019**. [CrossRef]
11. Zhu, M.; Wang, X.; Wang, Y. Human-like autonomous car-following model with deep reinforcement learning. *Transp. Res. Part C Emerg. Technol.* **2018**, *97*, 348–368. [CrossRef]
12. Yu, L.; Shao, X.; Wei, Y.; Zhou, K. Intelligent Land-Vehicle Model Transfer Trajectory Planning Method Based on Deep Reinforcement Learning. *Sensors* **2018**, *18*, 2905.
13. Deng, Y.; Bao, F.; Kong, Y.; Ren, Z.; Dai, Q. Deep Direct Reinforcement Learning for Financial Signal Representation and Trading. *IEEE Trans. Neural Netw. Learn. Syst.* **2017**, *28*, 653–664. [CrossRef]
14. Isele, D.; Rahimi, R.; Cosgun, A.; Subramanian, K.; Fujimura, K. Navigating occluded intersections with autonomous vehicles using deep reinforcement learning. In Proceedings of the 2018 IEEE International Conference on Robotics and Automation (ICRA), Brisbane, Australia, 21–25 May 2018; pp. 2034–2039.

15. Mnih, V.; Kavukcuoglu, K.; Silver, D.; Graves, A.; Antonoglou, I.; Wierstra, D.; Riedmiller, M. Playing atari with deep reinforcement learning. *arXiv* **2013**, arXiv:1312.5602.

16. Bahdanau, D.; Brakel, P.; Xu, K.; Goyal, A.; Lowe, G.; Pineau, J.; Courville, A.; Bengio, Y. An Actor-Critic Algorithm for Sequence Prediction. *arXiv* **2016**, arXiv:1607.07086.

17. Lillicrap, T.P.; Hunt, J.J.; Pritzel, A.; Heess, N.; Erez, T.; Tassa, Y.; Silver, D.; Wierstra, D. Continuous control with deep reinforcement learning. *Comput. Sci.* **2015**, *8*, A187.

18. Chen, C.; Chen, X.Q.; Ma, F.; Zeng, X.J.; Wang, J. A knowledge-free path planning approach for smart ships based on reinforcement learning. *Ocean Eng.* **2019**, *189*, 106299. [CrossRef]

19. Petres, C.; Romero-Ramirez, M.A.; Plumet, F. Reactive path planning for autonomous sailboat. In Proceedings of the 2011 IEEE International Conference on Advanced Robotics (ICAR), Tallinn, Estonia, 20–23 June 2011; pp. 112–117.

20. Mankabady, S. *The International Maritime Organization, Volume 1: International Shipping Rules*; Croom Helm Ltd.: London, UK, 1986; p. 450.

21. Kuwata, Y.; Wolf, M.T.; Zarzhitsky, D.; Huntsberger, T.L. Safe maritime autonomous navigation with COLREGS, using velocity obstacles. *IEEE J. Ocean. Eng.* **2013**, *39*, 110–119. [CrossRef]

22. Campbell, S.; Naeem, W. A rule-based heuristic method for colregs-compliant collision avoidance for an unmanned surface vehicle. *IFAC Proc. Vol.* **2012**, *45*, 386–391. [CrossRef]

23. Xue, Y.; Clelland, D.; Lee, B.S.; Han, D. Automatic simulation of ship navigation. *Ocean Eng.* **2011**, *38*, 2290–2305. [CrossRef]

24. Vettor, R.; Guedes Soares, C. Multi-objective evolutionary algorithm in ship route optimization. In *Maritime Technology and Engineering*; Taylor & Francis Group: London, UK, 2014; pp. 865–873.

25. Lazarowska, A. Ship's trajectory planning for collision avoidance at sea based on ant colony optimisation. *J. Navig.* **2015**, *68*, 291–307. [CrossRef]

26. Xin, J.; Zhong, J.; Yang, F.; Cui, Y.; Sheng, J. An Improved Genetic Algorithm for Path-Planning of Unmanned Surface Vehicle. *Sensors* **2019**, *19*, 2640. [CrossRef]

27. Xie, S.; Chu, X.; Zheng, M.; Liu, C. Ship predictive collision avoidance method based on an improved beetle antennae search algorithm. *Ocean Eng.* **2019**, *192*, 106542. [CrossRef]

28. Fu, K.; Li, Y.; Sun, H.; Yang, X.; Xu, G.; Li, Y.; Sun, X. A Ship Rotation Detection Model in Remote Sensing Images Based on Feature Fusion Pyramid Network and Deep Reinforcement Learning. *Remote Sens.* **2018**, *10*, 1922. [CrossRef]

29. Yang, J.; Liu, L.; Zhang, Q.; Liu, C. Research on Autonomous Navigation Control of Unmanned Ship Based on Unity3D. In Proceedings of the 2019 IEEE International Conference on Control, Automation and Robotics (ICCAR), Beijing, China, 19–22 April 2019; pp. 2251–2446.

30. Shen, H.Q.; Hashimoto, H.; Matsuda, A.; Taniguchi, Y.; Terada, D.; Guo, C. Automatic collision avoidance of multiple ships based on deep Q-learning. *Appl. Ocean Res.* **2019**, *86*, 268–288. [CrossRef]

31. Zhang, R.B.; Tang, P.; Su, Y.; Li, X.; Yang, G.; Shi, C. An adaptive obstacle avoidance algorithm for unmanned surface vehicle in complicated marine environments. *IEEE CAA J. Autom. Sin.* **2014**, *1*, 385–396.

32. Wang, Y.; Tong, J.; Song, T.Y.; Wan, Z.H. Unmanned Surface Vehicle Course Tracking Control Based on Neural Network and Deep Deterministic Policy Gradient Algorithm. In Proceedings of the OCEANS-MTS/IEEE Kobe Techno-Oceans (OTO), Kobe, Japan, 28–31 May 2018; pp. 1–5.

33. Zhang, X.; Wang, C.; Liu, Y.; Chen, X. Decision-Making for the Autonomous Navigation of Maritime Autonomous Surface Ships Based on Scene Division and Deep Reinforcement Learning. *Sensors* **2019**, *19*, 4055. [CrossRef] [PubMed]

34. Zhao, L.; Roh, M.I. COLREGs-compliant multiship collision avoidance based on deep reinforcement learning. *Ocean Eng.* **2019**, *191*, 106436. [CrossRef]

35. Lisowski, J.; Mohamed-Seghir, M. Comparison of Computational Intelligence Methods Based on Fuzzy Sets and Game Theory in the Synthesis of Safe Ship Control Based on Information from a Radar ARPA System. *Remote Sens.* **2019**, *11*, 82. [CrossRef]

36. Serrano, W. Deep Reinforcement Learning Algorithms in Intelligent Infrastructure. *Infrastructures* **2019**, *4*, 52. [CrossRef]

37. Ting-Ying, B.A.I. International Standards of Electronic Chart Display and Information System. *Hydrogr. Surv. Charting* **2004**, *2*, 67–70.

Article

A Generic Design of Driver Drowsiness and Stress Recognition Using MOGA Optimized Deep MKL-SVM

Kwok Tai Chui [1,*], Miltiadis D. Lytras [2,3] and Ryan Wen Liu [4]

1   Department of Technology, School of Science and Technology, The Open University of Hong Kong, Hong Kong
2   School of Business & Economics, Deree College—The American College of Greece, 153-42 Athens, Greece; mlytras@acg.edu
3   Effat College of Engineering, Effat University, Jeddah P.O. Box 34689, Saudi Arabia
4   Hubei Key Laboratory of Inland Shipping Technology, School of Navigation, Wuhan University of Technology, Wuhan 430063, China; wenliu@whut.edu.cn
*   Correspondence: jktchui@ouhk.edu.hk; Tel.: +852-2768-6883

Received: 27 December 2019; Accepted: 4 March 2020; Published: 7 March 2020

Abstract: Driver drowsiness and stress are major causes of traffic deaths and injuries, which ultimately wreak havoc on world economic loss. Researchers are in full swing to develop various algorithms for both drowsiness and stress recognition. In contrast to existing works, this paper proposes a generic model using multiple-objective genetic algorithm optimized deep multiple kernel learning support vector machine that is capable to recognize both driver drowsiness and stress. This algorithm simplifies the research formulations and model complexity that one model fits two applications. Results reveal that the proposed algorithm achieves an average sensitivity of 99%, specificity of 98.3% and area under the receiver operating characteristic curve (AUC) of 97.1% for driver drowsiness recognition. For driver stress recognition, the best performance is yielded with average sensitivity of 98.7%, specificity of 98.4% and AUC of 96.9%. Analysis also indicates that the proposed algorithm using multiple-objective genetic algorithm has better performance compared to the grid search method. Multiple kernel learning enhances the performance significantly compared to single typical kernel. Compared with existing works, the proposed algorithm not only achieves higher accuracy but also addressing the typical issues of dataset in simulated environment, no cross-validation and unreliable measurement stability of input signals.

Keywords: at-risk driving; deep support vector machine; driver drowsiness; driver stress; multi-objective genetic algorithm; multiple kernel learning

---

1. Introduction and Literature Review

The World Health Organization (WHO) has reported that the annual road traffic deaths and injuries remain unacceptably high as 1.35 million and 50 million respectively [1]. It has highlighted the road traffic is the 8th leading cause of death for people of all ages. More important, it ranks number one when it comes to the age group of 5–29 years old, which can wreak havoc on economic and social development. To be transformed into monetary measure, it accounts for 3% loss (equivalent to 2400 billion USD) in world gross domestic product (GDP). Injuries include minor cuts, whiplash, bruises to broken limbs, paralysis and spinal injuries. The report from the American Automobile Association (AAA) concluded that the poor driving behaviors leads to over 55% of fatal crashes [2]. With the ever-growing number of cars, the leading cause of death will soar from the 9th to the 7th position by 2030 if no action and prevention are carried out. In this paper, the issues of driver drowsiness and stress will be addressed.

Driver drowsiness is about the sleepiness of drivers that is an undesired condition in real-world driving. Surveys revealed that more than half of adult drivers felt sleepy while driving and even more severe that about 30% of them fell asleep [3]. This high prevalence can further estimate that drivers could experience accidents (affect or being influenced attributable to drowsy driving). It is worth mentioning that driver drowsiness is differed from driver fatigue [4]. Drivers drive unconsciously in the former condition but are with a conscious status in latter condition. In reality, people are usually fatigued in today's fast paced world.

Stress is body's way to react any kind of threat, challenge and demand. Research has revealed that stress has played a crucial role in adapting to driving and making decision [5]. The primary sources for driver stress are congestion and adverse driving condition as well as time pressure [6]. The stress can lead to poor and dangerous driving behaviors, for instance, flashing high beams, eliciting anger in drivers, road rage and aggressive driving, which are major causes of road traffic accidents [7]. Various recent studies have been carried out on the investigation between psychological factors and driving behaviors. In [8], results revealed that anger leads to stress, which is reflecting in the form of aggressive and negative cognitive driving behavior. On the other hand, research argued that risky drivers generally exhibited more antisocial and substance misuse, reward sensitive personality features and sensation seeking [9]. In addition, traffic penalties reported by public transport drivers are preceded by individual factors, personality and work-related when combined with driving anger. It may enhance negative results on traffic sanctions given they are preceded by risky road behaviors and affect overall road safety [10].

Repeated exposure of stressful conditions affects drivers' daily life and drivers have higher risks in suffering from stress-related health problems, for instance, accelerated aging [11], depression and anxiety [12], type 2 diabetes [13], asthma [14] and cardiovascular diseases [15]. These have indicated that driver stress can lead to long-term health issues.

1.1. Literature Review

Various research works have been carried out in driver drowsiness and driver fatigue detection. We believe that this paper is the first work that fully considers a generic model that can apply to recognize both driver drowsiness and stress. In this subsection, the literature review is divided into two parts, which firstly present the latest works on driver drowsiness recognition.

1.1.1. Existing Works of Driver Drowsiness Recognition

Typical algorithms for driver drowsiness recognition were based on three types of inputs: (i) the biometric-signal-based approach [16–19]; (ii) the vehicle-based approach [20–23] and (iii) the image-based approach [24–27]. Approach (i) is intrusive whereas approaches (ii) and (iii) are non-intrusive.

The first approach is illustrated as follows. In [16], driver drowsiness recognition was based on the input of respiratory signal, which measurement requires the tracking of the displacements of the diaphragm, abdominal and rib cage. A threshold was derived by analyzing the respiratory rate variability of the training dataset. The first work to adopt the electrocardiogram (ECG) signal was presented in [17], which is differed from traditional works that relied on partial information of the ECG signal, R wave or heart rate variability (HRV). The feature vector was formulated by cross-correlation coefficient between ECG signals and the classification problem was modeled by support vector machine (SVM). Another biometric-signal-based approach includes the electroencephalogram (EEG) signal, examples can be referred to [18] and [19] for the SVM model and long short-term memory (LSTM) model respectively.

When it comes to the vehicle-based approach, a lightweight threshold-based algorithm was proposed to analyze the status of drivers, with the continuous input of steering wheel angle [20]. With the same input, multilevel ordered logit model was implemented [21]. Some works utilized more measurements as inputs for driver drowsiness recognition. For instance, in [22], besides the steering wheel angle, pedal input, vehicle speed and acceleration were selected as features for classification model based on the dynamic Bayesian network. The deviation from the current lane could also be a useful indicator of driver drowsiness, as verified using an exponentially weighted moving average [23].

The third approach is based on the images of drivers, specifically the images are retrieved from continual video frames recorded by an in-vehicle camera. The textual and landmark information of the drivers' face were collected as inputs of the driver drowsiness recognition algorithm, implemented by the deep belief network [24]. Researchers in [25] constructed the features based on the detection of the face, eyes, nose and mouth of the drivers. A threshold was determined on the number of eye blinks per minute in order to deduce the percentage of drowsiness level. Mandal et al. proposed a fusion and reasoning method to measure the driver drowsiness, the following modules were included, head and shoulder detection, face detection based on the front view and oblique view analysis, eye detection based on Open Source Computer Vision Library (OpenCV) and Institute for Infocomm Research (I2R) as well as the eye openness estimation [26]. A multi-channel (3, 6 and 9 channels) second-order blind identification algorithm was proposed, which analyzed the yawn signals and eye blinks and yielded optimal thresholds for the drowsiness level [27].

1.1.2. Existing Works of Driver Stress Recognition

When it comes to driver stress recognition, there are less research publications compared to driver drowsiness recognition. However, both are of equal importance as life threatening sources. The approach for driver stress recognition was mainly based on the biometric-signal-based approach. Khattak et al. carried out statistical analysis (Wilcoxon Signed rank test, t-test and analysis of variance (ANOVA)) on the driver stress based on drivers' HRV [28]. Electrodermal activity (EDA) skin potential response (SPR) is another typical input for the driver stress detection, which an example was shown along with adaptive filtering and spike detection techniques [29]. Skin conductance and EEG served as inputs of the driver stress condition, which were further explored by the incremental association Markov blanket algorithm for feature extraction [30]. The least square support vector machine (LS-SVM) was applied to construct the classifier. In [31], researchers discussed the feasibility of the employment of HRV and photoplethysmogram (PPG) signal for driver stress recognition through ensemble learning of k-nearest neighbor (kNN), decision tree (DT) and linear discriminant analysis (LDA). Sparse Bayesian learning (SBL) and principal component analysis (PCA) were adopted to find out the optimal feature vector from a galvanic skin response (GSR), HRV and respiration [32]. Two kernel-based methods SVM and extreme learning machine (ELM) were applied and evaluated with typical kernels, sigmoid, radial basis function (RBF) and linear kernel.

It is noted that there were a few discussions on driver stress recognition via other approaches like steering wheel angle [33], steering wheel angle and road shape [34] and speech signal [35]. Besides, there is a recent work that detected the driver's status among normal, stress, fatigue and drowsiness [36]. However, it has been limited by assuming only one status. Differing from the authors work, we perform drowsiness and stress recognition in parallel to each other so that drivers will have multiple statuses, awake/drowsy and stress level.

1.2. Research Gaps and Motivation

Various methods have been proposed for driver drowsiness recognition [16–27] and driver stress recognition [28–35]. Nonetheless, there exist several challenges and limitations that require further research.

i.   There is no generic model for driver drowsiness and stress recognition, which is one model suits two applications. Existing studies considered the formulation separately and required the implementation of different models to serve two applications. This may increase the complexity of the recognition system as well as include more input signals.

ii.   The study in [37] analyzed the measurement reliability of several signals, ECG, EEG and video recording on real-world driving conditions (highway, urban and turning). EEG and video recording may not be reliable input signals as only 85% and 59% of the time the acquired signals are with good signal quality. The severe signal distortion in the rest of the period could alter the validity of the foundation of the problem formulations, which could be considered as misleading input and thus output. Therefore, EEG and video-based approaches experience challenges in signal acquisition.

iii.   There was limited utilization of a full ECG signal as signal input [17], which most of the existing works were focusing on HRV, which is a single point of an ECG signal. An in-depth analysis could be made on driver drowsiness and stress recognition via ECG signal.

iv.   Generally, shallow learning was adopted in existing works, except the discussion of a deep belief network in [24] for driver drowsiness recognition. Since there is room for improvement of accuracy of the existing classifiers, applying and investigating the deep learning technique could benefit on the classification performance.

To address the aforementioned challenges, the following measures are introduced in this paper:

i.   A generic model is proposed for the recognition of both driver drowsiness and stress.

ii.   The high measurement reliability ECG signal is adopted as the input signal.

iii.   The complete ECG signal is employed so that more representative features can be included in the feature extraction process.

iv.   The deep learning approach is selected to build the classifier. It is noted that there is a consideration of deep learning with small samples, which implies that typical deep neural network is not appropriate.

1.3. Research Contributions

The key contributions of this paper are summarized as follows.

i.   A generic model using the multiple-objective genetic algorithm (MOGA) optimized deep multiple kernel learning support vector machine (D-MKL-SVM) is proposed for the recognition of driver drowsiness and stress detection.

ii.   The ECG signal is newly applied for driver stress recognition in which the idea of using the ECG signal as an input was proposed in [17] for driver drowsiness recognition. The identical input signal and model serving two applications, i.e., driver stress and drowsiness recognition could reduce the complexity of the system theoretically and practically.

iii.   Deep support vector machine is employed, which takes the advantage in small samples problem.

iv.   The proposed generic model achieves highest accuracy for both driver drowsiness recognition and driver stress recognition compared to existing works. It also addresses typical issues of existing works, which are input signals of poor measurement stability, performance evaluation without cross-validation and collecting data using the simulated environment.

2. Materials and Methodology

In this section, the dataset for driver drowsiness and stress data was firstly presented. It was followed by data pre-processing techniques as well as beat segmentation for the ECG signal. Afterwards, the proposed generic model using MOGA optimized D-MKL-SVM would be discussed.

2.1. Driver Drowsiness and Stress Dataset

The dataset for driver drowsiness and stress data was retrieved from publicly available databases: the Stress Recognition in Automobile Drivers Database [38,39] and the cyclic alternating pattern (CAP) Sleep Database [39,40]. The first database includes 18 records of real-world driving in Boston, USA, which the route was started from the first baseline period, to the first city, to the first highway, to the second highway, to the second city and ended with the second baseline period, which is summarized in Figure 1a. The experiment was conducted in mid-morning or mid-afternoon to avoid heavy road traffic. Each participant was at rest (eyes closed and sitting inside the vehicle) for 15 minutes before and after the driving. This period of data provides a baseline of the participant. The medium stress environment was set-up by highway driving between a toll preceding the off-ramp and at the on-ramp. The high stress environment was side street and main street driving environment with a winding, narrow lamp. The ECG signal of the drivers were continually monitored and collected. The duration of records ranged from 50 to 90 min. Three stress levels: (i) high stress level; (ii) medium stress level and (iii) low stress level; were defined. To avoid the effect of traffic condition, the experiment was carrying out at mid-morning or mid-afternoon.

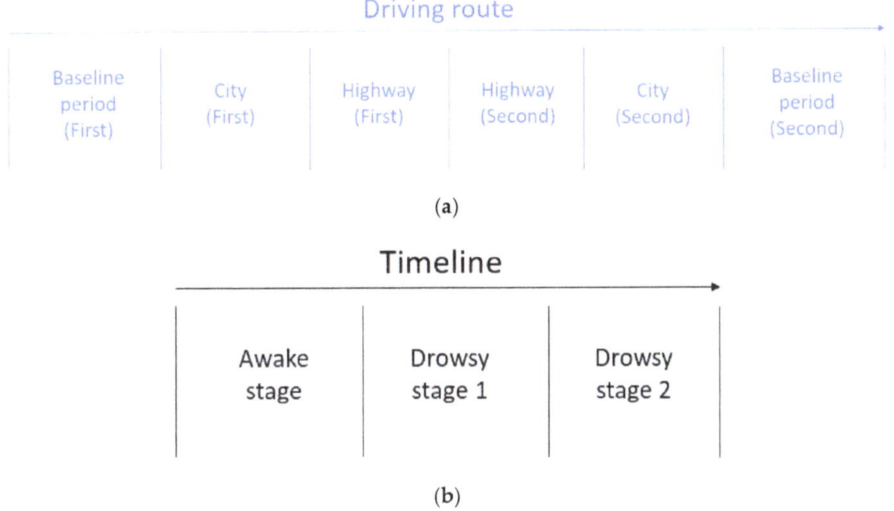

Figure 1. (**a**) Scenario setting in the Stress Recognition in Automobile Drivers Database and (**b**) the scenario setting in the cyclic alternating pattern (CAP) Sleep Database.

The second database had 108 records from sleep centers. These records contained six categories of sleep stages (i) awake stage; (ii) sleep stage 1; (iii) sleep stage 2; (iv) sleep stage 3; (v) sleep stage 4 and (vi) rapid eye movement (REM) stage. Only sleep stage 1 and sleep stage 2 were selected, which is drowsy stage 1 and drowsy stage 2 because they were the immediate stages after awake stage, which the timeline is shown in Figure 1b.

2.2. Data Pre-Processing

The data pre-processing of the dataset aimed at segmenting ECG beat into individual samples. We have adopted an existing Tompkins' method [41,42] as it achieves favorable performance in ECG beat segmentation. The major steps involved are dc offset elimination, digital band-pass filter, derivative filter, signal squaring, sliding window integration and locating Q, R and S waves. The detail was omitted as this was not the focus of this paper.

There are three classes in each of the driver drowsiness dataset and driver stress dataset. Their corresponding number of samples is tabulated in Table 1 after ECG beat segmentation. Totally, the driver drowsiness dataset and driver stress dataset contain 131,500 and 76,200 samples respectively.

Table 1. Definition of classes and number of samples in driver drowsiness and stress dataset.

Dataset	Class	Number of Samples
	Class 0: Awake stage	76,200
Driver drowsiness dataset	Class 1: Drowsy stage 1	35,300
	Class 2: Drowsy stage 2	20,000
	Class 0: High stress level	19,300
Driver stress dataset	Class 1: Medium stress level	45,000
	Class 2: Low stress level	11,900

2.3. Generic Model Using MOGA Optimized D-MKL-SVM

We have proposed a generic model using MOGA optimized D-MKL-SVM. Figure 2 shows the general flow of the model. Dataset and data pre-processing were discussed in Sections 2.1 and 2.2 respectively. For feature extraction, the rationale is explained as follows. Previous works had revealed that HRV or R waves of ECG signals could serve as an important feature for driver drowsiness recognition. In this paper, the consideration was further extended to a full utilization of the ECG signal. Cross-correlation and convolution techniques were applied between every two samples (as defined in Section 2.2), which resulted in cross-correlation coefficients and convolution coefficient and serving as feature inputs of driver drowsiness and stress data.

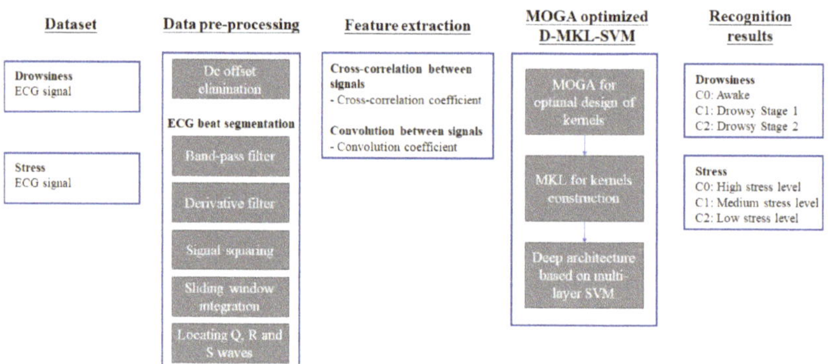

Figure 2. General flow of the proposed generic model multiple-objective genetic algorithm (MOGA) optimized deep multiple kernel learning support vector machine (D-MKL-SVM).

Cross-correlation between two ECG signals X_1 and X_2 with length of N = 100 (zero-padding when len{X}<100) is given in Equation (1).

$$\text{XCorr}_{X_1,X_2}[k] = \begin{cases} \sum_{n=k}^{N-1} X_1[n]X_2[n-k], & k \geq 0 \\ \sum_{n=0}^{N-|k|-1} X_1[n]X_2[n-k], & k < 0 \end{cases} \tag{1}$$

Convolution between X_1 and X_2 can be calculated by:

$$\text{Con}_{X_1,X_2}[k] = \sum_{k=0}^{N-1} X_1[n]X_2[n-k] \tag{2}$$

The model for driver drowsiness and stress recognition is built based on deep multi-layer SVM architecture. Generally, it consists of multiple hidden layers of SVM and an output layer of SVM. Compared to other deep learning architectures, multi-layer SVM takes several advantages like (i) the output layer SVM has strong regularization power to avoid over-fitting; (ii) able to handle problem of very large input vectors and few training samples and (iii) the design of kernel functions is more flexible.

The deep multi-layer SVM is structured as in Figure 3. Assume D, L, M and N are integers. Here the number of hidden layers had not been fixed, which is a general form of architecture. In this paper, experimental results indicate that the optimal setting for driver drowsiness and driver stress recognition is deep three-layered SVM and deep four-layered SVM respectively. In each of the SVM, the kernel function is customized by multiple kernel learning (MKL), which weighting factors $\omega_i = [0, 1] \; \forall i = 1, \ldots, N_k$ where there is N_k kernels, are optimally designed by MOGA.

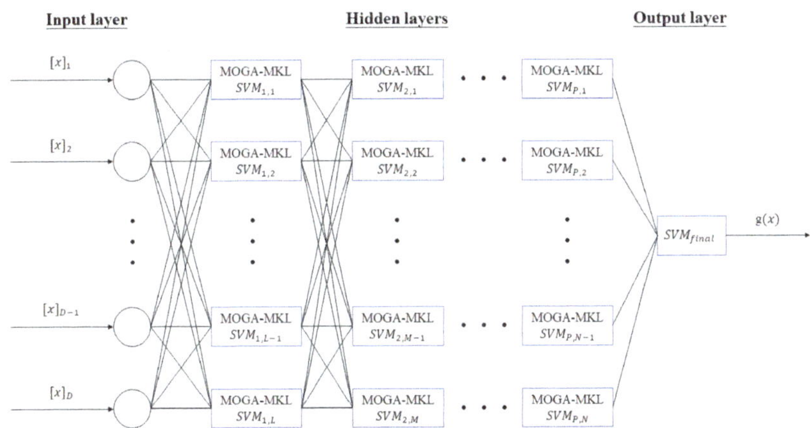

Figure 3. Architecture of MOGA optimized D-MKL-SVM.

The kernel function of every SVM is designed by the combinations of typical kernels, linear kernel, RBF kernel, polynomial kernel and sigmoid kernel. The classifier can achieve better performance by taking the advantages from each kernel. Four kernel properties are applied for joining these kernels. Weighting factors are also introduced between kernels and optimally solved by MOGA. It is worth mentioning that, each SVM in multi-layer SVM architecture may result in different sets of optimal weighting factors. The kernel properties are presented as follows [43]. Assume there are two Mercer's kernel $K_i(x_1, x_2) : \chi \cdot \chi \rightarrow R$ and $K_j(x_1, x_2) : \chi \cdot \chi \rightarrow R$, $K_e(x_1, x_2)$ is the resultant Mercer's kernel. It is trivial to derive that joining properties between P_1, P_2, P_3 and P_4 in Equations (3)–(6) is also governed by Mercer's theorem.

$$P_1 : K_e(x_1, x_2) = K_i(x_1, x_2) + K_j(x_1, x_2) \tag{3}$$

$$P_2 : K_e(x_1, x_2) = b \cdot K_i(x_1, x_2) \tag{4}$$

$$P_3 : K_e(x_1, x_2) = K_i(x_1, x_2) + b \tag{5}$$

$$P_4 : K_e(x_1, x_2) = K_i(x_1, x_2) \cdot K_j(x_1, x_2) \tag{6}$$

The optimal design of kernel functions for each SVM is considered as a multi-objective optimization (MOO) problem, which is solved by MOGA. MOO is practically important because daily life applications are generally with multi-objective and those objectives are conflicted with each other [44,45]. Traditional single objective optimization often normalizes and combines objectives into single objective; whereas MOO provides trade-off optimal solutions, and these compromise of the solutions and can increase the satisfaction of the decision-makers. In addition, MOO is characterized by different search spaces, multiple objectives and cardinalities-optimal solutions [46]. Here are quick review of definitions of important concepts: (i) Objective space is defined as the multidimensional space of the objective functions; (ii) Pareto optimal solution is defined as the optimal solution in the objective space and (iii) Pareto front is defined as the set of Pareto optimal solutions.

Attributed to multiple conflicting objectives, there exists a challenge in obtaining single optimal objective. As a result, the technique of domination was introduced. Consider a minimization problem of the N-objective, we say candidate solution a is dominated by another candidate solution b if and only if function values of a is partially less than b, its mathematical expression is:

$$\begin{cases} f_n(a) \geq f_n(b), & \forall n = 1,2,\ldots,N \\ f_n(a) > f_n(b), & \exists n = 1,2,\ldots,N \end{cases} \tag{7}$$

Hence, non-dominated Pareto optimal solutions are desired. The flow of the MOGA for the optimal design of MKL for each SVM is shown in Figure 4. The key steps are summarized:

i.   Initialize the values of objective function as well as the population size.
ii.   Based on the first step, compute the values of the objective function for all individuals within the population. This creates a list of values.
iii. Rank the individuals based on the list of values in Step (ii).
iv.   For a defined population size, the nature of stochastic selection errors governs the convergency of the population, which depends on a small group of Pareto optimal solutions, rather than all optimal solutions.
v.   Lengthen the distance between Pareto optimal solutions along the axis of objective functions. This could increase the diversity of the population in order to lower the tendency of the convergence to small group solutions. Thus, the niche count is utilized and defined.
vi.   Generate a new offspring.
vii. Evaluate the values of objective functions.
viii. Calculate ranks assignment and niche count repeatedly in the new offspring.
ix.   MOGA can be terminated in two situations: (i) the output reaches the Pareto front; or (ii) generation (number of iterations) reaches the maximum number of generations.

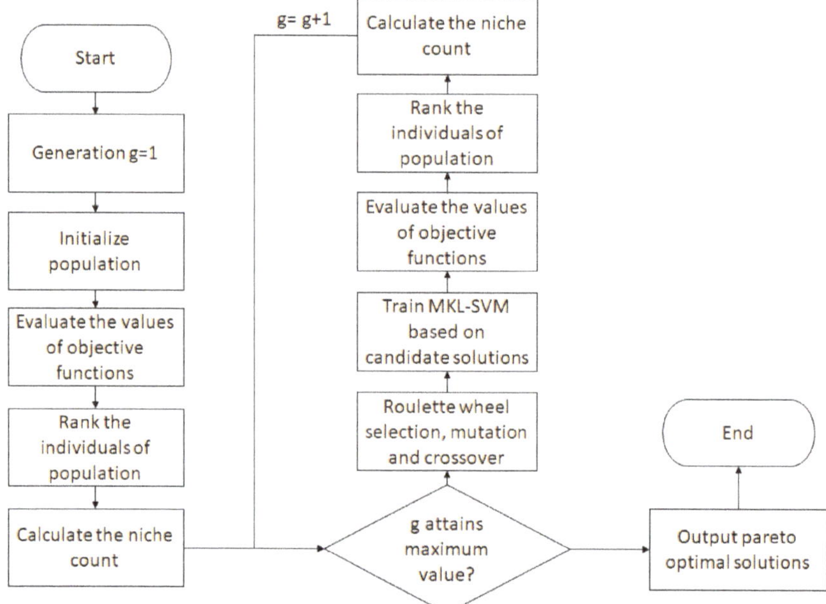

Figure 4. MOGA process for the optimal design of MKL-SVM.

The MOO for driver drowsiness and stress recognition is formulated as follows, with objectives O_1, O_2 and O_3:

$$O_1 : \max \sum_{i=1}^{N} \alpha_i - 0.5 \sum_{i=1}^{N} \sum_{j=1}^{N} \alpha_i \alpha_j y_i y_j K_e(x_i, x_j) \tag{8}$$

$$O_2 : \max TN/N_n \tag{9}$$

$$O_3 : \max TP/N_p \tag{10}$$

where O_1 aims at maximizing the margin, defining α_i as the Lagrange multiplier, $y_i \in \{-1, +1\}$ as the output class and $K_e(x_i, x_j)$ is the resultant Mercer's kernel defined by applying kernel properties in Equations (3)–(6) on typical kernels: linear kernel, RBF kernel, polynomial kernel and sigmoid kernel. O_2 maximizes the specificity, which is the ratio of true negative (TN) and number of negative samples (N_n). O_3 maximizes the sensitivity, which is the ratio of true positive (TP) and number of positive samples (N_p).

The binary classifier SVM is extended to multi-class SVM by the 1-against-1 method as it outperforms the one-against-all approach in general applications [47–49]. When it comes to performance evaluation, 10-fold cross-validation is selected [43,49,50].

3. Analysis and Results

Various scenarios were analyzed to reveal the effectiveness of proposed generic model using MOGA optimized D-MKL-SVM for driver drowsiness and stress recognition. Four parts would be discussed in detail: (i) the performance of the proposed generic model MOGA optimized D-MKL-SVM was evaluated; (ii) study of the effectiveness of MOGA comparing with pure D-MKL-SVM; (iii) comparing the performance between MKL and single typical kernel and (iv) comparing the performance of proposed algorithm and related works.

3.1. Performance Evaluation of MOGA Optimized D-MKL-SVM

The evaluation criteria for driver drowsiness and stress recognition via the MOGA optimized D-MKL-SVM algorithm are the specificity, sensitivity and area under the receiver operating characteristic curve (AUC). Figure 5 shows the average of sensitivity, specificity and AUC (based on 10-fold cross-validation) of driver drowsiness and driver stress versus the number of layers in MOGA optimized D-MKL-SVM. As a trade-off to computation time, the number of layers was limited to 5.

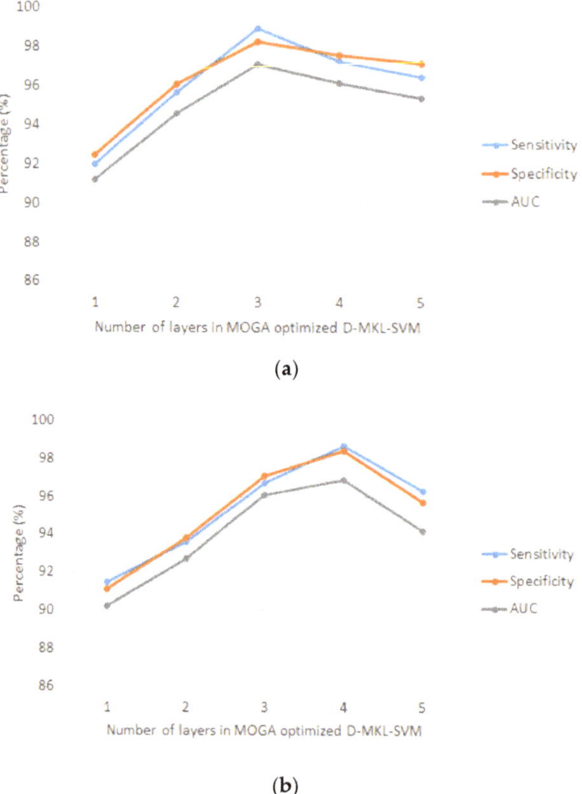

(**a**)

(**b**)

Figure 5. Average sensitivity, specificity and area under the receiver operating characteristic curve (AUC) versus number of layers in MOGA optimized D-MKL-SVM under 10-fold cross-validation: (**a**) driver drowsiness recognition and (**b**) driver stress recognition.

Results revealed that the proposed algorithm achieved favorable performance in average sensitivity, specificity and AUC in both driver drowsiness and stress recognition. For driver drowsiness recognition, the best performance was obtained using deep three-layered SVM, with average sensitivity of 99%, specificity of 98.3% and AUC of 97.1%. For driver stress recognition, the best performance was yielded when deep four-layered SVM was adopted, with average sensitivity of 98.7%, specificity of 98.4% and AUC of 96.9%. On the other hand, the average number of generations (rounded to nearest integer) in each layer for driver drowsiness and stress recognition is shown in Figure 6. As a result, the proposed approach was generic, which suited both applications.

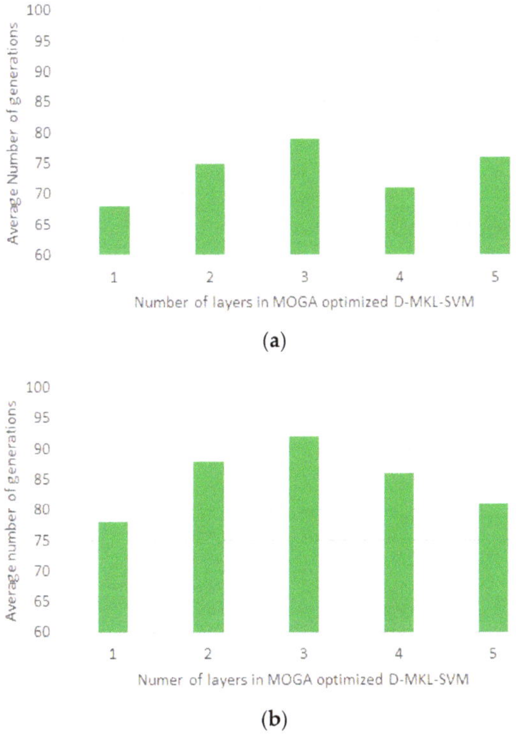

(a)

(b)

Figure 6. Average number of generations versus the number of layers in MOGA optimized D-MKL-SVM under 10-fold cross-validation: (**a**) driver drowsiness recognition and (**b**) driver stress recognition.

3.2. Study on the Benefits of MOGA

For parameter selection, grid search is a typical method that would carry out a simple trial and error on the values parameters, given a range and certain step size between successive test values. The grid search method requires one to test all the scenarios with the step size that requires excessive computation power when it comes to a small step size and large difference in boundaries. To avoid excessive iterations, the step size for weighting factors using grid search was selected to be 0.05 and 0.1.

Figure 7 shows the average sensitivity, specificity and AUC of the proposed algorithm and grid search method with a step size of 0.05 and 0.1, in driver drowsiness and stress recognition. The results were summarized as follows. For driver drowsiness recognition as in Figure 7a, the performance indicators of average sensitivity, specificity and AUC of the proposed algorithm were 99%, 98.3% and 97.1%. For the grid search approach with a step size of 0.05, the indicators were 90.3%, 91.5% and 89.2%. When the step size was doubled the indicators were 89.4%, 88.7% and 86.7%. Hence, the average improvement by MOGA was 8.64% and 11.2% comparing with grid search with a step size of 0.05 and 0.1 respectively.

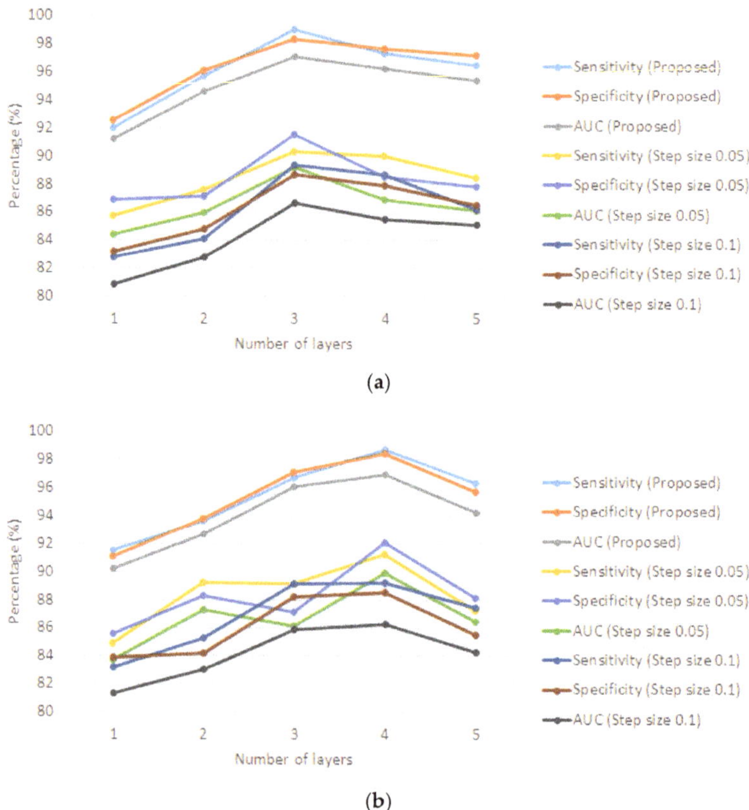

(a)

(b)

Figure 7. Average sensitivity, specificity and AUC of the proposed algorithm using MOGA and a traditional grid search with a step size of 0.05 and a step size of 0.1, versus the number of layers under a 10-fold cross-validation: (**a**) driver drowsiness recognition and (**b**) driver stress recognition.

When it comes to driver stress recognition, referring to Figure 7b, the proposed algorithm achieved 98.7%, 98.4% and 96.9%, the grid search approach with a step size of 0.05 achieved 91.2%, 92.1% and 88.9% and that for a step size of 0.1 was 89.2%, 88.5% and 86.3%. Thus, the average improvement by MOGA was 8.01% and 11.4% comparing with a grid search with a step size of 0.05 and 0.1 respectively.

It was concluded that the proposed algorithm outperformed the grid search method. All the methods agreed that the optimal numbers of deep layers for driver drowsiness and stress recognition were three and four respectively. A step size of 0.05 was better than that of 0.1 as the number of considered solutions was significantly increased. MOGA had better searching in the solution space whereas a grid search only bounds to limited solutions. One may argue that the step size can be further reduced so that all the possible solutions can be approximately analyzed, however, it requires extensive computation power, which is normally not a generic and feasible approach practically. Here, the proposed algorithm was a deep learning approach, which involved many optimization problems (optimal SVM).

3.3. Study on the Benefits of MKL

To reveal the improvement of MKL, analysis is made between MKL and single typical kernel: linear kernel, RBF kernel, polynomial kernel and sigmoid kernel. Figure 8 shows the results of MKL, linear kernel, RBF kernel, polynomial kernel and sigmoid kernel for driver drowsiness and stress recognition. To better visualize the results, each subfigure compares MKL with two of the kernels.

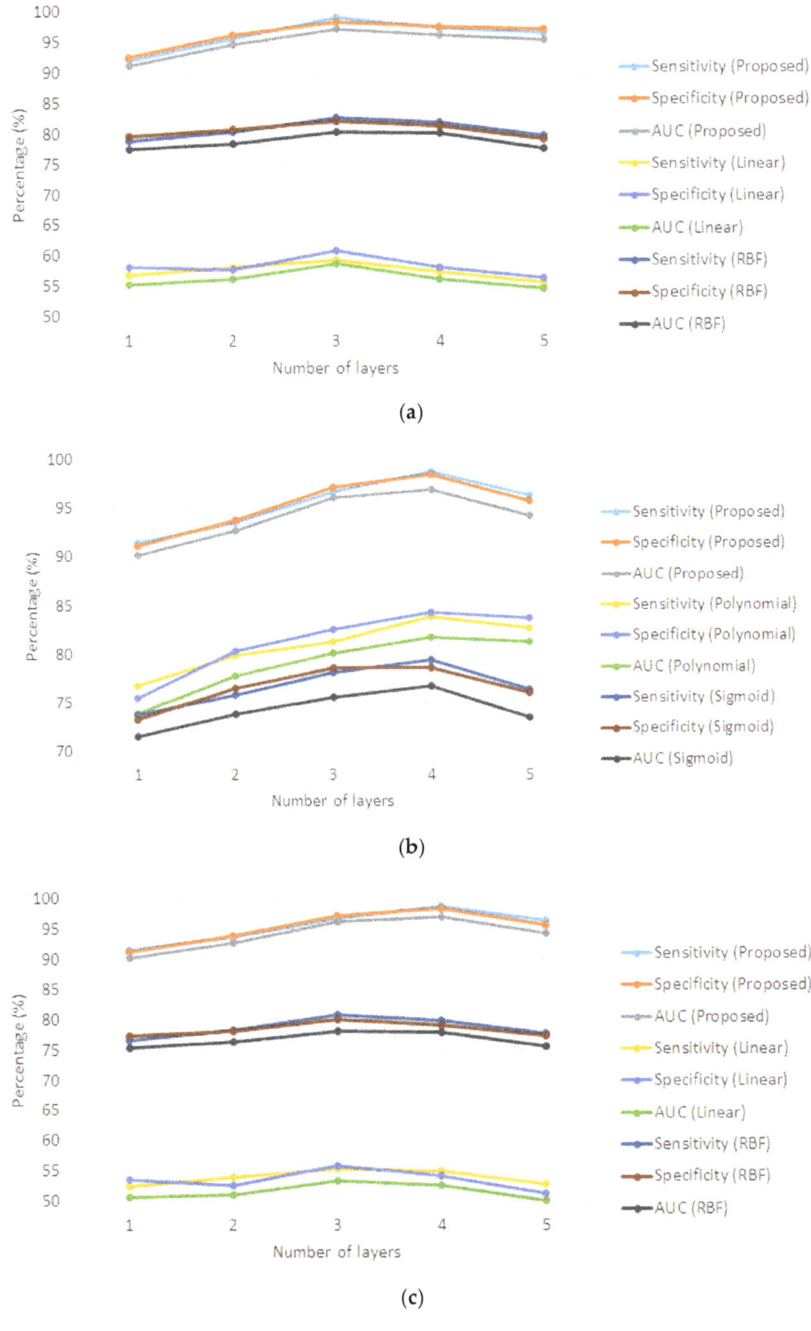

Figure 8. *Cont.*

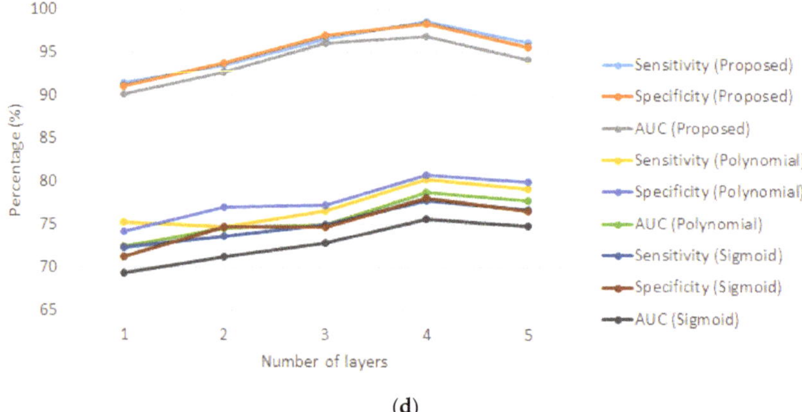

Figure 8. Average sensitivity, specificity and AUC of the proposed algorithm using multiple kernel learning (MKL) and typical single kernel versus number of layers under 10-fold cross-validation: (**a**) the MKL approach versus linear and radial basis function (RBF) kernels for driver drowsiness recognition; (**b**) the MKL approach versus polynomial and sigmoid kernels for driver drowsiness recognition; (**c**) the MKL approach versus linear and RBF kernels for driver stress recognition and (**d**) the MKL approach versus polynomial and sigmoid kernels for driver stress recognition.

Driver drowsiness recognition was analyzed based on Figure 8a,b. It can be concluded that the proposed algorithm using MKL significantly improved the performance indicators compared with all single typical kernel. Results between different kernel agreed with the optimal number of layers of three. The average performance improvement by the proposed algorithm was 64.6%, 20.1%, 17.7% and 25.2% compared to linear kernel, RBF kernel, polynomial kernel and sigmoid kernel, respectively.

The same finding was observed for driver stress recognition in Figure 8c,d. The optimal number of layers matches (which is four) in all approaches. On average, the performance improvement by the proposed algorithm was 82.0%, 24.3%, 22.5% and 26.9% respectively.

MKL combines the advantages and properties of various kernels so that the resultant kernel function outperforms any single typical kernel. One of the other key reasons is typical kernels are not customized to a specific application. Therefore, the proposed approach using MKL, which the weighting factors were optimally designed by MOGA, is a customized approach for designing the optimal kernel function for driver drowsiness and stress recognition.

3.4. Comparisons to Related Works

First, the comparison was carried out on driver drowsiness recognition. Table 2 summarizes the performance between proposed MOGA optimized D-MKL-SVM and existing works [16–27]. It summarizes the category of input signals, dataset, methodology, types of cross-validation and performance. Existing works may experience the following issues, which lead to lower reliability (and thus performance) when the methodology is implemented practically.

Table 2. Performance comparison between the proposed method and existing works for driver drowsiness recognition.

Work	Category	Dataset	Methodology	Cross-Validation	Performance
[16]	Biometric-signal-based (respiratory signal)	20 volunteers (simulated environment) Samples: 2246 awake; 1035 drowsy	Threshold-based approach by the tracking of the displacements of diaphragm, abdominal and rib cage	Leave-one-subject-out	Specificity: 96.6% Sensitivity: 90.3%
[17]	Biometric-signal-based (ECG signal)	18 volunteers (real-world driving environment) Samples: unknown	SVM using polynomial kernel using cross-correlation coefficient	10-fold cross-validation	Sensitivity: 77.4% Specificity: 76.5% Overall accuracy: 76.9%
[18]	Biometric-signal-based (EEG signal)	17 volunteers (simulated environment) Samples: 255 awake; 477 slightly drowsy; 167 moderate drowsy; 98 significant drowsy; 20 extremely drowsy	SVM using RBF kernel using RBP (α) and movement power	Leave-one-subject-out	Overall accuracy: 93.7%
[19]	Biometric-signal-based (EEG signal)	16 volunteers (simulated environment) Samples: unknown	LSTM using spectral entropy and instantaneous frequency	10-fold cross-validation	Overall accuracy: 94.3%
[20]	Vehicle-based (steering wheel angle)	6 volunteers (real-world driving environment) Samples: 92 awake; 99 drowsy	Threshold-based approach by analyzing steering wheel angle	No	Accuracy: 78.0%
[21]	Vehicle-based (steering wheel angle)	10 volunteers (simulated environment) Samples: Total 7020	Multilevel ordered logit model	No	Accuracy: 72.9%
[22]	Vehicle-based (steering wheel angle, pedal input, vehicle speed and acceleration)	72 volunteers (simulated environment) Samples: 840 awake; 21 drowsy	Dynamic Bayesian Network algorithm	No	Specificity: 85% AUC: 77%
[23]	Vehicle-based (deviation from the current lane)	Unknown number of volunteers (simulated environment) Samples: 4000 awake; 4000 drowsy	Exponentially weighted moving average	No	Sensitivity: 76% Accuracy: 86%
[24]	Image-based (image of driver's head)	30 volunteers (real-world driving environment) Samples: 11568 awake; 4224 moderate drowsy; 4560 severe drowsy	Deep belief network	10-fold cross-validation	Average accuracy: 96.7%
[25]	Image-based (eyes)	Unknown number of volunteers (simulated environment) Samples: 2500 images	Threshold-based approach by analyzing eye blinking frequency	No	Accuracy: 89%
[26]	Image-based (eyes)	15 volunteers (simulated environment) Samples: 1068 images in total	Fusion and reasoning method including head and shoulder detection, face detection based on front view and oblique view analysis, eye detection	No	Average accuracy: 90.1%
[27]	Image-based (yawns and eyes)	15 volunteers (simulated environment) Samples: unknown	Threshold-based approach using second-order blind identification algorithm	No	Accuracy: From 27.2% to 95.3% to under different scenarios
Proposed algorithm	Biometric-signal-based	126 volunteers (real-world driving environment) Samples: 76,200 awake; 35,300 sleep stage 1, 20,000 sleep stage 2	MOGA optimized D-MKL-SVM with cross-correlation and convolution coefficients	10-fold cross-validation	Sensitivity: 99% Specificity: 98.3% AUC: 97.1%

Area under the curve (AUC); Receiver operating characteristic (ROC); Deep multiple kernel learning support vector machine (D-MKL-SVM); Electrocardiogram (ECG); Electroencephalography (EEG); Long short-term memory (LSTM); Multiple-objective genetic algorithm (MOGA); Radial basis function (RBF); Relative band power (RBP); Support vector machine (SVM).

Generally, the first part of any methodology is the source of the input signal in which study in [37] analyzed the measurement reliability of several signals, ECG, EEG and video recording on real-world driving conditions (highway, urban and turning). EEG and a video recording may not be reliable input signals as only 85% and 59% of the time the acquired signals were with good signal quality. The severe signal distortion in the rest of the period could alter the validity of the foundation of the problem formulations, which could be considered as misleading input and thus output. Therefore, EEG and video-based approaches experience challenges in signal acquisition. As a result, the performance in related works [18,19,24–27] is reduced in a practical situation, which the methodology is governed by a maximum accuracy of 85% and 59% for an EEG-based and image-based approach.

It can be seen from Table 2 that some works [20–23,25–27] did not employ cross-validation. The bias training dataset (lack of generalization) may be selected to obtain high accuracy. In addition, many existing works [16,18,19,21–23,25–27] utilized simulation driving datasets. These were not convincing to reflect actual driver's status.

The proposed MOGA optimized D-MKL-SVM algorithm utilizes the ECG signal, which provides a reliable measurement of stability. The real-world driving dataset and 10-fold cross-validation were adopted. Therefore, the performance of proposed work was more reliable and robust.

Consider driver stress recognition, Table 3 compares the performance between the proposed algorithm and related works [28–35], towards a category of input signals, dataset, methodology, types of cross-validation and performance. Similar to driver drowsiness recognition, related works in driver stress recognition experienced similar issues.

Table 3. Performance comparison between the proposed method and existing works for driver stress recognition.

Work	Category	Dataset	Methodology	Cross-Validation	Performance
[28]	Biometric-signal-based (HRV)	22 volunteers (real-world driving environment) Samples: Unknown	Wilcoxon Signed rank test, *t*-test and ANOVA	No	No (statistical analysis between HRV and stress level)
[29]	Biometric-signal-based (EDA SPR)	15 volunteers (simulated environment) Samples: 510	Adaptive filtering and spike detection	No	Accuracy: 83.9%
[30]	Biometric-signal-based (Skin conductance and EEG)	30 volunteers (real-world driving environment) Samples: 713 normal; 430 low anger; 315 medium anger; 161 high anger	Incremental association Markov blanket and least square SVM	10-fold cross-validation	Accuracy: 82.2%
[31]	Biometric-signal-based (HRV and PPG)	21 volunteers (real-world driving environment) Samples: Unknown	Ensemble learning of kNN, DT and LDA	10-fold cross-validation	Accuracy: 86.9%
[32]	Biometric-signal-based (GSR), HRV and respiration)	18 volunteers (real-world driving environment) Samples: 588 high stress level; 588 medium stress level; 588 low stress level	SVM and ELM	Leave-one-subject-out	SVM Sensitivity: 88.5% Specificity: 94.2% ELM Sensitivity: 88.2% Specificity: 94.1%
[33]	Vehicle-based (steering wheel angle)	8 volunteers (simulated environment) Samples: 2154 normal; 2287 stress	SVM	No	Accuracy: 82.5%

Table 3. *Cont.*

Work	Category	Dataset	Methodology	Cross-Validation	Performance
[34]	Vehicle-based (steering wheel angle and road shape)	4 volunteers (real-world driving environment) Samples: 220	Multilayer perceptron	10-fold cross-validation	Accuracy: 46.9%
[35]	Speech-based (speech and GSR)	N/A	SVM	10-fold cross-validation	92.4%
Proposed algorithm	Biometric-signal-based	18 volunteers (real-world driving environment) Samples: 19,300 high stress level; 45,000 medium stress level; 11,900 low stress level	MOGA optimized D-MKL-SVM with cross-correlation and convolution coefficients	10-fold cross-validation	Sensitivity: 98.7% Specificity: 98.4% AUC: 96.9%

Analysis of variance (ANOVA); Decision tree (DT); Electrodermal activity (EDA); Extreme learning machine (ELM); Galvanic skin response (GSR); Heart rate variability (HRV); k-nearest neighbor (kNN); Linear discriminant analysis (LDA); Photoplethysmogram (PPG); Signal potential response (SPR).

The EEG signal was adopted in [30], which the input signal has a measurement stability of 85% [37]. Some of the works [28,29,33] did not analyze the method using cross-validation. Data was collected using the simulated environment in [29,35]. It is noted that there is an unspecified dataset in [35]. To compare the rest of the existing works, the proposed algorithm outperforms [30–32,34] significantly.

As a result, the proposed MOGA optimized D-MKL-SVM outperforms existing works and serves as a generic approach for both driver drowsiness and stress recognition.

4. Conclusions

Owning to the fact that existing works on driver drowsiness and stress recognition have room for improvement, which suffer from the following concerns: (i) input signals of poor measurement stability, (ii) performance evaluation without cross-validation and (iii) collecting data using a simulated environment, this paper proposed a generic model using the multiple-objective genetic algorithm (MOGA) optimized deep multiple kernel learning support vector machine (D-MKL-SVM). For driver drowsiness recognition, the average sensitivity, specificity and AUC were 99%, 98.3% and 97.1% respectively. For driver stress recognition, the average sensitivity, specificity and AUC were 98.7%, 98.4% and 96.9% respectively. For performance evaluation, we firstly analyzed the importance of MOGA versus the traditional grid search method and MKL versus the single typical kernel. The proposed generic model achieved the highest accuracy for both driver drowsiness recognition and driver stress recognition compared to existing works.

The proposed approach can improve road safety as follows. Once drowsy events have been concluded, a sound could be emitted to awake drivers. In contrast to drowsiness recognition, stress recognition cannot directly prevent a stressed driving accident. The decision should come along with some relaxation techniques in order to lower the stress level of drivers. There are some measures to relieve stress while driving: (i) turn off disturbing noise, (ii) take deep breaths at stoplights, (iii) say what you see, (iv) try not to have long haul driving, (v) drive in slow and safe areas and (vi) embrace mindfulness. It is extremely important to reduce the stress level while driving. It not only helps at preventing accidents but also increases the happiness after driving, since stress can be accumulated as a snowball effect.

The future work will be suggested in two ways: (i) analyze the performance of proposed work with a real-world driving dataset, towards volunteers of different countries, so that the proposed algorithm can be adopted worldwide and (ii) test if the proposed generic model is applicable to cardiovascular diseases recognition because the proposed approach utilizes the ECG signal as an input, which is also the key input for cardiovascular diseases recognition.

Author Contributions: Formal analysis, K.T.C., M.D.L. and R.W.L.; Investigation, K.T.C., M.D.L. and R.W.L.; Methodology, K.T.C.; Validation, K.T.C., M.D.L. and R.W.L.; Visualization, K.T.C.; Writing—original draft, K.T.C., M.D.L. and R.W.L. Writing—review and editing, K.T.C., M.D.L. and R.W.L. All authors have read and agreed to the published version of the manuscript.

Funding: This research received no external funding.

Conflicts of Interest: The authors declare no conflict of interest.

References

1. World Health Organization. *Global Status Report on Road Safety 2018*; World Health Organization: Geneva, Switzerland, 2018.

2. Du, X.; Shen, Y.; Chang, R.; Ma, J. The exceptionists of Chinese roads: The effect of road situations and ethical positions on driver aggression. *Transp. Res. Part F Traffic Psychol. Behav.* **2018**, *58*, 719–729. [CrossRef]

3. Rosekind, M.R. Underestimating the societal costs of impaired alertness: Safety, health and productivity risks. *Sleep Med.* **2005**, *6*, S21–S25. [CrossRef]

4. Sikander, G.S.; Anwar, S. Driver fatigue detection systems: A review. *IEEE Trans. Intell. Transp. Syst.* **2019**, *20*, 2339–2352. [CrossRef]

5. Useche, S.A.; Ortiz, V.G.; Cendales, B.E. Stress-related psychosocial factors at work, fatigue, and risky driving behavior in bus rapid transport (BRT) drivers. *Accid. Anal. Prev.* **2017**, *104*, 106–114. [CrossRef] [PubMed]

6. Qu, W.; Zhang, Q.; Zhao, W.; Zhang, K.; Ge, Y. Validation of the Driver Stress Inventory in China: Relationship with dangerous driving behaviors. *Accid. Anal. Prev.* **2016**, *87*, 50–58. [CrossRef]

7. Precht, L.; Keinath, A.; Krems, J.F. Effects of driving anger on driver behavior–Results from naturalistic driving data. *Transp. Res. Part F Traffic Psychol. Behav.* **2017**, *45*, 75–92. [CrossRef]

8. Ge, Y.; Qu, W.; Jiang, C.; Du, F.; Sun, X.; Zhang, K. The effect of stress and personality on dangerous driving behavior among Chinese drivers. *Accid. Anal. Prev.* **2014**, *73*, 34–40. [CrossRef]

9. Brown, T.G.; Ouimet, M.C.; Eldeb, M.; Tremblay, J.; Vingilis, E.; Nadeau, L.; Pruessner, J.; Bechara, A. Personality, executive control, and neurobiological characteristics associated with different forms of risky driving. *PLoS ONE* **2016**, *11*, e0150227. [CrossRef]

10. Montoro, L.; Useche, S.; Alonso, F.; Cendales, B. Work environment, stress, and driving anger: A structural equation model for predicting traffic sanctions of public transport drivers. *Int. J. Environ. Res. Public Health* **2018**, *15*, 497. [CrossRef]

11. Maurya, P.K.; Noto, C.; Rizzo, L.B.; Rios, A.C.; Nunes, S.O.V.; Barbosa, D.S.; Sethi, S.; Zeni, M.; Mansur, R.B.; Maes, M.; et al. The role of oxidative and nitrosative stress in accelerated aging and major depressive disorder. *Prog. Neuropsychopharmacol. Biol. Psychiatry* **2016**, *65*, 134–144. [CrossRef]

12. Smoller, J.W. The genetics of stress-related disorders: PTSD, depression, and anxiety disorders. *Neuropsychopharmacology* **2016**, *41*, 297. [CrossRef] [PubMed]

13. Hackett, R.A.; Steptoe, A. Type 2 diabetes mellitus and psychological stress—A modifiable risk factor. *Nat. Rev. Endocrinol.* **2017**, *13*, 547. [CrossRef] [PubMed]

14. Marshall, G.D., Jr. Psychological stress, immunity, and asthma: Developing a paradigm for effective therapy and prevention. *Curr. Opin. Behav. Sci.* **2019**, *28*, 14–19. [CrossRef]

15. Esler, M. Mental stress and human cardiovascular disease. *Neurosci. Biobehav. Rev.* **2017**, *74*, 269–276. [CrossRef] [PubMed]

16. Guede-Fernández, F.; Fernández-Chimeno, M.; Ramos-Castro, J.; García-González, M.A. Driver drowsiness detection based on respiratory signal analysis. *IEEE Access* **2019**, *7*, 81826–81838. [CrossRef]

17. Chui, K.T.; Tsang, K.F.; Chi, H.R.; Wu, C.K.; Ling, B.W.K. Electrocardiogram Based Classifier for Driver Drowsiness Detection. In Proceedings of the 2015 IEEE 13th International Conference on Industrial Informatics, (INDIN '15), Cambridge, UK, 22–24 July 2015; IEEE: Piscataway, NJ, USA, 2015; pp. 600–603.

18. Li, G.; Chung, W.Y. Combined EEG-gyroscope-tDCS brain machine interface system for early management of driver drowsiness. *IEEE Trans. Hum. syst.* **2018**, *48*, 50–62. [CrossRef]

19. Budak, U.; Bajaj, V.; Akbulut, Y.; Atilla, O.; Sengur, A. An Effective Hybrid Model for EEG-Based Drowsiness Detection. *IEEE Sens. J.* **2019**, *19*, 7624–7631. [CrossRef]

20. Li, Z.; Li, S.; Li, R.; Cheng, B.; Shi, J. Online detection of driver fatigue using steering wheel angles for real driving conditions. *Sensors* **2017**, *17*, 495. [CrossRef]

21. Chai, M. Drowsiness monitoring based on steering wheel status. *Transp. Res. Part D Transp. Environ.* **2019**, *66*, 95–103. [CrossRef]
22. McDonald, A.D.; Lee, J.D.; Schwarz, C.; Brown, T.L. A contextual and temporal algorithm for driver drowsiness detection. *Accid. Anal. Prev.* **2018**, *113*, 25–37. [CrossRef]
23. Assuncao, A.N.; Aquino, A.L.; Santos, C.D.M.; Ricardo, C.; Guimaraes, R.L.; Oliveira, R.A. Vehicle Driver Monitoring through the Statistical Process Control. *Sensors* **2019**, *19*, 3059. [CrossRef] [PubMed]
24. Zhao, L.; Wang, Z.; Wang, X.; Liu, Q. Driver drowsiness detection using facial dynamic fusion information and a DBN. *IET Intell. Transp. Syst.* **2018**, *12*, 127–133. [CrossRef]
25. Ahmad, R.; Borole, J.N. Drowsy Driver Identification Using Eye Blink Detection. *Int. J. Comp. Sci. Inf. Technol.* **2015**, *6*, 270–274.
26. Mandal, B.; Li, L.; Wang, G.S.; Lin, J. Towards detection of bus driver fatigue based on robust visual analysis of eye state. *IEEE Trans. Intell. Transp.* **2016**, *18*, 545–557. [CrossRef]
27. Zhang, C.; Wu, X.; Zheng, X.; Yu, S. Driver drowsiness detection using multi-channel second order blind identifications. *IEEE Access* **2019**, *7*, 11829–11843. [CrossRef]
28. Khattak, Z.H.; Fontaine, M.D.; Boateng, R.A. Evaluating the impact of adaptive signal control technology on driver stress and behavior using real-world experimental data. *Trans. Res. Part F Traffic Psychol. Behav.* **2018**, *58*, 133–144. [CrossRef]
29. Affanni, A.; Bernardini, R.; Piras, A.; Rinaldo, R.; Zontone, P. Driver's stress detection using skin potential response signals. *Measurement* **2018**, *122*, 264–274. [CrossRef]
30. Wan, P.; Wu, C.; Lin, Y.; Ma, X. Driving Anger States Detection Based on Incremental Association Markov Blanket and Least Square Support Vector Machine. *Discrete Dyn. Nat. Soc.* **2019**, *2019*, 2745381. [CrossRef]
31. Dobbins, C.; Fairclough, S. Signal Processing of Multimodal Mobile Lifelogging Data towards Detecting Stress in Real-World Driving. *IEEE Trans. Mob. Comput.* **2019**, *18*, 632–644. [CrossRef]
32. Chen, L.L.; Zhao, Y.; Ye, P.F.; Zhang, J.; Zou, J.Z. Detecting driving stress in physiological signals based on multimodal feature analysis and kernel classifiers. *Expert Syst. Appl.* **2017**, *85*, 279–291. [CrossRef]
33. Lee, B.G.; Chung, W.Y. Wearable glove-type driver stress detection using a motion sensor. *IEEE Trans. Intell. Transp.* **2017**, *18*, 1835–1844. [CrossRef]
34. Muñoz-Organero, M.; Corcoba-Magaña, V. Predicting upcoming values of stress while driving. *IEEE Trans. Intell. Transp.* **2017**, *18*, 1802–1811. [CrossRef]
35. Kurniawan, H.; Maslov, A.V.; Pechenizkiy, M. Stress Detection from Speech and Galvanic Skin Response Signals. In Proceedings of the 26th IEEE International Symposium on Computer-Based Medical Systems, Porto, Portugal, 20–22 June 2013; IEEE: Piscataway, NJ, USA, 2013; pp. 209–214.
36. Choi, M.; Koo, G.; Seo, M.; Kim, S.W. Wearable device-based system to monitor a driver's stress, fatigue, and drowsiness. *IEEE T. Instrum. Meas.* **2017**, *67*, 634–645. [CrossRef]
37. Sun, Y.; Yu, X.B. An Innovative Nonintrusive Driver Assistance System for Vital Signal Monitoring. *IEEE J. Biomed. Health Inform.* **2014**, *18*, 1932–1939. [CrossRef] [PubMed]
38. Healey, J.A.; Picard, R.W. Detecting stress during real-world driving tasks using physiological sensors. *IEEE Trans. Intell. Transp.* **2005**, *6*, 156–166. [CrossRef]
39. Goldberger, A.L.; Amaral, L.A.N.; Glass, L.; Hausdorff, J.M.; Ivanov, P.C.H.; Mark, R.G.; Mietus, J.E.; Moody, G.B.; Peng, C.K.; Stanley, H.E. PhysioBank, PhysioToolkit, and PhysioNet: Components of a New Research Resource for Complex Physiologic Signals. *Circulation* **2003**, *101*, e215–e220. [CrossRef]
40. Terzano, M.G.; Parrino, L.; Sherieri, A.; Chervin, R.; Chokroverty, S.; Guilleminault, C.; Hirshkowitz, M.; Mahowald, M.; Moldofsky, H.; Rosa, A.; et al. Atlas, rules, and recording techniques for the scoring of cyclic alternating pattern (CAP) in human sleep. *Sleep Med.* **2001**, *2*, 537–553. [CrossRef]
41. Kohler, B.U.; Hennig, C.; Orglmeister, R. The principles of software QRS detection. *IEEE Eng. Med. Biol.* **2002**, *21*, 42–57. [CrossRef]
42. Tompkins, W.J. *Biomedical Digital Signal Processing C-Language Examples and Laboratory Experiments for the IBM® PC*; Prentice Hall: Upper Saddle River, NJ, USA, 2000; pp. 236–264.
43. Herbrich, R. *Learning Kernel Classifiers Theory and Algorithms*; The MIT Press: London, UK, 2002.
44. Thokala, P.; Devlin, N.; Marsh, K.; Baltussen, R.; Boysen, M.; Kalo, Z.; Longrenn, T.; Mussen, F.; Peacock, S.; Watkins, J.; et al. Multiple criteria decision analysis for health care decision making—An introduction: Report 1 of the ISPOR MCDA Emerging Good Practices Task Force. *Value Health* **2016**, *19*, 1–13. [CrossRef]

45. Mardani, A.; Zavadskas, E.K.; Khalifah, Z.; Jusoh, A.; Nor, K.M. Multiple criteria decision-making techniques in transportation systems: A systematic review of the state of the art literature. *Transport* **2016**, *31*, 359–385. [CrossRef]

46. Ruiz-Torrubiano, R.; Suárez, A. A memetic algorithm for cardinality-constrained portfolio optimization with transaction costs. *Appl. Soft. Comput.* **2015**, *36*, 125–142. [CrossRef]

47. Tavakoli, M.; Shokridehaki, F.; Akorede, M.F.; Marzband, M.; Vechiu, I.; Pouresmaeil, E. CVaR-based energy management scheme for optimal resilience and operational cost in commercial building microgrids. *Int. J. Electr. Power Energy Syst.* **2018**, *100*, 1–9. [CrossRef]

48. Marzband, M.; Azarinejadian, F.; Savaghebi, M.; Pouresmaeil, E.; Guerrero, J.M.; Lightbody, G. Smart transactive energy framework in grid-connected multiple home microgrids under independent and coalition operations. *Renew. Energ.* **2018**, *126*, 95–106. [CrossRef]

49. Hsu, C.W.; Lin, C.J. A comparison of methods for multiclass support vector machines. *IEEE Trans. Neural Netw.* **2002**, *13*, 415–425.

50. Epstein, J.I.; Zelefsky, M.J.; Sjoberg, D.D.; Nelson, J.B.; Egevad, L.; Magi-Galluzzi, C.; Vickers, A.J.; Parwani, A.V.; Reuter, V.E.; Fine, S.W.; et al. A contemporary prostate cancer grading system: A validated alternative to the Gleason score. *Eur. Urol.* **2016**, *69*, 428–435. [CrossRef]

 *sensors*

Article

Hybrid Dynamic Traffic Model for Freeway Flow Analysis Using a Switched Reduced-Order Unknown-Input State Observer

Yuqi Guo [1], Bin Li [1], Matthew Daniel Christie [2], Zongzhi Li [3], Miguel Angel Sotelo [4], Yulin Ma [5,*], Dongmei Liu [1] and Zhixiong Li [5]

[1]   Research Institute of Highway, Ministry of Transport of China, Beijing 100088, China; gyq@itsc.cn (Y.G.); libin@itsc.cn (B.L.); ldm@itsc.cn (D.L.)
[2]   School of Mechanical, Materials, Mechatronic and Biomedical Engineering, University of Wollongong, Wollongong, NSW 2522, Australia; mdc769@uowmail.edu.au
[3]   Department of Civil, Architectural and Environmental Engineering, Illinois Institute of Technology, Chicago, IL 60616, USA; lizz@iit.edu
[4]   Department of Computer Engineering, University of Alcalá, 28801 Alcalá de Henares (Madrid), Spain; miguel.sotelo@uah.es
[5]   Suzhou Automotive Research Institute, Tsinghua University, Suzhou 215134, China; zhixiong_li@uow.edu.au
*   Correspondence: mayulin@tsari.tsinghua.edu.cn; Tel.: +86-13051025205

Received: 17 December 2019; Accepted: 5 March 2020; Published: 13 March 2020

Abstract: This paper introduces a new methodology for reconstructing vehicle densities of freeway segments by utilizing the limited data collected by traffic-counting sensors and developing a macroscopic traffic stream model formulated as a switched reduced-order state observer design problem with unknown or partially known inputs. Specifically, the traffic network is modeled as a hybrid dynamic system in a state space that incorporates unknown inputs. For freeway segments with traffic-counting sensors installed, vehicle densities are directly computed using field traffic count data. A reduced-order state observer is designed to analyze traffic state transitions for freeway segments without field traffic count data to indirectly estimate the vehicle densities for each freeway segment. A simulation-based experiment is performed applying the methodology and using data of a segment of Beijing Jingtong freeway in Beijing, China. The model execution results are compared with the field data associated with the same freeway segment, and highly consistent results are achieved. The proposed methodology is expected to be adopted by traffic engineers to evaluate freeway operations and develop effective management strategies.

Keywords: urban freeway; hybrid dynamic system; state transition; unknown inputs observer; vehicle density

---

1. Introduction

The estimation of vehicle densities on highway segments has been of considerable interest in recent decades. Research has been underway in developing practical methods in the context of the macroscopic traffic flow dynamic model for vehicle density estimation using different types of estimators. In particular, the state observer method [1,2] has been rapidly adopted by researchers and practitioners for traffic state estimation. In a study [3], traffic state was estimated by using an adaptive observer. Based on the cell transmission model, a centralized observer was considered in another work [4], and the estimation was further improved in a subsequent study [5]. In reference [6], the switched distributed observer was studied using the consensus theory. Based on the piecewise affine (PWA) system model of traffic network, various switched-state observers were also designed, such as the centralized observer [7–11], decentralized observer [8], and distributed observer [9,10], to estimate vehicle densities associated with segments of highways with different functional classifications.

It should be noted that the above-mentioned state observers were designed based on the ideal traffic flow dynamic model. That is, both the system disturbance and the measurement noise were not taken into consideration in the modeling procedure, and all the inputs were treated as measurable signals. Hence, these types of observers are termed as known-inputs state observers. As a practical matter, disturbances, especially unknown input signals, cannot be ignored in actual traffic networks. Otherwise, the proposed model could not reflect the real traffic flow transmission rule.

With advancements of sensing and positioning technologies, measurement accuracy and precision of traffic sensors have significantly improved, leading to the possibility of ignoring measurement noises. However, unknown inputs must be explicitly accounted for owing to their impacts on the predictability of the proposed models. The existing literature largely considers dynamic models with unknown inputs, while measurement noises are ignored. In the context of unknown-input observer models, dynamic models are classified as linear and nonlinear systems [12–24], continuous and discrete systems [19,21, 25,26], and time-varying and time-invariant systems [27,28]. For instance, the unknown-input observer was studied based on the optimal data fusion approach [29]. Both full-order and reduced-order observers for continuous dynamical systems with unknown inputs were investigated [29]. A design strategy of the nonlinear unknown-input state observer for a nonlinear system was reported [30]. A novel full-order unknown-input observer for the continuous system along with the existence of the necessary conditions was proposed [31]. Additional unknown-input observers for different types of systems were also reported [32–35].

In addition, as an effective and practical tool, deep learning methods are also popular in estimating traffic flows. In a study [36], spatio-temporal factors were considered in traffic prediction, and a multi-attention network was proposed to predict traffic conditions at different locations on a road network graph. In order to deeply capture the high-order spatial–temporal correlations among the road links, Zhang et al. [37] performed a road network-level prediction using a network-scale deep traffic prediction model, called TrafficGAN, where the generative adversarial nets were utilized to predict traffic flows under an adversarial learning framework. In another study [38], a deep learning framework was presented to solve the traffic forecasting problem; the traffic flow was modelled as a diffusion process on a direct graph, and a diffusion convolutional recurrent neural network was introduced to predict the traffic flow.

Estimation of traffic flow density is the basis of many transportation applications, such as route planning and vehicle routing [39–41]. For example, based on the estimation of the traffic flow density, the distribution of traffic congestion in a road network was identified, and then, alternative routes were assigned for selected vehicles to avoid congested roads [39]. Typically, traffic network can be modeled as a hybrid dynamic system by means of multi-mode switching [11]. However, most of the existing state observers were designed on the basis of the ideal traffic flow dynamic model, where both the system disturbance and the measurement noise were not taken into consideration in the modeling procedure. Especially, the inputs were treated as measurable signals and were known in advance. Hence, these types of observers are termed known-input state observers. From a practical viewpoint, system disturbance and measurement noise always exist in the actual traffic flow network, and these input signals are often unknown but cannot be ignored. In order to estimate vehicle density for real-world freeway segments, a traffic flow model should include system disturbance and measurement noise. However, few studies have systematically addressed the vehicle density estimation problem with unknown inputs. It is crucial to explore a more effective way to solve the design challenge of the unknown-input state observer on the basis of the hybrid dynamic traffic network model. To this avail, this paper introduces a switched unknown-input state observer to reconstruct vehicle densities of segments of a highway system which maintains field data measurements by using traffic sensors for only a fraction of segments and limited known inputs. Compared with other existing methods, the proposed dynamic model has the following advantages: (1) it is more realistic in representing the actual traffic networks and (2) the actual vehicle densities can be readily reconstructed using the unknown-input state observer.

The remainder of the paper is organized as follows. Section 2 provides the background information on the traffic flow dynamic model and elaborates the proposed observer model with unknown inputs state. Section 3 evaluates the proposed model using real-world traffic data. Section 4 summarizes the findings of this study.

2. Proposed Method

An overview of the proposed traffic density estimation method is given in Figure 1. The details of constructing the predictive model are described in the following sub-sections.

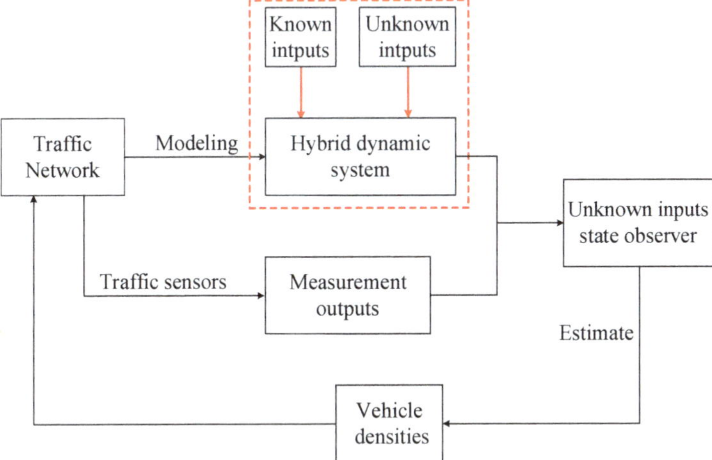

Figure 1. An overview of the proposed method.

2.1. Hybrid Dynamic System

This section first reviews the hybrid dynamic traffic flow model that combines the dynamic graph hybrid automata with the cell transmission model (CTM) and then briefly describes the problem of vehicle density estimation. The hybrid dynamic traffic flow model is described by

$$\begin{cases} x(t+1) = A_{\sigma(t)}x(t) + B_{\sigma(t)}u(t) \\ y(t) = Cx(t) \end{cases} \tag{1}$$

where $x = [\rho_1, \cdots, \rho_n]^T \in \mathbf{R}^n$ represents the vehicle density vector, $u \in \mathbf{R}^m$ is the input vector, $y \in \mathbf{R}^q$ is the measured output vector, A_σ, B_σ, and C are the system matrix, the input matrix, and the output matrix, respectively, $\sigma : [0, +\infty) \rightarrow \{1, 2, \cdots, s\}$ is the switching function that maps the index time stage into an index set $\{1, 2, \cdots, s\}$, and each index corresponds to a different mode of the system.

Based on the dynamic model shown above, different types of state observers can be designed to estimate the vehicle densities of a traffic stream [7–11]. However, it is difficult to apply this method to a real-world traffic network. This is because all sources of disturbances associated with unknown-input signals and measurement errors are ignored in the modeling process, making it difficult to apply the actual traffic flow transmission rule. Further, the exclusion of measurement errors in the design of state observers renders the estimated data incapable of reflecting the real traffic states. Therefore, disturbances need to be included in the base model, which creates the augmented model

$$
\begin{cases}
x(t+1) = A_{\sigma(t)}x(t) + B_{\sigma(t)}u(t) + D_{\sigma(t)}v(t) \\[2mm]
y(t) = Cx(t)
\end{cases}
\tag{2}
$$

where $v \in \mathbf{R}^p$ is the unknown-input signal, $u \in \mathbf{R}^m$ is the known-input signal, $D_\sigma \in \mathbf{R}^{p \times n}$ is the known noise matrix with appropriate dimensions. The others are the same as in the model described by (1).

Remark 1. *With the development of sensor and positioning technologies, the accuracy and precision of both mobile and fixed traffic detectors have been improving. The influence of measurement errors can be marginally neglected in most cases. However, system noise cannot be completely eliminated. As such, the augmented dynamic model ignores the measurement errors, while it incorporates unknown inputs to capture the system noise.*

The augmented model deals with situations where traffic data are partially available for some highway segments equipped with sensors. For the remaining highway segments without traffic sensors, a state observer needs to be designed to reconstruct the traffic states. To reduce the structural complexity of the traffic state observer, it is advantageous to design a reduced-order state observer.

2.2. Unknown-Input State Observer

This section introduces the design of the reduced-order unknown-input state observer for the augmented dynamic traffic model of vehicle density estimation. The essence of the state observer design is to accurately reconstruct vehicle densities $\hat{x}$ and estimate the unknown inputs $\hat{v}$ for the system $\{u, v, y, A, B, C, D\}$, such that the following conditions are satisfied:

$$
\begin{cases}
e(t) = x(t) - \hat{x}(t) \to 0 \\[2mm]
\widetilde{v}(t) = v(t) - \hat{v}(t) \to 0
\end{cases}
\tag{3}
$$

where x is the vehicle density vector, $\hat{x}$ is the estimated density vector, v is the input vector, $\hat{v}$ is the estimated input vector, e is the estimation error between the actual and the estimated vehicle densities, and $\widetilde{v}$ is the estimation error between the real and the estimated inputs.

Definition 1. Unknown-Input State Observer: *A state observer is defined as an unknown-input observer if its estimation error approaches zero asymptotically, regardless of the presence of unknown inputs or disturbances in a dynamic system.*

Definition 2. Switched Unknown-Input State Observer: *A state observer is termed a switched unknown-input observer for a dynamic system if, and only if, its state estimation error system is asymptotically stable for any switching sequence, regardless of the existence of unknown inputs in the system.*

As the preparation for the unknown-input state observer design, the following three assumptions were made:

(i) $\text{rank}D_\sigma = p$ and $\text{rank}C = q$.
(ii)   The pair (A_σ, C) is observable or detectable.
(iii) $q \geq p$, $\text{rank}(CD_\sigma) = p$.

Remark 2. *The above assumptions imply that the matrices C and D_σ are full row rank and full column rank, respectively. These characteristics can always be met by optimizing the configuration of the matrix C and redefining the noise matrix D_σ.*

Since $\mathrm{rank}\,C = q$, $R \in \mathbf{R}^{(n-q)\times n}$ can be arbitrarily chosen, and the following transformation matrix $P \in \mathbf{R}^{n\times n}$ is nonsingular.

$$P \triangleq \begin{bmatrix} C \\ R \end{bmatrix} \tag{4}$$

The inverse matrix of P is denoted as

$$Q \triangleq P^{-1} = \begin{bmatrix} C \\ R \end{bmatrix}^{-1} = \begin{bmatrix} Q_1 & Q_2 \end{bmatrix} \tag{5}$$

The matrix R is not unique but needs to be chosen to ensure that the matrix P is invertible. By using linear nonsingular transformation, the corresponding matrices can be rewritten as follows

$$\begin{cases} \overline{A} = QAQ^{-1} = \begin{bmatrix} \overline{A}_{11} & \overline{A}_{12} \\ \overline{A}_{21} & \overline{A}_{22} \end{bmatrix} \\[2mm] \overline{B} = QB = \begin{bmatrix} \overline{B}_1 \\ \overline{B}_2 \end{bmatrix} \\[2mm] \overline{C} = CQ = \begin{bmatrix} I_q & 0 \end{bmatrix} \\[2mm] \overline{D} = QD = \begin{bmatrix} \overline{D}_1 \\ \overline{D}_2 \end{bmatrix} \end{cases} \tag{6}$$

Subsequently, by using the transformation $\overline{x} = Q^{-1}x$ and $y = CQ\overline{x}$, the dynamic system (2) can be re-constructed with the following specification:

$$\begin{cases} \begin{bmatrix} \overline{x}_1(t+1) \\ \overline{x}_2(t+1) \end{bmatrix} = \begin{bmatrix} \overline{A}_{11} & \overline{A}_{12} \\ \overline{A}_{21} & \overline{A}_{22} \end{bmatrix}_{\sigma(t)} \begin{bmatrix} \overline{x}_1(t) \\ \overline{x}_2(t) \end{bmatrix} + \begin{bmatrix} \overline{B}_1 \\ \overline{B}_2 \end{bmatrix}_{\sigma(t)} u(t) + \begin{bmatrix} \overline{D}_1 \\ \overline{D}_2 \end{bmatrix}_{\sigma(t)} v(t) \\[4mm] y(t) = \begin{bmatrix} I_q & 0 \end{bmatrix} \begin{bmatrix} \overline{x}_1(t) \\ \overline{x}_2(t) \end{bmatrix} \end{cases} \tag{7}$$

where $\overline{x}_1 = y \in \mathbf{R}^q$, $\overline{x}_2 \in \mathbf{R}^{n-q}$.

The above analysis shows that vehicle densities $\overline{x}_1$ can be obtained by the measurement output y, while only partial vehicle densities $\overline{x}_2$ need an estimate. Hence, the reduced-order state observer needs to be designed to complete the density reconstruction.

Remark 3. *For segments of a highway network with field traffic-counting sensors installed, traffic states can be directly assessed by traffic data collected by the sensors. Conversely, for highway segments without field traffic-counting sensors, a state observer needs to be designed to estimate the traffic states. The joint use of field traffic count data and traffic state estimates by the state observer could help derive highly accurate and precise values of vehicle densities. As such, vehicle density estimation boils down to the design of an effective state observer. Practically, designing a reduced-order state observer becomes the key to solving the vehicle density estimation problem.*

Based on Remark 3, the dynamic system (7) can be formulated as the following:

$$\begin{cases} \overline{x}_2(t+1) = \overline{A}_{22_{\sigma(t)}} \overline{x}_2(t) + \overline{A}_{21_{\sigma(t)}} y(t) + \overline{B}_2 u(t) + \overline{D}_{2_{\sigma(t)}} v(t) \\[3mm] y(t+1) - \overline{A}_{11_{\sigma(t)}} y(t) = \overline{A}_{12_{\sigma(t)}} \overline{x}_2(t) + \overline{B}_1 u(t) + \overline{D}_{1_{\sigma(t)}} v(t) \end{cases} \tag{8}$$

Theorem 1. *In the presence of an invertible matrix$P$, such that Equation (8) is satisfied, and the pair$(A_\sigma, C)$is observable or detectable, there must exist a reduced-order state observer in the form (9), such that the vehicle densities of the system (2) can be estimated.*

The reduced-order state observer can be obtained by

$$\begin{cases} z(t+1) = F_\sigma z(t) + G_\sigma y(t) + H_\sigma u(t) \\ \hat{x}(t) = Q_1 y(t) + Q_2(z + L_\sigma y) \end{cases} \tag{9}$$

The proof for the state observer begins with (8), which can be further rewritten as

$$\begin{cases} \overline{x}_2(t+1) = \overline{A}_{22_{\sigma(t)}} \overline{x}_2(t) + \left[\overline{A}_{21_{\sigma(t)}} y(t) + \overline{B}_2 u(t) + \overline{D}_{2_{\sigma(t)}} v(t) \right] \\ \\ y(t+1) - \overline{A}_{11_{\sigma(t)}} y(t) - \overline{B}_1 u(t) - \overline{D}_{1_{\sigma(t)}} v(t) = \overline{A}_{12_{\sigma(t)}} \overline{x}_2(t) \end{cases} \tag{10}$$

Based on the following equivalent substitution

$$\begin{cases} \overline{u}(t) \triangleq \overline{A}_{21_{\sigma(t)}} y(t) + \overline{B}_2 u(t) + \overline{D}_{2_{\sigma(t)}} v(t) \\ \\ w(t) \triangleq y(t+1) - \overline{A}_{11_{\sigma(t)}} y(t) - \overline{B}_1 u(t) - \overline{D}_{1_{\sigma(t)}} v(t) \end{cases} \tag{11}$$

the following normative form is obtained:

$$\begin{cases} \overline{x}_2(t+1) = \overline{A}_{22_{\sigma(t)}} \overline{x}_2(t) + \overline{u}(t) \\ \\ w(t) = \overline{A}_{12_{\sigma(t)}} \overline{x}_2(t) \end{cases} \tag{12}$$

It should be noted that observability or detectability of the pair (A_σ, C) implies that (A_{22}, A_{12}) is also observable or detectable. Therefore, for state $\overline{x}_2$, a full-order observer can be designed as:

$$\hat{\overline{x}}_2(t+1) = \left[\overline{A}_{22_{\sigma(t)}} - \overline{L}_{\sigma(t)} \overline{A}_{12_{\sigma(t)}} \right] \overline{x}_2(t) + \overline{L}_{\sigma(t)} w(t) + \overline{u}(t) \tag{13}$$

Combining (11) with (13), the following expression is constructed:

$$\hat{\overline{x}}_2(t+1) = \left[\overline{A}_{22_{\sigma(t)}} - \overline{L}_{\sigma(t)} \overline{A}_{12_{\sigma(t)}} \right] \overline{x}_2(t) + \overline{L}_{\sigma(t)} \Psi + \overline{A}_{21_{\sigma(t)}} y(t) + \overline{B}_2 u(t) + \overline{D}_{2_{\sigma(t)}} v(t) \tag{14}$$

where $\Psi = y(t+1) - \overline{A}_{11_{\sigma(t)}} y(t) - \overline{B}_1 u(t) - \overline{D}_{1_{\sigma(t)}} v(t)$.

Note that $z = \hat{\overline{x}}_2 - \overline{L}_\sigma y$, so we obtain

$$\begin{aligned} z(t+1) &= \hat{\overline{x}}_2(t+1) - \overline{L}_{\sigma(t)} y(t+1) \\ &= \left[\overline{A}_{22_{\sigma(t)}} - \overline{L}_{\sigma(t)} \overline{A}_{12_{\sigma(t)}} \right] \overline{x}_2(t) + \left(\overline{A}_{21_{\sigma(t)}} - \overline{L}_{\sigma(t)} \overline{A}_{11_{\sigma(t)}} \right) y(t) \\ &\quad + \left(\overline{B}_2 - \overline{L}_{\sigma(t)} \overline{B}_1 \right) u(t) + \left(\overline{D}_{2_{\sigma(t)}} - \overline{L}_{\sigma(t)} \overline{D}_{1_{\sigma(t)}} \right) v(t) \\ &= F_{\sigma(t)} z(t) + G_{\sigma(t)} y(t) + H u(t) + J_{\sigma(t)} v(t) \end{aligned} \tag{15}$$

where

$$\begin{cases} F_{\sigma(t)} = A_{22_{\sigma(t)}} - L_{\sigma(t)} A_{12_{\sigma(t)}} \\ G_{\sigma(t)} = F_{\sigma(t)} L_{\sigma(t)} + A_{21_{\sigma(t)}} - L_{\sigma(t)} A_{11_{\sigma(t)}} \\ H_{\sigma(t)} = B_{2_{\sigma(t)}} - L_{\sigma(t)} B_{1_{\sigma(t)}} \\ J_{\sigma(t)} = D_{2_{\sigma(t)}} - L_{\sigma(t)} D_{1_{\sigma(t)}} \end{cases}$$

The traffic states of $\hat{\bar{x}}_2$ can be reconstructed by

$$\hat{\bar{x}}_2(t) = z(t) + \bar{L}_{\sigma(t)} y(t) \tag{16}$$

where $\bar{L}_{\sigma(t)}$ is the feedback matrix of the state observer.

The reconstructed traffic states $\bar{x}_1$ and $\bar{x}_2$ can be denoted as

$$\begin{cases} \hat{\bar{x}}_1 = \bar{x}_1 = y \\[2mm] \hat{\bar{x}}_2 = z + L_\sigma y \end{cases} \tag{17}$$

Correspondingly, the vehicle density estimates are obtained in the following equation:

$$\hat{\bar{x}} = \begin{bmatrix} \hat{\bar{x}}_1 \\ \hat{\bar{x}}_2 \end{bmatrix} = \begin{bmatrix} y \\ z + L_\sigma y \end{bmatrix} \tag{18}$$

With $\bar{x} = Px$ being held to be true, we obtain $x = P^{-1}\bar{x} = Q\bar{x}$ and $\hat{x} = Q\hat{\bar{x}}$. This further leads to

$$\hat{x} = \begin{bmatrix} Q_1 & Q_2 \end{bmatrix} \begin{bmatrix} y \\ z + L_\sigma y \end{bmatrix} = Q_1 y + Q_2(z + L_\sigma y) \tag{19}$$

Finally, the reduced-order state observer can be derived according to (9). Figure 2 illustrates the structure of the reduced-order unknown-input state observer.

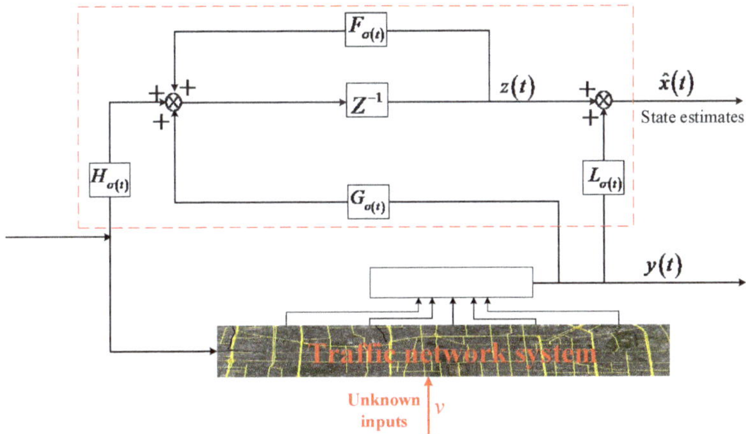

Figure 2. The structure of the reduced-order unknown-input state observer.

2.3. Estimation of Observer Parameters

Further to establishing the structure of the reduced-order unknown-input state observer as (9), the parameters of its unknown input vector $v(t)$ need to be estimated for real-world implementation. To initiate this process, the matrix $L_{\sigma(t)}$ in (16) can be computed by imposing the following condition:

$$J_{\sigma(t)} = D2_{\sigma(t)} - L_{\sigma(t)} D1_{\sigma(t)} = 0 \tag{20}$$

Meanwhile, the matrix $F_{\sigma(t)}$ must possess Schur stability, so that the existence of the state observer as described by Expression (9) is guaranteed. This implies that the matrix $A_{22_{\sigma(t)}} - L_{\sigma(t)} A_{12_{\sigma(t)}}$ has stable eigenvalues. Also, it is essential to compute the feedback matrix $L_{\sigma(t)}$ in order to determine the matrix $\bar{D}_{\sigma(t)}$. The procedure for deriving the matrix $L_{\sigma(t)}$ is given below.

First, with the above assumptions in place, the following conditions are satisfied:

$$\text{rank}D_\sigma = \text{rank}(CD_\sigma) = \text{rank}D_{1_\sigma} = p \tag{21}$$

Then, the matrix $L_{\sigma(t)}$ can be computed by combining (20) and (21):

$$L_\sigma = D_{2_\sigma}D_{1_\sigma}^+ + K_\sigma(I_q - D_{1_\sigma}D_{1_\sigma}^+) \tag{22}$$

where $D_{1_{\sigma(t)}}^+$ is the generalized inverse of $D_{1_{\sigma(t)}}$.

Because the matrix D_{1_σ} is of full column rank, the matrix $D_{1_{\sigma(t)}}^+$ can be calculated by:

$$D_{1_\sigma}^+ = (D_{1_\sigma}^T D_{1_\sigma})^{-1}D_{1_\sigma}^T, \ \forall K \in \mathbf{R}^{(n-q)\times q} \tag{23}$$

Also, there exists an orthogonal matrix S_σ with the following conditions satisfied:

$$\begin{cases} S_\sigma D_{1_\sigma} = \begin{bmatrix} \overline{D}_{1_\sigma} \\ 0 \end{bmatrix} \\[2em] S_\sigma A_{12_\sigma} = \begin{bmatrix} \overline{A}_{12_\sigma}^{1} \\ \overline{A}_{12_\sigma}^{2} \end{bmatrix} \\[2em] K_\sigma S_\sigma^T = \begin{bmatrix} \overline{K}_{1_\sigma} & \overline{K}_{2_\sigma} \end{bmatrix} \end{cases} \tag{24}$$

where $\overline{D}_{1_\sigma} \in \mathbf{R}^{p\times p}$ is a nonsingular matrix, $\overline{A}_{12_\sigma}^{1} \in \mathbf{R}^{p\times 1}, \overline{K}_{1_\sigma} \in \mathbf{R}^{1\times p}$.

Next, the matrices F_σ, G_σ, and H_σ can be computed by:

$$\begin{aligned} F_\sigma &= A_{22_\sigma} - L_\sigma A_{12_\sigma} \\ &= A_{22_\sigma} - D_{2_\sigma}D_{1_\sigma}^+ A_{12_\sigma} + K_\sigma(I_q - D_{1_\sigma}D_{1_\sigma}^+)A_{12_\sigma} \\ &= A_{22_\sigma} - D_{2_\sigma}\overline{D}_{1_\sigma}^{-1}\overline{A}_{12_\sigma}^{1} - \overline{K}_{2_\sigma}\overline{A}_{12_\sigma}^{2} \\ &= F_{1_\sigma} - \overline{K}_{2_\sigma}\overline{A}_{12_\sigma}^{2} \end{aligned} \tag{25}$$

$$G_{\sigma(t)} = F_{\sigma(t)}\overline{L}_{\sigma(t)} + \overline{A}_{21_{\sigma(t)}} - \overline{L}_{\sigma(t)}\overline{A}_{11_{\sigma(t)}} \tag{26}$$

$$H_{\sigma(t)} = \overline{B}_2 - \overline{L}_{\sigma(t)}\overline{B}_1 \tag{27}$$

2.4. Design of the State Observer

After developing the procedure for estimating the parameters for the reduced-order unknown-input state observer, the subsequent effort is centered on proposing a design procedure of the state observer for switched systems. The design comprises a generic process with the following steps:

Step 1: Compute matrices $A_{\sigma(t)}$, B, and $D_{\sigma(t)}$ and construct output matrix C using data collected by traffic-counting sensors for the preparation of traffic flow modeling;

Step 2: Verify the validity of the conditions $\text{rank}C = q, \text{rank}D_\sigma = \text{rank}(CD_\sigma) = q, q \geq p$ and iteratively reconfigure the matrices C and D_σ until the above conditions are satisfied;

Step 3: Estimate the state transformation matrix $P \triangleq \begin{bmatrix} C \\ R \end{bmatrix}$; let $\overline{x} = P^{-1}x$ and calculate the values of (6);

Step 4: Determine the feedback matrix $L_\sigma = D_{2_\sigma}D_{1_\sigma}^+ + K(I_q - D_{1_\sigma}D_{1_\sigma}^+)$;

Step 5: Develop an $(n \times n)$ orthogonal matrix S_σ that satisfies the conditions in (24);

Step 6: Generate the matrix $\overline{K}_{2_\sigma}$ to satisfy Schur stability for F_σ;

Step 7: Derive the matrices G_σ and H_σ;

Step 8: Establish the reduced-order unknown-input state observer for switched systems in accordance with (9).

3. Case Study: Beijing Jingtong Freeway

In this section, an experiment example will be presented to demonstrate the validity and the practicability of the proposed approach by applying the designed state-jump observer to the Beijing Jingtong freeway. The selected road section is approximately 3.5 km long and is comprises three lanes. In accordance with the segment partition rules mentioned in reference [17], the road section was divided into 10 cells, as shown in Table 1.

Table 1. Cell length.

Cell Number	Length (m)	Cell Number	Length (m)
1	300	6	275
2	160	7	435
3	460	8	400
4	430	9	450
5	400	10	406

3.1. Data Collection and Processing

Figure 3 presents the segment of Jingtong Freeway, Beijing, China used for methodology application, particularly the design of an unknown-input state observer for analyzing traffic states to reconstruct vehicle densities essential to traffic flow modeling. The west–east directional freeway segment is labeled as segment AB and encompasses four on-ramps and four off-ramps, respectively labeled in ascending order from 1 to 4. The directional segment AB is partitioned into 10 cells marked from 1 to 10 correspondingly. Tables 1 and 2 list the details of the cell lengths and pertinent parameters of vehicles' operational characteristics.

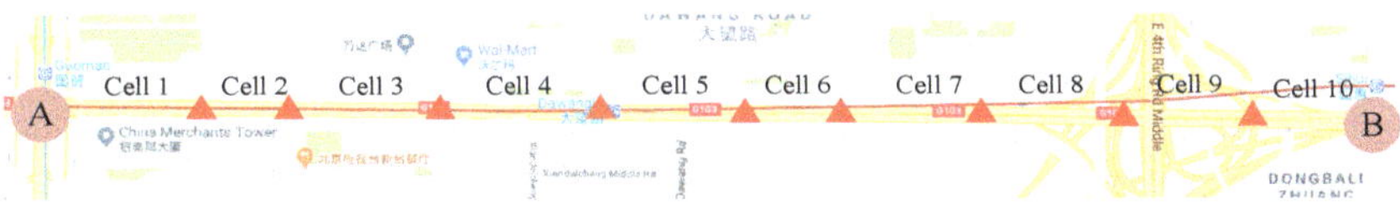

Figure 3. Illustration of the Jingtong freeway segment for the methodology application.

Table 2. Cell-specific parameters of vehicles' operational characteristics.

Cell Number	V (km/h)	W (km/h)	C (veh/h)	ρ^0 (veh/km)	ρ^m (veh/km)
1–10	65	20	2800	46	185

3.2. Analysis Results

Figure 4 depicts a VISSIM-based simulation model developed for the directional segment AB to generate simulated traffic stream data [42]. The simulation execution period was 3 h in a typical weekday from noon to 3:00 p.m., with a data reporting interval of 5 s. Virtual traffic-counting sensors were placed in each cell to collect data on vehicle densities used to verify the accuracy and precision of traffic state estimates from the reduced-order unknown-input state observer. In addition, virtual traffic-counting sensors were installed in the on-ramp Sections 2 and 3 to collect traffic data as known inputs. Conversely, on-ramp Sections 1 and 4 were not equipped with traffic-counting sensors and could be treated as roadway sections with unknown inputs.

Figure 4. Topological representation of the Jingtong freeway segment in the VISSIM model.

Remark 4. *Although many traffic simulation software have been developed for traffic system modelling, VISSIM is one of the most practical traffic simulation software to model urban traffic and public transit operations. Compared with other traffic simulation software, such as SParamics, VISSIM reserves many data interfaces, which enables users to easily redevelop the model of interest in accordance with their own needs by incorporating some new algorithms. As a result, a VISSIM model can be modified to continue previous research, which provides significant opportunities for sustainable and comprehensive traffic simulation research.*

As sufficient and necessary conditions for the existence of the unknown-input state observer, the pair (A_σ, C) must be observable or detectable. In this respect, traffic-counting sensors were installed in cells 2, 3, 4, 5, 6, 8, 9, and 10, which facilitated collecting data on vehicle densities that were subsequently used to design the observer. Therefore, the output matrix $C = [c_{i,j}]$ was a 7×10 matrix, and $c_{1,2} = c_{2,3} = c_{3,4} = c_{4,6} = c_{5,8} = c_{6,9} = c_{7,10} = 1$, with all other entries of the matrix set to zero. With virtual traffic-counting sensors installed in cells 2, 3, 4, 5, 6, 8, 9, and 10, data on vehicle densities of those cells could be directly collected. For the remaining cells including cells 1, 5, 7 without virtual sensors, vehicle densities associated with them needed to be derived based on traffic states estimated by the reduced-order unknown-input state observer. Figures 5 and 6 show the simulated and estimated vehicle densities for multiple cells evolving over different time points of the analysis. Using a color-coding mechanism of green, yellowand red, vehicle densities would increase from green to yellow and then to red, representing free-flow to capacity and then to congested traffic stream conditions. As the red color becomes darker, it shows more severe traffic congestion.

Further, vehicle densities for cells 1, 5, and 7 were reconstructed using traffic state estimates by the reduced-order unknown-input state observer, as shown in Figure 7.

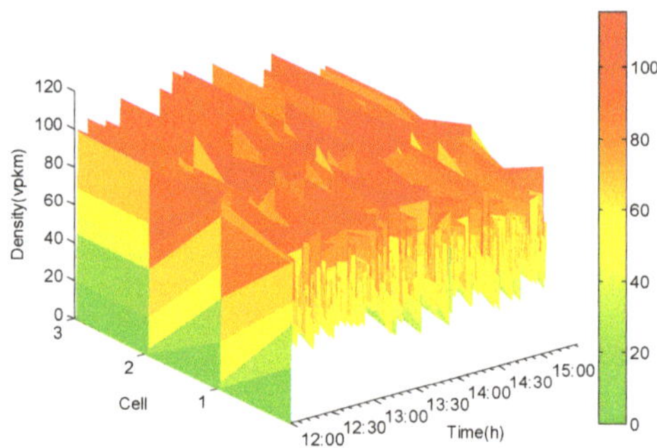

Figure 5. Simulated data for vehicle densities.

Vehicle densities could be reconstructed by the designed reduced-order unknown-input state observer withmodest accuracy. Compared with the results of the known-input observer [11], the estimation needs to be more precise. Once this issue gets resolved, the proposed methodology will become highly practical for deployment to a large traffic network.

Congested road segments could be further identified from the estimation results. Thus, driving-route planning could be optimized effectively for GPS systems, greatly enhancing travel efficiency.

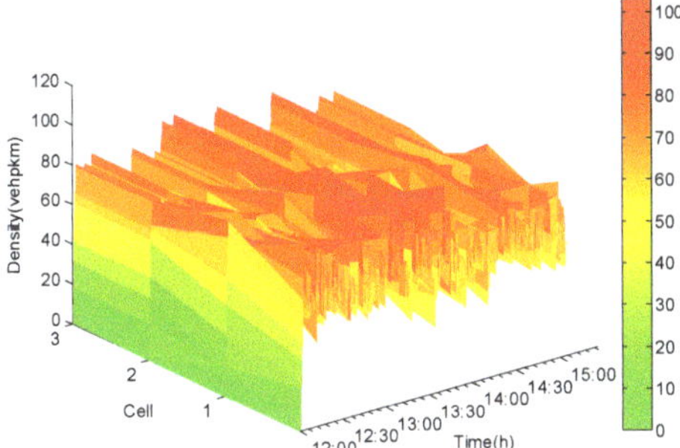

Figure 6. Estimated vehicle densities.

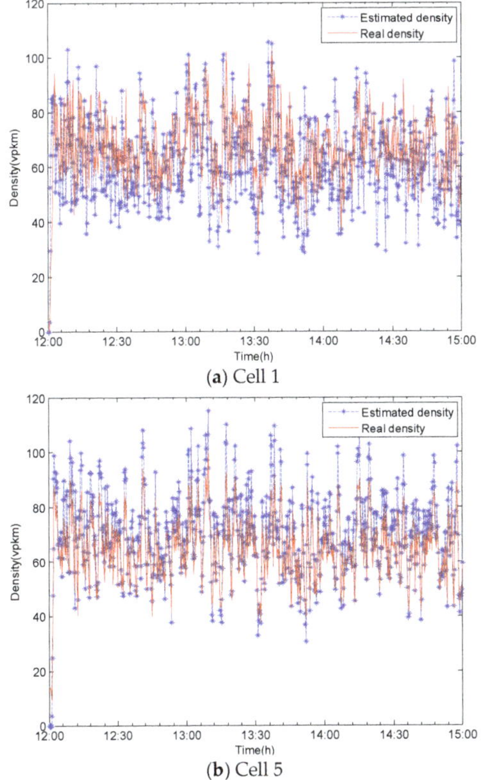

(**a**) Cell 1

(**b**) Cell 5

Figure 7. *Cont.*

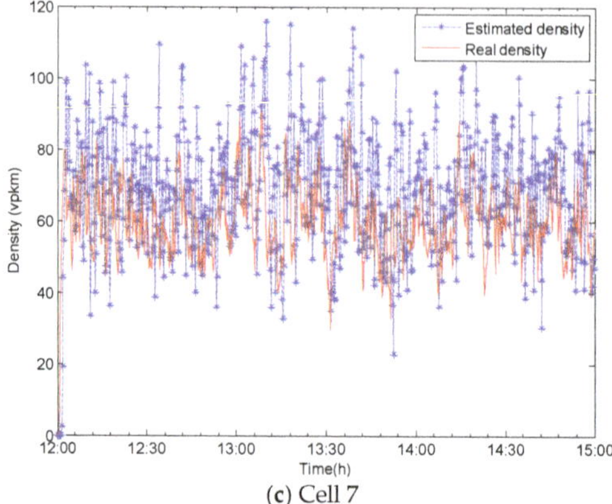

(c) Cell 7

Figure 7. Comparison of simulated and estimated vehicle densities for (**a**) Cell 1, (**b**) Cell 5, and (**c**) Cell 7.

Remark 5. *It is well known that for the most reliable validation, the proposed model should be evaluated using real-world traffic data. However, it is always difficult to obtain real-world data. We have tried our best to measure traffic data from the real world and will present the analysis result in our future work. Nevertheless, computer simulation has already been recognized as an effective tool to verify various theoretical models; in this study, the effectiveness of the proposed method was verified using simulated data.*

Remark 6. *In this study, only one street was considered in the method validation. It is very important to consider the influence of adjunction streets to analyze the effects of other cells and streets on the traffic network. To this end, as a first step, this study introduced on-ramp traffic into the analyzed street to investigate its effect on the main road.*

In our case, on-ramp traffic flowed into cell 1. The on-ramp traffic volume was 10% of the cell 1 traffic. The prediction results for cell re ias shown in Figure 8. As can be seen, the proposed model was still effective for traffic density prediction. However, compared to Figure 7a, because the traffic flow on the main road was affected by the on-ramp vehicles, the estimated accuracy of vehicle density determination was reduced by 15%. As a result, it was crucial to consider the influence of other streets on traffic flow and develop related observer functions to reduce it and improve the accuracy of the prediction model. This coupled-street issue will be addressed in our next work.

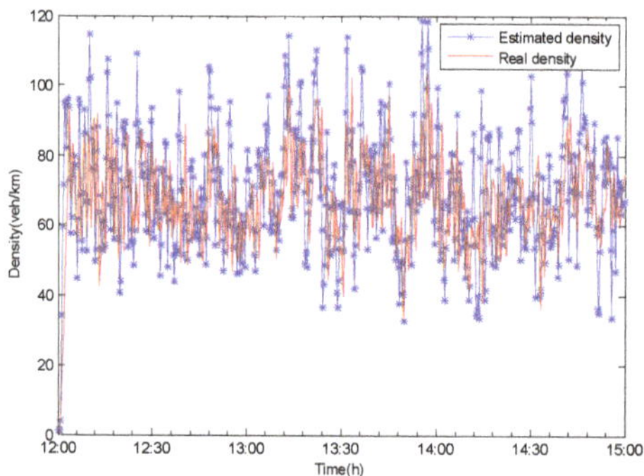

Figure 8. Cell 1 with on-ramp traffic.

4. Conclusions

Based on the hybrid dynamic traffic system with unknown inputs, a switched unknown-input state observer was designed, and the issues of vehicle density estimation and congestion identification were investigated. We showed that the unknown-inputsobserver was able to reconstruct the vehicle densities of road sections which were not equipped with traffic sensors. This strategy for vehicle density estimation was applied to The Beijing Jingtong freeway. Experimental results demonstrated that the estimated densities matched the actual densities reasonably well, and thus congestion can be identified effectively using ths model.

However, in this study, only simulation data obtained by VISSIM were used to verify the performance of the observer, thus the results have some limitations. Meanwhile, the design of the model parameters did not consider the coupled-street effect. In future work, we will choose a practical road network and collect real data to evaluate and optimize the model parameters. The coupled-street issue will also be addressed.

Author Contributions: Y.M. and Z.L. (Zhixiong Li) conceived and designed the experiments; Y.G., B.L., D.L. and M.D.C. performed the experiments and wrote the paper; M.A.S. and Z.L. (Zongzhi Li) revised the paper; M.A.S., D.L., and Z.L. (Zongzhi Li) analyzed the data. All authors have read and agreed to the published version of the manuscript.

Funding: This work was funded by the National Key Research and Development Program of China (2018YFB1601300 & 2016YFB0100903), JITRI Suzhou Automotive Research Institute Project (CEC20190404), Basic Scientific Research Business Expenses Special Funds from National Treasury (2019-0019, 2019-0092, 2018-9067, 2018-9060, 2017-9038, zx-2017-014), and Australia Research Council (No. DE190100931).

Acknowledgments: We would also like to thank the anonymous reviewers, whose meticulous reading and thoughtful comments helped improve this paper.

Conflicts of Interest: The authors declare no conflict of interest.

References

1. Luenberger, D.G. Observers for multivariable systems. *IEEE Trans. Autom. Control* **2003**, *11*, 190–197. [CrossRef]
2. Luenberger, D.G. An introduction to observers. *IEEE Trans. Autom. Control* **1971**, *16*, 596–602. [CrossRef]
3. Alvarez-Icaza, L.; Munoz, L.; Sun, X.; Horowitz, R. Adaptive observer for traffic density estimation. In Proceedings of the IEEE American Control Conference (ACC), Boston, MA, USA, 30 June–2 July 2004; pp. 2705–2710.
4. Canudas-de-Wit, C.; Ojeda, L.; Kibangou, A. Graph constrained-CTM observer design for the Grenoble south ring. *IFAC Symp. Control Transp. Syst.* **2012**, *45*, 197–202. [CrossRef]
5. Morbidi, F.; Ojeda, L.; Canudas-de-Wit, C.; Bellicot, I. A new robust approach for highway traffic density estimation. In Proceedings of the European Control Conference (ECC), Strasbourg, France, 24–27 June 2014; pp. 2575–2580.
6. Vivas, C.; Siri, S.; Ferrara, A. Distributed consensus-based switched observers for freeway traffic density estimation. In Proceedings of the 54th IEEE Decision and Control Conference (CDC), Osaka, Japan, 15–18 December 2015; pp. 3445–3450.
7. Chen, Y.; Guo, Y.; Wang, Y. Modeling and Density Estimation of an Urban Freeway Network Based on Dynamic Graph Hybrid Automata. *Sensors* **2017**, *17*, 716. [CrossRef] [PubMed]
8. Guo, Y.; Chen, Y.; Zhang, C. Decentralized state-observer-based traffic density estimation of large-scale urban freeway network by dynamic model. *Information* **2017**, *8*, 95. [CrossRef]
9. Guo, Y.; Chen, Y.; Li, W. Traffic density estimation of urban freeway by dynamic model based distributed observer. In Proceedings of the 17th COTA International Conference of Transportation Professionals (CICTP), Shanghai, China, 7–9 July 2017; pp. 603–612.
10. Guo, Y.; Chen, Y.; Li, W.; Zhang, C. Distributed State-Observer-Based Traffic Density Estimation of Urban Freeway Network. In Proceedings of the 20th IEEE International Conference on Intelligent Transportation Systems (ITSC), Yokohama, Japan, 16–19 October 2017; pp. 1177–1182.
11. Guo, Y. Dynamic-model-based switched proportional-integral state observer design and traffic density estimation for urban freeway. *Eur. J. Control* **2018**, *44*, 103–113. [CrossRef]

12. Kudva, P.; Viswanadham, N.; Ramakrishna, A. Observers for linear systems with unknown inputs. *IEEE Trans. Autom. Control* **1980**, *25*, 113–115. [CrossRef]

13. Guan, Y.; Saif, M. A Novel Approach to the Design of Unknown Input Observers. *IEEE Trans. Autom. Control* **1991**, *36*, 632–635. [CrossRef]

14. Darouach, M.; Zasadzinski, M.; Xu, S. Full-order observers for linear systems with unknown inputs. *IEEE Trans. Autom. Control* **1994**, *39*, 607–609. [CrossRef]

15. Saif, M. A disturbance accommodating estimator for bilinear systems. In Proceedings of the 1993 IEEE American Control Conference (ACC), San Francisco, CA, USA, 2–4 June 1993; pp. 945–949.

16. Gao, N.; Darouach, M.; Voos, H. New unified H∞ dynamic observer design for linear systems with unknown inputs. *Automatica* **2016**, *65*, 43–52. [CrossRef]

17. Lungu, M.; Lungu, R. Full-order observer design for linear systems with unknown inputs. *Int. J. Control* **2012**, *85*, 1602–1615. [CrossRef]

18. Bhattacharyya, S. Observer design for linear systems with unknown inputs. *IEEE Trans. Autom. Control* **1978**, *23*, 483–484. [CrossRef]

19. Li, Z.; Guo, Z.; Hu, C.; Li, A. Full-order observers for linear systems with unknown inputs. *Comput. Electr. Eng.* **2017**, *60*, 100–115.

20. Zasadzinski, M.; Rafaralahy, H.; Mechmeche, C.; Darouach, M. On disturbance decoupled observer for a class of bilinear systems. *Int. J. Control* **1998**, *120*, 371–377. [CrossRef]

21. Pertew, A.; Marquez, H.; Zhao, Q. Design of unknown input observers for Lipschitz nonlinear systems. In Proceedings of the 2005 IEEE American Control Conference (ACC), Portland, OR, USA, 8–10 June 2005; pp. 4198–4203.

22. Seliger, R.; Frank, P. Fault diagnosis by disturbance decoupled nonlinear observers. In Proceedings of the 30th IEEE Conference on Decision and Control (CDC), Brighton, UK, 11–13 December 1991; pp. 2248–2253.

23. Seliger, R.; Frank, P. Robust component fault detection and isolation in nonlinear dynamic systems using nonlinear unknown input observers. In Proceedings of the IFAC/IMACS Symposium SAFEPROCESS, Baden-Baden, Germany, 10–13 September 1991; pp. 313–318.

24. Yang, H.; Saif, M. Monitoring and Diagnosis of a Class of Nonlinear Systems Using Nonlinear Unknown Input Observer. In Proceedings of the IEEE Conference on Control Applications, Dearborn, MI, USA; 1996; pp. 1006–1011.

25. Liu, Y.; Wang, Z.; He, X.; Zhou, D. Observer design for systems with unknown inputs and missing measurements. In Proceedings of the 35th Chinese Control Conference (CCC), Chengdu, China, 27–29 July 2016; pp. 1799–1803.

26. Sharma, V.; Sharma, B.; Nath, R. Reduced order unknown input observer for discrete time system. In Proceedings of the 2016 IEEE Region 10 Conference (TENCON), Singapore, 22–25 November 2016; pp. 3443–3446.

27. Mihai, L.; Romulus, L. Reduced Order Observer for Linear Time-Invariant Multivariable Systems with Unknown Inputs. *Circuits Syst. Signal Process.* **2013**, *32*, 2883–2898.

28. Stefen, H.; Stanislaw, H. Observer design for systems with unknown inputs. *J. Appl. Math. Comput. Sci.* **2005**, *15*, 431–446.

29. Lyubchyk, L. Optimal data fusion in decentralized stochastic unknown input observers. In Proceedings of the 7th IEEE International Conference on Intelligent Data Acquisition and Advanced Computing Systems (IDAACS), Berlin, Germany, 12–14 September 2013; pp. 358–362.

30. Chen, W.; Saif, M. Design of a TS based fuzzy nonlinear unknown input observer with fault diagnosis applications. In Proceedings of the 2007 IEEE American Control Conference, New York, NY, USA, 9–13 July 2007; pp. 2545–2550.

31. Chen, J.; Patton, R.; Zhang, H. Design of unknown input observers and robust fault detection filters. *Int. J. Control* **1996**, *63*, 85–105. [CrossRef]

32. Gonzalez, J.; Sueur, C. Unknown Input Observer with stability: A Structural Analysis Approach in Bond Graph. *Eur. J. Control* **2018**, *41*, 25–43. [CrossRef]

33. Ifqir, S.; Ichalal, D.; Oufroukh, N.A. Robust interval observer for switched systems with unknown inputs: Application to vehicle dynamics estimation. *Eur. J. Control* **2018**, *44*, 3–14. [CrossRef]

34. Edwards, C.; Chee, P. A Comparison of Sliding Mode and Unknown Input Observers for Fault Reconstruction. *Eur. J. Control* **2006**, *12*, 245–260. [CrossRef]

35. Osoriogordillo, G.; Darouach, M.; Astorgazaragoza, C. H∞ dynamical observers design for linear descriptor systems. Application to state and unknown input estimation. *Eur. J. Control* **2015**, *26*, 35–43. [CrossRef]

36. Zheng, C.; Fan, X.; Wang, C.; Qi, J. GMAN: A Graph Multi-Attention Network for Traffic Prediction. In Proceedings of the Thirty-Fourth Conference on Artificial Intelligence (AAAI 2020), New York, NY, USA, 7–12 February 2020.

37. Zhang, Y.; Wang, S.; Chen, B.; Cao, J.; Huang, Z. TrafficGAN: Network-Scale Deep Traffic Prediction with Generative Adversarial Nets. *IEEE Trans. Intell. Transp. Syst.* **2019**, 1–12. [CrossRef]

38. Li, Y.; Shahabi, R.; Yu, C.; Liu, Y. Diffusion Convolutional Recurrent Neural Network:Data-Driven Traffic Forecasting. *arXiv* **2017**, arXiv:1707.01926.

39. Cao, Z.; Jiang, S.; Zhang, J.; Guo, H. A Unified Framework for Vehicle Rerouting and Traffic Light Control to Reduce Traffic Congestion. *IEEE Trans. Intell. Transp. Syst.* **2017**, *18*, 1958–1973. [CrossRef]

40. Cao, Z.; Guo, H.; Zhang, J.; Fastenrath, U. Multiagent-Based Route Guidance for Increasing the Chance of Arrival on Time. In Proceedings of the Thirtieth AAAI Conference on Artificial Intelligence (AAAI 2016), Phoenix, AZ, USA, 12–17 February 2016; pp. 3817–3820.

41. Cao, Z.; Guo, H.; Zhang, J. A Multiagent-Based Approach for Vehicle Routing by Considering Both Arriving on Time and Total Travel Time. *ACM Trans. Intell. Syst. Technol.* **2017**, *19*, 1–21. [CrossRef]

42. PTV. *VISSIM 5.20 User Manual*; Planung Transport Verkeher AG: Karlsruhe, Germany, 2009.

Article

Using Deep Learning to Forecast Maritime Vessel Flows

Xiangyu Zhou [1,2,*], **Zhengjiang Liu** [1], **Fengwu Wang** [1], **Yajuan Xie** [2,3] **and Xuexi Zhang** [4]

[1] Navigation College, Dalian Maritime University, Dalian 116026, China; liuzhengjiang@dlmu.edu.cn (Z.L.); wangfw650105@163.com (F.W.)
[2] Centre for Maritime Studies, National University of Singapore, Singapore 118414, Singapore; isexy@nus.edu.sg
[3] Department of Industrial Systems Engineering and Management, National University of Singapore, Singapore 117576, Singapore
[4] School of Automation, Guangdong University of Technology, Guangzhou 510006, China; zxxnet@gdut.edu.cn
* Correspondence: zhou.x.y@outlook.com

Received: 13 February 2020; Accepted: 18 March 2020; Published: 22 March 2020

Abstract: Forecasting vessel flows is important to the development of intelligent transportation systems in the maritime field, as real-time and accurate traffic information has favorable potential in helping a maritime authority to alleviate congestion, mitigate emission of GHG (greenhouse gases) and enhance public safety, as well as assisting individual vessel users to plan better routes and reduce additional costs due to delays. In this paper, we propose three deep learning-based solutions to forecast the inflow and outflow of vessels within a given region, including a convolutional neural network (CNN), a long short-term memory (LSTM) network, and the integration of a bidirectional LSTM network with a CNN (BDLSTM-CNN). To apply those solutions, we first divide the given maritime region into $M \times N$ grids, then we forecast the inflow and outflow for all the grids. Experimental results based on the real AIS (Automatic Identification System) data of marine vessels in Singapore demonstrate that the three deep learning-based solutions significantly outperform the conventional method in terms of mean absolute error and root mean square error, with the performance of the BDLSTM-CNN-based hybrid solution being the best.

Keywords: maritime vessel flows; intelligent transportation systems; deep learning

---

1. Introduction

Forecasting traffic flows of vessels has been recognized as a challenging task in the maritime intelligent transportation system, since it could be affected by various complex factors [1,2]. Accurate and timely traffic information is of significant importance to both maritime managers and individual vessels, as it not only helps the former to better conduct port planning [3,4], alleviate congestion [5,6], mitigate emission of GHG (greenhouse gases) and improve public security [7,8], but also enables the latter to better operate the ship navigation system and plan a route [9,10], so as to avoid collisions and reduce the potential cost due to late arrival [11].

Various solutions have been studied for forecasting maritime traffic flow. Wang et al. [12] developed a vessel flow prediction model using a back propagation (BP) neural network, in which training and data sampling were conducted based on a residual analysis. Haiyan and Youzhen [13] came up with a hybrid scheme for vessel flow forecasting, which integrated an RBF (radial basis function) neural network, grey forecasting and auto-regression into a framework of support vector regression (SVR). The experimental results based on a real dataset demonstrated that the combined method significantly outperformed each individual. Li et al. [14] conceived a hybrid method centering around a robust support vector regression (RSVR) model. In that method, a chaotic cloud simulated annealing genetic algorithm was first developed to optimize the parameters of RSVR, then the kernel principal component analysis (KPCA) algorithm was utilized to finalize the input vectors. Afterwards, the hybrid framework was established, and the test based on a real dataset justified its advantages over others. Yang et al. [15] developed an enhanced hybrid RBF neural network (EHRBF-NN) to forecast the traffic flow of vessels. The method was robust in that it incorporated a regression tree and a particle swarm optimization (PSO) algorithm into the RBF neural network. More specifically, the regression tree was exploited to calculate the parameters (i.e., centers and radius) of RBF, and the PSO was designated to address the issue of potential over-fitting and update the weights of the neural network. Liu et al. [16] proposed a two-step method to predict vessel flows. Particularly, in the first step, they separated vessel traffic flow into low-rank and sparse components by adopting a non-convex model. Then in the second step, the ARIMA (auto-regressive integrated moving average) and WNN (wavelet neural network) were explored to predict the flows based on the low-rank and sparse components from the fist step. He et al. [17] proposed an improved Kalman model for short-term vessel traffic flow prediction. In that model, a preliminary prediction was first done using a regression analysis, which was then adopted to replace the transfer equation of the Kalman filter. Xiao et al. [11] came up with a probabilistic method for maritime traffic forecasting. The method was novel in that it took both waterway pattern knowledge and vessel motion stability into account. In particular, the waterway pattern was learnt via a lattice-based DBSCAN algorithm which also helped narrow down the problem scale, and the motion behavior was modeled through a kernel density estimation, thus, the combination of them was supposed to successfully predict the traffic flow.

Although some successes have been achieved in maritime traffic flow forecasting, most of the methods above are shallow and outdated to some extent, given the fact that the deep learning approach has been unarguably deemed as the state-of-the-art in many areas, such as computer vision, natural language processing, and transportation. The promising advantages brought by deep learning can be summarized as two aspects: (1) deep neural networks can be designed to solve end-to-end tasks, which directly takes the raw data as input without elaborating the intermediate feature extraction; (2) it usually achieves significantly higher accuracy in comparison with conventional methods. Despite the fact that numerous deep learning approaches have been applied in traffic flow or traffic condition forecasting, most of them focus on the land transportation instead of maritime. Lv et al. [18] proposed a deep learning framework to predict the freeway traffic flow, by exploring the big data from road infrastructure sensors. At the heart of that approach was a stacked auto-encoder (SAE) model, which was adopted to learn the latent traffic flow features by inherently taking spatial and temporal correlations into account. Chen et al. [19] implemented a typical stacked LSTM (long short-term memory) network to forecast the traffic congestion, which was applied to a realistic navigation map. The results demonstrated that the proposed approach was able to effectively learn the hidden patterns of traffic flows. Cui et al. [20] proposed a deep stacked bidirectional and unidirectional LSTM (SBU-LSTM) network to predict the traffic speed, which considered both forward and backward dependencies of the traffic data. More specifically, in that approach, a bidirectional LSTM (BDLSM) layer was developed to learn the spatial features and bidirectional temporal relationships by exploiting the historical data, as such it was able to grasp the dynamic nature of the traffic. Polson and Sokolov [21] came up with a deep learning architecture to predict the short-term traffic flow. Particularly, that architecture integrated a linear model with ℓ_1 regularization and a branch of tanh layers, the first layer of which was designed to identify the spatio-temporal relations among predictors and other layers. As a consequence, it was able to capture the sharp nonlinear patterns of the traffic flow. Yu et al. [22] designed a spatio-temporal deep neural network to predict the short-term traffic speeds. Enlightened by the research outcomes of motion detection in the field of computer vision, they first segmented the road network into grids, then the traffic information inside each grid was input to a convolutional neural network (CNN) to learn the spatial relationship. After that, the output of which was further taken as the input of a LSTM network, to capture the temporal relationship. Zhang et al. [23] proposed a deep spatio-temporal residual network (ST-ResNet) to forecast the citywide crowd inflow and outflow. In that approach, a residual neural network framework was first adopted to model the temporal closeness, period, and trend properties of the crowd flows, then a sequence of residual convolutional units were designed to model the spatial relationships of crowd flows, and lastly, the output of those residual neural networks were aggregated as the preceding layer to the final output layer of real crowd flows. Guo et al. [24] presented a framework for predictor fusion to forecast the short-term traffic condition. In the framework, three strategies were employed to evaluate the fusion performance, i.e., average fusion, weighted fusion and kNN fusion. Experimental results based on real dataset verified the significant advantages of the fusion method over the stand-alone ones. Liu and Chen [25] developed a novel prediction model for passenger flow using deep learning approach. The core of the model was the combination of a pre-trained unsupervised SAE with a supervised deep neural network, which was supposed to well extract the hierarchical features, so that the passenger flow for any periods from Monday to Sunday could be successfully forecast. Ke et al. [26] devised a deep learning approach to predict the road traffic congestion. The approach was distinguished in that it utilized visual signals to learn the traffic flow, and moving object detection and moving speed calculation were done by integrating a CNN and a Gaussian mixture model (GMM).

Obviously, the applications of deep learning in land transportation have been investigated far more extensively than that of maritime. On one hand, large-scale data for the land transportation is not that hard to obtain compared with maritime transportation, which is the foundation of applying the deep learning techniques. On the other hand, unlike the vessels in the sea, the mobilities of cars or people on land are mostly restricted to streets or rails, therefore, the traffic pattern for which is comparatively easy to be revealed. Moreover, we would like to note that, a variety of research with respects to the deep learning in maritime have been conducted, however most of them focus on vessel type identification [27], trajectory prediction or reconstruction [28], anomaly detection [29,30] and collision avoidance [31–33]. The problem of how to adopt deep learning to forecast the traffic flows for maritime has not yet been studied.

Therefore, in this paper, centering around the deep learning approach, we propose three different solutions to forecast the inflow and outflow of vessels within a given marine region. More specifically, the three solutions are featured by a CNN, a LSTM network, and the integration of a bidirectional LSTM network with a CNN (BDLSTM-CNN), respectively. In particular, with respects to the BDLSTM-CNN based solution, each input will be first fed into the convolutional layers, then they will go through the forward layer and backward layer of a bidirectional LSTM network. As such, this hybrid solution is supposed to coherently learn the spatial and temporal dependencies pertaining to the vessel flows. Moreover, to apply the three deep learning based solutions into practice, we divide the given marine areas into M×N grids. Then all the solutions are implemented to forecast the inflow and outflow for each grid. Experimental results based on the real AIS data for a given area in Singapore show that the three deep learning based solutions significantly outperform the conventional method, in which the hybrid solution BDLSTM-CNN achieves the best performance.

The remainder of the paper is organized as follows. Section 2 introduces the preliminaries first and then elaborates the structures and logic of the three deep learning based solutions. Section 3 presents the comprehensive experimental results and analysis. Section 4 concludes the paper and states the future works.

2. Deep Learning Based Solutions

In this section, we first introduce the preliminaries regarding the vessel flow forecasting problem. Then we present three deep learning based solutions, i.e., CNN, RNN, and BDLSTM-CNN, respectively.

2.1. Preliminaries

Our target is to forecast the inflow and outflow of vessels within a given area. To this end, we first divide this area into $M \times N$ grids. Thus, the inflow and outflow at a time step can be represented by two matrix, respectively, and the element of which refers to the counts of vessels entering or leaving the corresponding grid during that time step. In this regard, our task comes down to predict the inflow matrix and outflow matrix using the historical traffic flow data. Formally, given the historical observations $\{X_t | t = 0, 1, \ldots, n-1\}$, our task is to predict $X_{t+\tau}$, where $\tau \geq 1$ and is an integer. In particular, $X_t = [I_t, O_t]$, where I_t and O_t represent the inflow matrix and outflow matrix at time step t, respectively. Taking Figure 1 as an example, at time step t, two vessels will enter grid g_2, and one vessel will leave, then $I_t(g_2) = 2$, and $O_t(g_2) = 1$, respectively.

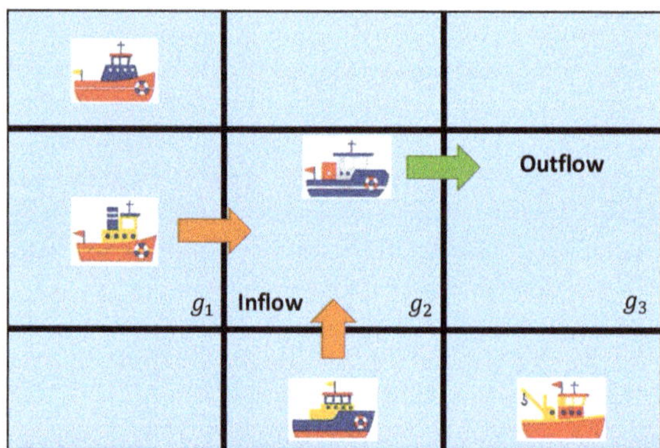

Figure 1. Illustration of the Inflow and Outflow.

2.2. CNN Based Solution

Convolutional neural network has been successfully applied in many areas, such as video analysis [34] and transportation [35], which is characterized by its strong capability of capturing the spacial representation. CNN can be employed here to handle the traffic flow prediction problem for two reasons. On one hand, the flow of vessels in a grid could be affected by the neighbor or distant grids, and CNN has the potential to capture this dependency, as the convolution operation inside CNN somehow can help predict the movement of vessels from the spatial perspective. On the other hand, the inflow and outflow for the whole given area are represented as matrix in our problem, which is the most inherent format of inputs to CNN. And the structure of the proposed CNN-based solution is depicted in Figure 2.

Accordingly, this CNN based solution has m channels, which correspond to the m historical observations as the inputs. Meanwhile, all the channels share the same weights. In particular, each input contains both the inflow and outflow matrix, and each input channel includes two convolutional layers, two ReLu layers, one batch normalization (BN) layer, and one dropout layer. Among them, the convolutional layers play the most important role, which are used to learn spacial features of different levels [22]. The underlying logic of convolutional layer is expressed as follows:

$$o_r^l = \sum_k W_{kr}^l * o_k^{l-1} + b_k^l, \tag{1}$$

where o_r^l refers to the output of the r_{th} filter in the l_{th} layer; o_k^{l-1} refers to the output of the k_{th} filter of the preceding layer; W_{kr}^l and b_k^l refers to the weights and bias; $*$ refers to the convolution operation. Besides, the BN layer is used to scale the range of feature values; the ReLu layer acts as the activation function; the Dropout layer is used to select the salient features from the receptive region, so as to avoid redundant features and reduce the scale of the computation; and the output layer generates both inflow and outflow matrix that we would like to predict. To adopt the CNN based solution to solve the vessel flow forecasting problem, we take X_{t-m+1} to X_t as the inputs at time step t and $X_{t+\tau}$ as the output Y_t.

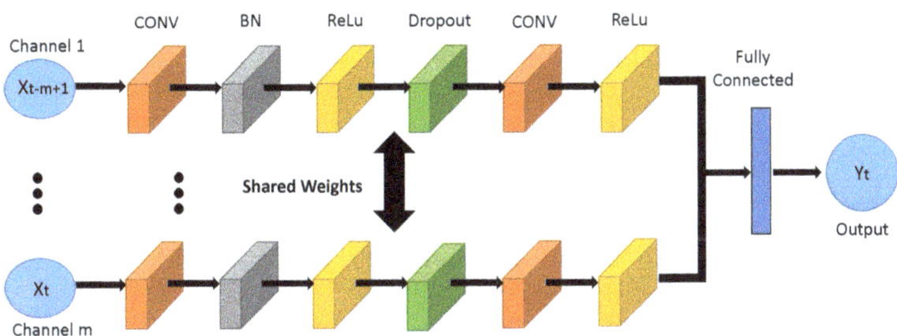

Figure 2. Illustration of the CNN based Solution.

2.3. LSTM Based Solution

Due to the capability of sequential and temporal modeling, recurrent neural network (RNN) also has been successfully applied to many challenging practices, such as natural language processing, stock forecasting and crowd density prediction [36]. However, traditional RNN sometimes suffers from the issue of vanishing and exploding gradient if the learning sequence is long. To address this issue, a variant of RNN, termed LSTM (Long short-term memory) model, was devised, which utilized memory cells with various gates to preserve useful information for long-term dependencies [37]. In view of the fact that the vessel flow for a given area can be considered as the classical time series with temporal dependency, the LSTM network is supposed to well capture the temporal correlations in the flow forecasting problem.

More specifically, the structure of a single layer LSTM network is depicted in Figure 3a. And the LSTM network updates itself at time step t as follows [38]:

$$
\begin{aligned}
f_t &= \sigma(W^f[h_{t-1}, x_t] + b_f) \\
i_t &= \sigma(W^i[h_{t-1}, x_t] + b_i) \\
\tilde{C}_t &= tanh(W^C[h_{t-1}, x_t] + b_C) \\
C_t &= f_t * C_{t-1} + i_t * \tilde{C}_t \\
o_t &= \sigma(W^o[h_{t-1}, x_t] + b_o) \\
h_t &= o_t * C_t,
\end{aligned}
\tag{2}
$$

where h_t is the hidden state; i_t, f_t and o_t refer to the input gate, forget gate and output gate, respectively; $\tilde{C}_t$ and C_t refer to the input modulation gate and memory gate, respectively. $\{W^f, W^i, W^C, W^o\}$ are the weights, and $\{b^f, b^i, b^c, b^o\}$ are the biases for the corresponding gates; $\sigma(\cdot)$ and $tanh(\cdot)$ are sigmoid and hyperbolic tangent activation functions, respectively. The memory cell unit C_t contains two components, i.e., previous memory cell unit C_{t-1} modulated by f_t and $\tilde{C}$, which is modeled by the current input, and previous hidden state, modulated by the input gate i_t [38]. The essence of sigmoidal operation for i_t and f_t normalizes themselves into the scope of [0,1]. Particularly, they could be deemed as knobs that LSTM learns to selectively forget its previous memory or consider its current input. In a similar way, the output gate o_t models the transfer from memory cells to hidden states. On the basis of these mechanisms, the LSTM network is supposed to learn complex and temporal dynamics that exist in sequential vessel movement measurements, engendering a satisfactory performance for vessel flow forecasting.

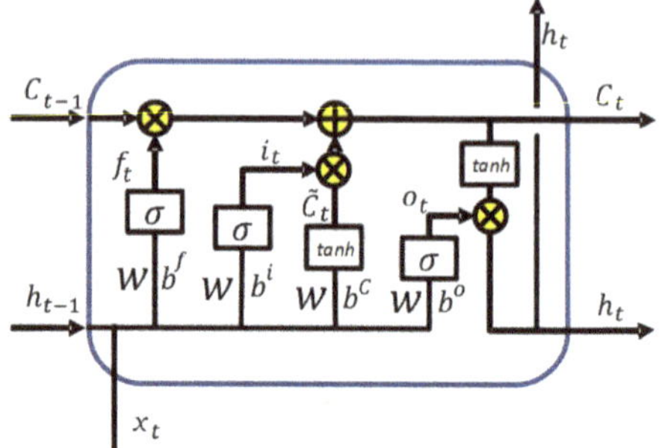

(**a**) Structure of LSTM Network

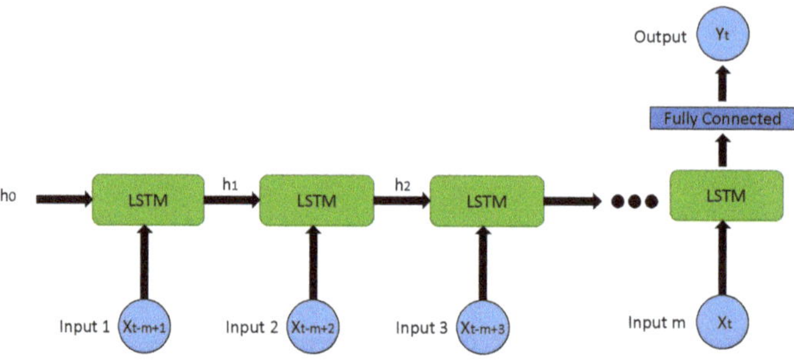

(**b**) Structure of LSTM based Solution

Figure 3. Structure of the LSTM network, and the LSTM based Solution.

With the fundamental logic of the LSTM network, we build up the complete structure of the LSTM based solution, which is depicted in Figure 3b. In this solution, there are m inputs at the time step t, i.e., $X_t, X_{t-1}, \ldots, X_{t-m+1}$, all of which are fed into the LSTM networks in order. Afterwards, the outputs of the LSTM network are connected to a fully connected network layer, then Y_t is regarded as the final output, which is set as $X_{t+\tau}$.

2.4. BDLSTM-CNN Based Hybrid Solution

As we have stated previously, the vessel flow forecasting relies on both spatial and temporal dependencies. Therefore, it would be desirable to integrate the CNN and the LSTM network into a comprehensive framework, to coherently learn the spatiotemporal relationships for the vessel flow forecasting. However, with respect to the temporal feature learning, we exploit a more powerful recurrent neural network, namely, bidirectional LSTM, to replace the traditional unidirectional LSTM.

2.4.1. Bidirectional LSTM

The idea of BDLSTMs is derived from the bidirectional RNN, which is used to address a crucial issue that, the conventional RNN is only able to make use of the previous context, thus they only learn representations from previous time steps [39]. However, we might have to learn representations from future time steps to better understand the context and eliminate ambiguity sometimes. As a consequence, the bidirectional RNN was developed to achieve this goal, which processed sequence data in both forward and backward directions with two separate hidden layers. And both of them are connected to the same output layer, as depicted in Figure 4a. More specifically, the bidirectional RNN separates the hidden layer into two parts, forward state sequence $\overrightarrow{h}$ and backward state sequence $\overleftarrow{h}$, and they are computed as follows [40]:

$$
\begin{aligned}
\overrightarrow{h}_t &= \mathcal{H}(W_{x\overrightarrow{h}}x_t + W_{\overrightarrow{h}\overrightarrow{h}}\overrightarrow{h}_{t-1} + b_{\overrightarrow{h}}) \\
\overleftarrow{h}_t &= \mathcal{H}(W_{x\overleftarrow{h}}x_t + W_{\overleftarrow{h}\overleftarrow{h}}\overleftarrow{h}_{t+1} + b_{\overleftarrow{h}}) \\
y_t &= W_{\overrightarrow{h}y}\overrightarrow{h}_t + W_{\overleftarrow{h}y}\overleftarrow{h}_t + b_y.
\end{aligned}
\tag{3}
$$

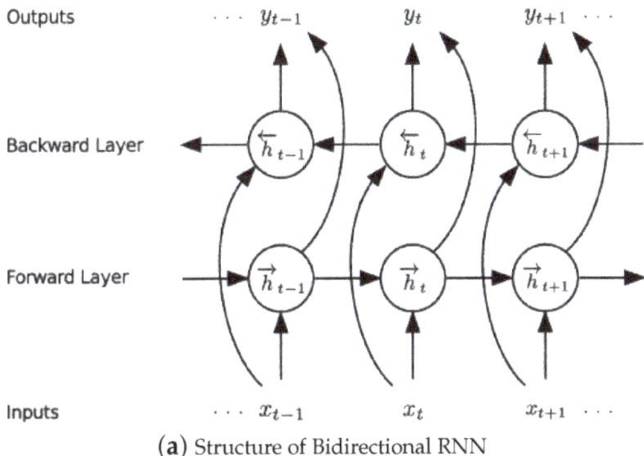

(**a**) Structure of Bidirectional RNN

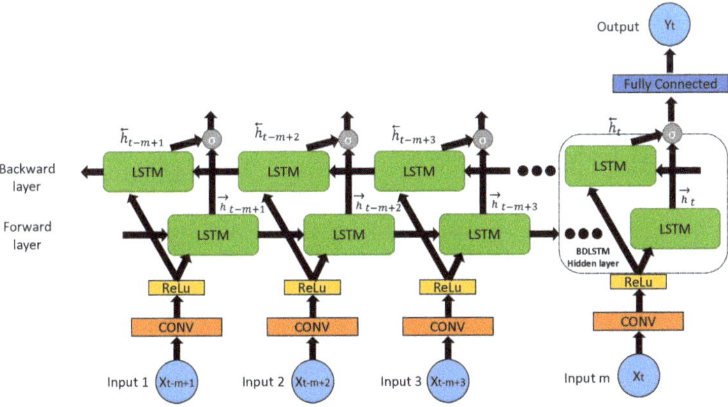

(**b**) Structure of the BDLSTM-CNN based Hybrid Solution

Figure 4. Structure of Bidirectional RNN, and the Hybrid Solution.

Accordingly, the deep bidirectional RNN can be established by stacking multiple RNN hidden layers on top of each other. Each hidden state sequence, h^n, is replaced by the forward and backward, $\overrightarrow{h}^n$ and $\overleftarrow{h}^n$. On the other hand, deep bidirectional LSTM (BDLSTM) can be achieved based on the integration of deep bidirectional RNN and the LSTM network. Thus, the BDLSTM is supposed to have the ability to model the deep representation of long-span features [39].

2.4.2. The Hybrid Solution

With the framework of bidirectional LSTM, CNN is further integrated to build up a hybrid solution, which is depicted in Figure 4b. In this framework, each input first goes through a convolution layer and a Relu layer, to learn the spatial dependency. Then the output will be fed into the bidirectional LSTM network (i.e, the dotted box in Figure 4b), as the input for both forward sequence and backward sequence. Afterwards, the outputs of the bidirectional LSTM hidden layers are further connected to a sigmoid activation function, which is followed by a fully connected layer. Lastly, the final output is regarded as the predicted vessel flow.

To summarize from a high level perspective, at each time step t, there are m inputs that will be fed into the synthesized networks, i.e., X_{t-m+1}, X_{t-m+2}, ..., and X_t, each of which consists of both inflow matrix and outflow matrix. At the same time, the output Y_t is set as $X_{t+\tau}$. Normally, τ is equal to 1, however it can also take much larger integer values. On the other hand, since the vessel flow forecasting is a regression task in nature, root mean square error or mean absolute error are usually adopted as the loss function. However, in this hybrid solution, we consider a new loss function for final objective optimization, namely, smooth ℓ_1 loss [41], which is expressed as follows:

$$loss(x,y) = \begin{cases} 0.5(x-y)^2, & if\ |x-y| < 1; \\ |x| - 0.5, & otherwise. \end{cases} \tag{4}$$

Smooth ℓ_1 loss, also called Huber loss, is usually less sensitive to abnormal inputs, and also helps the networks prevent gradient exploding to some extent [41].

3. Experimentation and Evaluation

In this section, we conduct experimentation in different settings to test the proposed solutions, and demonstrate their advantages over the baseline. Particularly, we first introduce the data processing and experimental settings, then we compare and evaluate the three deep learning based solutions. Finally, we compare the three solutions with a conventional method, i.e., support vector regression (SVR).

3.1. Data Processing and Experimentation Setup

We use the AIS (Automatic Identification System) data of maritime vessels to perform the forecasting task, which is an automatic tracking system that uses transponders on ships to generate trajectory-related information [30]. The AIS data for a vessel contains many useful attributes regarding its movement and mobility, which can be obtained from https://www.marinetraffic.com. Here we mainly exploit the information of vessel ID, coordinates (i.e., longitude and latitude) and time stamp. The testing filed we selected is an rectangle marine area southeast to Singapore, which is shown as the red rectangle in Figure 5. The location of the left upper point is $(1.287979°, 103.892723°)$, and that of the right lower point is $(1.235027°, 103.996817°)$. In all the experimentation we conducted, we uniformly divided this area into 7×7 grids (i.e., $M = N = 7$), which means that both inflow matrix I_t and outflow matrix O_t have a size of 7×7.

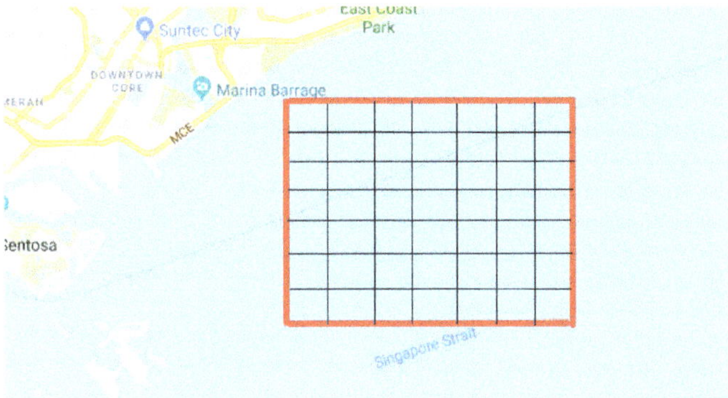

Figure 5. The Selected Marine Area: 7×7 Grids.

Then we process and analyze the AIS data, and identify all the vessels entering and leaving those grids accordingly. Particularly, the AIS data we collected for the given area lasts about 31 days, from 01-10-2013 to 31-10-2013. We set the duration for each time step as 5 min. Then we have 8525 samples for each grid (with some data missed), which include the amount of vessels entering and leaving the given grid. We divide them into training dataset and testing dataset according to the time order, i.e., from 01-10-2013 to 25-10-2013, and from 26-10-2013 to 31-10-2013, which include 6305 and 2220 samples, respectively. In addition, we implement all the experimentation using pytorch, in a laptop with Intel i7 CPU, 8G RAM, and Nvidia GTX 1060.

3.2. Error Performance for the Deep Learning Based Solutions

We set $Y_t = X_{t+1}$ and $Y_t = X_{t+2}$, respectively, and conduct experimentation according to the above configurations. We use two types of errors to measure the performance of the proposed solutions, i.e., mean absolute error (MAE) and root mean square error (RMSE), since they are the two most important metrics for regression problems. We also use their normalized forms to further evaluate the performance, i.e., NMAE and NRMSE. All the results for the three deep learning based solutions are shown in Figures 6 and 7. Before looking into them, we would like to note that the errors in those figures are calculated by considering inflow and outflow together. Additionally, all the results below are obtained based on the testing data.

From Figure 6a,b we can see that, as the training iteration increases, the MAE for testing data of the three deep learning based solutions drop quickly. With respect to both $Y_t = X_{t+1}$ and $Y_t = X_{t+2}$, all the solutions seem to converge after about 85 iterations. Pertaining to the two different Y_t settings, the BDLSTM-CNN based solution always achieves the lowest MAE, which are around 1.11 and 1.14, respectively. In contrast, the LSTM based solution achieves the second lowest MAE of 1.15 and 1.25, and the CNN based solution achieves the third lowest MAE of 1.29 and 1.35. It makes sense that the BDLSTM-CNN based hybrid solution engenders the best results as it coherently explores the spatial and temporal dependencies regarding the vessel flows, by integrating both CNN and LSTM. Moreover, with respect to the temporal expression, it exploits the bidirectional LSTM to take advantage of both past and future information, which is supposed to get better results than the unidirectional LSTM. Comparing the CNN based solution and the LSTM based solution, we can observe that, the latter presents a better performance, which is probably due to that the temporal relationship is much relatively important in comparison with the spatial relationship in this case. Although that the former solution takes every five preceding flow matrix to generate the succeeding one, and the convolution layer also has the potential to predict the movement of vessels from the spatial perspective, the structure of CNN itself does not have a scheme to capture the temporal correlation. This also might be able to explain the oscillation in the curve of CNN based solution, because it dose not have any mechanism to capture the seasonality pattern in the vessel flows. On the other hand, there are only slight oscillations for LSTM based solution and BDLSTM-CNN based solution, which justifies their favorable capability of handling the seasonality in sequence. Comparing Figure 6a,b, we can observe that, the performance of the three solutions slightly deteriorates for $Y_t = X_{t+2}$. It is reasonable as one can always make a more accurate forecast for a near future than for a far future. However, the hybrid solution still achieves the lowest error. Additionally, we would like to highlight that, the difference between the LSTM based solution and the hybrid solution is larger for $Y_t = X_{t+2}$ than that of $Y_t = X_{t+1}$. It happened because in the case of $Y_t = X_{t+2}$, the bidirectional LSTM may bring more useful information from the backward layer, in comparison with the solo forward layer in the LSTM based solution.

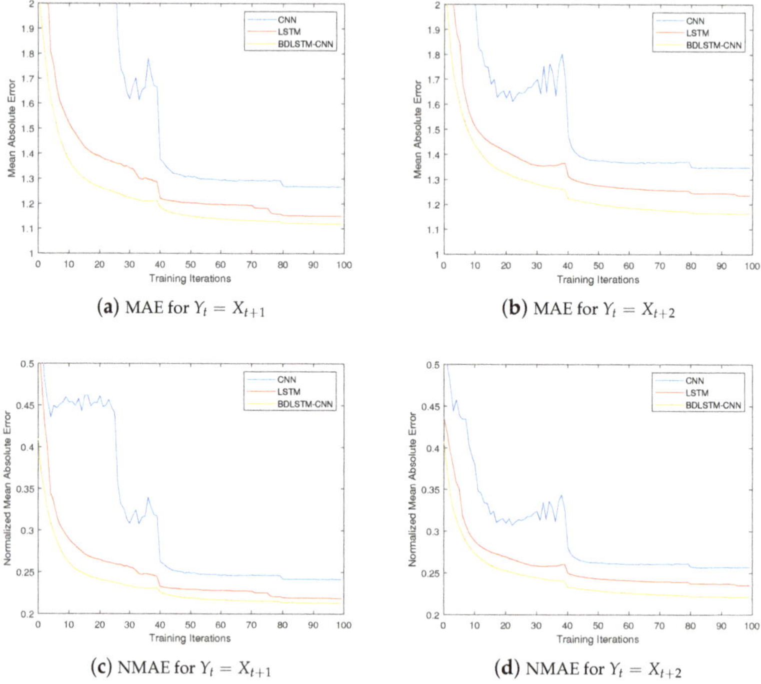

(**a**) MAE for $Y_t = X_{t+1}$ (**b**) MAE for $Y_t = X_{t+2}$

(**c**) NMAE for $Y_t = X_{t+1}$ (**d**) NMAE for $Y_t = X_{t+2}$

Figure 6. Performance Regarding MAE and NMAE.

From Figure 6c,d, we can see that they share similar pattern with Figure 6a,b. This is normal because they are simply the normalized version of the latter, which can be regarded as a kind of error rate. The difference in the curve shape comes from the fact that we only show part of errors in Figure 6a,b, in order to highlight the region of interests. Looking into Figure 6c,d, we can observe that, the three solutions achieve error rates (in terms of MAE) of around 24.5%, 22.5% and 22.0%, respectively, for $Y_t = X_{t+1}$, and around 26%, 24% and 22.5%, respectively, for $Y_t = X_{t+2}$, with the hybrid solution being the lowest. The error rates in terms of MAE would be decreased about 1.5%, 1.5% and 0.5%. So it can be concluded that the results of three deep learning based solutions considerably are better than the traditional approach expressed in MAE, with the performance of the BDLSTM-CNN based hybrid solution being the smallest.

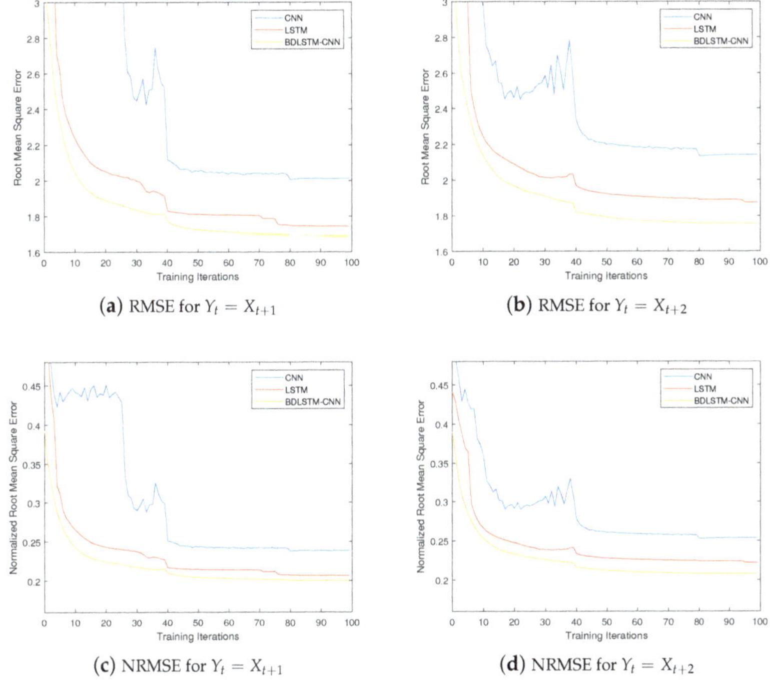

(**a**) RMSE for $Y_t = X_{t+1}$

(**b**) RMSE for $Y_t = X_{t+2}$

(**c**) NRMSE for $Y_t = X_{t+1}$

(**d**) NRMSE for $Y_t = X_{t+2}$

Figure 7. Performance Regarding RMSE and NRMSE.

Likewise, we also use the RMSE (root mean square error) to further evaluate the proposed solutions, and the results are depicted in Figure 7, which share similar pattern with that of Figure 6. From Figure 7a,b, we can observe that, the three solutions achieve a RMSE of around 2.01, 1.75 and 1.68 for $Y_t = X_{t+1}$, respectively, and around 2.18, 1.88 and 1.75 for $Y_t = X_{t+2}$, respectively, with the hybrid solution being the lowest for both cases. From Figure 7c,d, we can see that, the three solutions achieve error rates (in terms of RMSE) of around 24%, 21% and 20%, respectively, for $Y_t = X_{t+1}$, and around 26%, 23.5% and 21%, respectively, for $Y_t = X_{t+2}$, with the hybrid solution being the lowest for both cases. Similarly, the error rates in terms of RMSE will be lowered by 2%, 2.5% and 1%. Consequently, it can be found that the outcomes of three deep learning based solutions noticeably outperformed than the traditional approach in terms of RMSE, with the performance of the BDLSTM-CNN based hybrid solution being the lowest.

Combining the performance of both MAE and RMSE in Figures 6 and 7, it can be possibly proving that the BDLSTM-CNN based hybrid solution outperforms the CNN based solution and the LSTM based solution. However, considering that, (1) all the solutions only adopt five historical data to predict a new one, (2) all the errors are derived based on integrating inflow and outflow together, we believe that the performance of all the solutions are sufficiently satisfactory, although superiority and inferiority exist among them.

3.3. Breakdown Performance for the Hybrid Solution

Since the BDLSTM-CNN based solution has the best overall performance, we look into this hybrid solution and analyze its prediction capability in a breakdown perspective. To this end, we plot the curves of the average forecasting value of the inflow and outflow for the whole given region, and the forecasting value of inflow and outflow for a given grid, respectively. We would like to note that, all the results below are obtained based on the testing data.

We first plot the above forecasting values against the ground truth for $Y_t = X_{t+1}$ in Figure 8. From Figure 8a we can see that, the ground truth value of the average by considering inflow and outflow together, changes dramatically as time goes on. However, the BDLSTM-CNN based solution can well capture those trends, such as the sharp changes at time step 100, 1300, 1700 and 2200, respectively, because this hybrid solution combines the advantages of the capability of CNN and BDLSTM, to learn the spatial and temporal features in a unified way. Nevertheless, we observe some imperfect tracking, such as time step 250, which does not catch the crest. However, considering that the ground truth is the average of both inflow and outflow for the whole given area, some minor errors are tolerable. We also plot the forecasting curves of a single grid for inflow and outflow in Figure 8b,c, respectively, by taking grid (6,6) as an example. From Figure 8b, we observe that, the hybrid solution can catch most of the crests and troughs for the inflow curve, no matter how sharp or smooth they are, such as the crest at time step 400 and the trough at time step 1250. An unsatisfactory forecasting is found as well at time step 1800, where a sharp crest is missed. From Figure 8c, we observe that the hybrid solution can also successfully follow both sharp and smooth crests and troughs for the outflow curve, such as the crest at time step 1050, and trough at time step 1450. Although the hybrid solution mismatched the ground truth at some points, such as time step 1800, most of the failures are tolerable.

We then continue to plot the forecasting values against the ground truth for $Y_t = X_{t+2}$ in Figure 9. From Figure 9a we can see that, the hybrid solution is still able to basically track the ground truth value, although slight deterioration is observed in comparison with the performance in Figure 8a. This can be explained by the fact that, a near future is comparatively easier to be forecast than a distant future. Despite the slight deterioration in the average forecasting, the hybrid solution still shows competitive performance for inflow and outflow forecasting pertaining to grid (6,6), which mostly well captures all the sharp or smooth crests and troughs.

Combining the results in both Figures 8 and 9, we can conclude that the hybrid solution has strong capability of forecasting the inflow and the outflow of vessels. Though the inflow and outflow changed dramatically, the trend can be well captured by the hybrid solution.

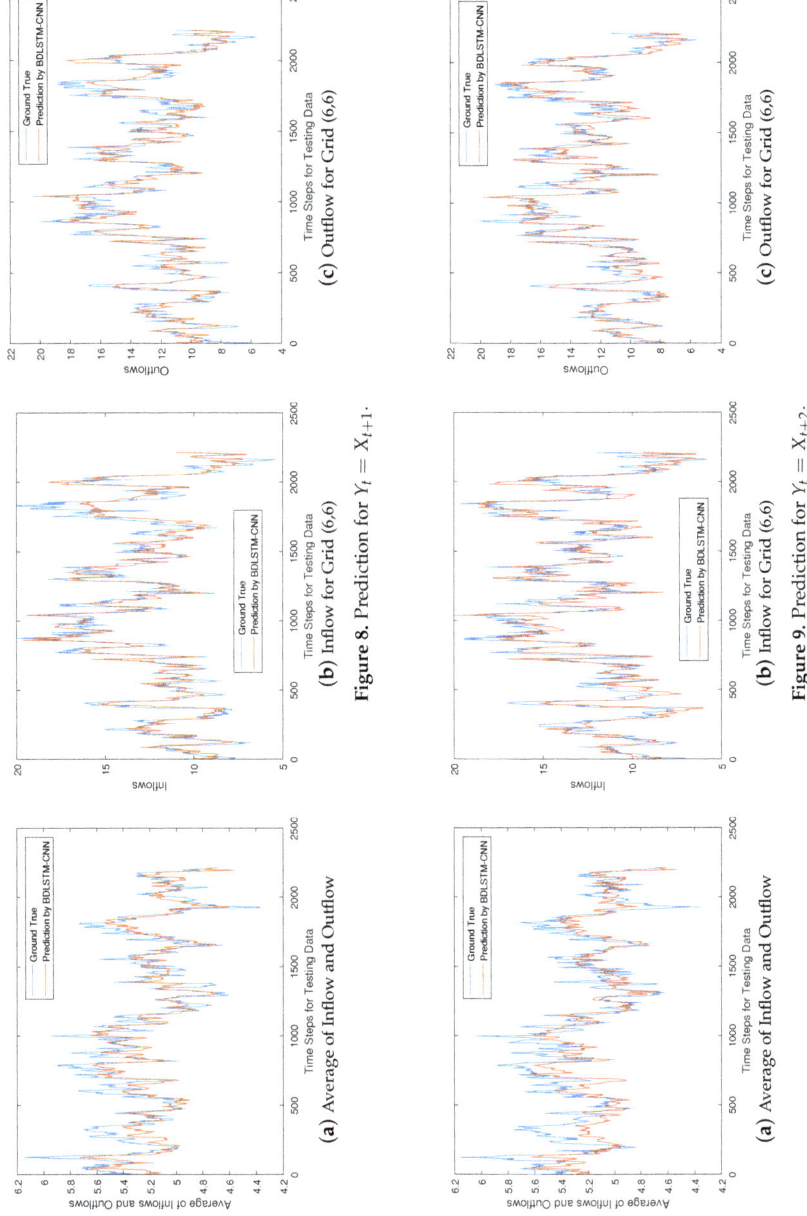

Figure 8. Prediction for $Y_t = X_{t+1}$.

Figure 9. Prediction for $Y_t = X_{t+2}$.

3.4. Comparison with the Conventional Method

In this subsection, we compare the performance of the three deep learning based solutions with a conventional method, namely the support vector regression (SVR) based approach. In particular, we use the python sklearn package to implement this function, and utilize the same training dataset to optimize its parameters, and then do the evaluation using the same testing dataset, the results of which are recorded in Table 1. Since the normalized errors reflect a kind of error rate in forecasting, we mainly concentrate on the normalized forms of the two errors, i.e., NMAE and NRMSE.

Table 1. Comparison for Error Rates (%)

	SVR	CNN	LSTM	BDLSTM-CNN
$Y_t = X_{t+1}$, NMAE	50.1	24.5	22.5	22.0
$Y_t = X_{t+2}$, NMAE	50.5	26.0	24.0	22.5
$Y_t = X_{t+1}$, NRMSE	51.2	24.0	21.0	20.0
$Y_t = X_{t+2}$, NRMSE	51.4	26.0	23.5	21.0

From Table 1 we can see that, the error rates of the SVR method are around 51% for both measurements of MAE and RMSE, almost twice as much as the three deep learning based solutions. The inferiority comes from the fact that, the SVR method does not have a sophisticated scheme to learn the underlying spatial dependency, or the long-term temporal dependency. In contrast, CNN has a convolution layer, and LSTM or BDLSTM has a memory and gate to handle the two situations [42,43], respectively. On the other hand, the BDLSTM-CNN based hybrid solution always achieves the lowest error rates of 20% to 22.5%, which implies a forecasting accuracy of 77.5% to 80%. Considering that we only use five historical data to predict a new one, and both inflow and outflow of vessels change dramatically, the performance achieved by the hybrid solution is sufficiently satisfactory.

4. Conclusions and Future Work

In this paper, we propose three deep learning based solutions to forecast the flow of maritime vessels. To apply the deep learning approach, we first divide the given marine area into $M \times N$ grids, then we predict both the inflow and outflow of vessels for each grid. In particular, the three solutions are characterized by a CNN, a LSTM, and the integration of BDLSTM and CNN, respectively. When testing them with the real AIS data of vessels, the hybrid solution based on BDLSTM-CNN achieves the best performance in terms of mean absolute error (MAE) and root mean square error (RMSE), both error rates will be decreased by 1–2% compared with other methods, as it is able to coherently learn the spatial and temporal representations for both the inflow and outflow. On the other hand, when further comparing them with a conventional approach, the three deep learning based solutions significantly outperform the SVR method. We would like to note that, $N = 7$ is only an empirical value for the given area in this paper, and it can vary with different scenarios. Moreover, the number of columns does not need to be the same with the rows.

However, the proposed approach still needs to be improved further and tested more extensively. In future, we will work on the following directions: (1) we will consider an attention model in the BDLSTM-CNN based solution to further improve the performance; (2) we will explore more relevant features, such as weather, date, kinematics and kinetics of ship, variable speeds of ship movement, collision avoidance maneuvers; (3) we will use more than five data points to predict a new one, and also forecast the flow longer-time away, such as $Y_t = X_{t+3}, X_{t+4}, \ldots$; (4) we will investigate multi-agent based methods and reinforcement learning based methods to solve the route planning for maritime vessel flows; (5) we will try to apply integral of multiplied by absolute error (ITAE) and integral of time multiplied by square error (ITSE) performance criteria to assess the quality of vessel traffic forecasting.

Author Contributions: Conceptualization, X.Z. (Xiangyu Zhou), Z.L.; methodology, X.Z. (Xiangyu Zhou), F.W.; software, X.Z. (Xiangyu Zhou), Y.X., X.Z. (Xuexi Zhang), writing–original draft preparation, X.Z. (Xiangyu Zhou); writing–review and editing, Z.L., F.W., Y.X. All authors have read and agreed to the published version of the manuscript.

Funding: This research was funded by China-APEC Cooperation Fund/Research on integration of education, training and employment of seafarers in APEC region (grant number 01871901), National Key Research and Development Program of China (grant number 2018YFC0810402).

Conflicts of Interest: The authors declare no conflict of interest.

References

1. Zissis, D.; Xidias, E.K.; Lekkas, D. Real-time vessel behavior prediction. *Evol. Syst.* **2016**, *7*, 29–40. [CrossRef]
2. Xiao, Z.; Fu, X.; Zhang, L.; Ponnambalam, L.; Goh, R.S.M. Data-driven multi-agent system for maritime traffic safety management. In Proceedings of the IEEE 20th International Conference on Intelligent Transportation Systems (ITSC), Yokohama, Japan, 16–19 October 2017; pp. 1–6.
3. Fazi, S.; Roodbergen, K.J. Effects of demurrage and detention regimes on dry-port-based inland container transport. *Transp. Res. A* **2018**, *89*, 1–18. [CrossRef]
4. Zhou, C.; Li, H.; Lee, B.K.; Qiu, Z. A simulation-based vessel-truck coordination strategy for lighterage terminals. *Transp. Res. A* **2018**, *95*, 149–164. [CrossRef]
5. Chen, L.; Hopman, H.; Negenborn, R.R. Distributed model predictive control for vessel train formations of cooperative multi-vessel systems. *Transp. Res. A* **2018**, *92*, 101–118. [CrossRef]
6. Kazemi, S.; Abghari, S.; Lavesson, N.; Johnson, H.; Ryman, P. Open data for anomaly detection in maritime surveillance. *Expert Syst. Appl.* **2013**, *40*, 5719–5729. [CrossRef]
7. Ruiz-Aguilar, J.J.; Turias, I.; Moscoso-López, J.A.; Jiménez-Come, M.J.; Cerbán, M. Forecasting of short-term flow freight congestion: A study case of Algeciras Bay Port (Spain). *DYNA* **2016**, *83*, 163–172. [CrossRef]
8. Hänninen, M.; Banda, O.A.V.; Kujala, P. Bayesian network model of maritime safety management. *Expert Syst. Appl.* **2014**, *41*, 7837–7846. [CrossRef]
9. Chen, C.H.; Khoo, L.P.; Chong, Y.T.; Yin, X.F. Knowledge discovery using genetic algorithm for maritime situational awareness. *Expert Syst. Appl.* **2014**, *41*, 2742–2753. [CrossRef]
10. Kisialiou, Y.; Gribkovskaia, I.; Laporte, G. Robust supply vessel routing and scheduling. *Transp. Res. A* **2018**, *90*, 366–378. [CrossRef]
11. Xiao, Z.; Ponnambalam, L.; Fu, X.; Zhang, W. Maritime traffic probabilistic forecasting based on vessels' waterway patterns and motion behaviors. *IEEE Trans. Intell. Transp. Syst.* **2017**, *18*, 3122–3134. [CrossRef]
12. Wang, C.; Zhang, X.; Chen, X.; Li, R.; Li, G. Vessel traffic flow forecasting based on BP neural network and residual analysis. In Proceedings of the 4th International Conference on Information, Cybernetics and Computational Social Systems (ICCSS), Dalian, China, 24–26 July 2017; pp. 350–354.
13. Haiyan, W.; Youzhen, W. Vessel traffic flow forecasting with the combined model based on support vector machine. In Proceedings of the International Conference on Transportation Information and Safety (ICTIS), Wuhan, China, 25–28 June 2015; pp. 695–698.
14. Li, M.W.; Han, D.F.; Wang, W.L. Vessel traffic flow forecasting by RSVR with chaotic cloud simulated annealing genetic algorithm and KPCA. *Neurocomputing* **2015**, *157*, 243–255. [CrossRef]
15. Yang, L.; Hao, Y.; Liu, Q.; Zhu, X. Ship traffic volume forecast in bridge area based on enhanced hybrid radial basis function neural networks. In Proceedings of the International Conference on Transportation Information and Safety (ICTIS), Wuhan, China, 25–28 June 2015; pp. 38–43.
16. Liu, R.W.; Chen, J.; Liu, Z.; Li, Y.; Liu, Y.; Liu, J. Vessel traffic flow separation-prediction using low-rank and sparse decomposition. In Proceedings of the IEEE 20th International Conference on Intelligent Transportation Systems (ITSC), Yokohama, Japan, 16–19 October 2017; pp. 1–6.
17. He, W.; Zhong, C.; Sotelo, M.A.; Chu, X.; Liu, X.; Li, Z. Short-term vessel traffic flow forecasting by using an improved Kalman model. *Clust. Comput.* **2017**; pp. 1–10. [CrossRef]
18. Lv, Y.; Duan, Y.; Kang, W.; Li, Z.; Wang, F.Y.; others. Traffic flow prediction with big data: A deep learning approach. *IEEE Trans. Intell. Transp. Syst.* **2015**, *16*, 865–873. [CrossRef]
19. Chen, Y.Y.; Lv, Y.; Li, Z.; Wang, F.Y. Long short-term memory model for traffic congestion prediction with online open data. In Proceedings of the IEEE 19th International Conference on Intelligent Transportation Systems (ITSC), Rio de Janeiro, Brazil, 1–4 November 2016; pp. 132–137.

20. Cui, Z.; Ke, R.; Wang, Y. Deep Stacked Bidirectional and Unidirectional LSTM Recurrent Neural Network for Network-wide Traffic Speed Prediction. In Proceedings of the 6th International Workshop on Urban Computing (UrbComp), Halifax, Nova Scotia, Canada, 14 August 2016.

21. Polson, N.G.; Sokolov, V.O. Deep learning for short-term traffic flow prediction. *Transp. Res. A* **2017**, *79*, 1–17. [CrossRef]

22. Yu, H.; Wu, Z.; Wang, S.; Wang, Y.; Ma, X. Spatiotemporal recurrent convolutional networks for traffic prediction in transportation networks. *Sensors* **2017**, *17*, 1501. [CrossRef] [PubMed]

23. Zhang, J.; Zheng, Y.; Qi, D. Deep Spatio-Temporal Residual Networks for Citywide Crowd Flows Prediction. In Proceedings of the 31st AAAI Conference on Artificial Intelligence (AAAI), San Francisco, CA, USA, 4–9 February 2017; pp. 1655–1661.

24. Guo, F.; Polak, J.W.; Krishnan, R. Predictor fusion for short-term traffic forecasting. *Transp. Res. A* **2018**, *92*, 90–100. [CrossRef]

25. Liu, L.; Chen, R.C. A novel passenger flow prediction model using deep learning methods. *Transp. Res. A* **2017**, *84*, 74–91. [CrossRef]

26. Ke, X.; Shi, L.; Guo, W.; Chen, D. Multi-Dimensional Traffic Congestion Detection Based on Fusion of Visual Features and Convolutional Neural Network. *IEEE Trans. Intell. Transp. Syst.* **2018**, *20*, 2157–2170. [CrossRef]

27. Nguyen, D.; Vadaine, R.; Hajduch, G.; Garello, R.; Fablet, R. A Multi-task Deep Learning Architecture for Maritime Surveillance using AIS Data Streams. In Proceedings of the 2018 IEEE International Conference on Data Science and Advanced Analytics (DSAA), Turin, Italy, 1–3 October 2018.

28. Jiang, X.; Liu, X.; de Souza, E.N.; Hu, B.; Silver, D.L.; Matwin, S. Improving point-based AIS trajectory classification with partition-wise gated recurrent units. In Proceedings of the International Joint Conference on Neural Networks (IJCNN), Anchorage, AK, USA, 14–19 May 2017, pp. 4044–4051.

29. Mazzarella, F.; Vespe, M.; Alessandrini, A.; Tarchi, D.; Aulicino, G.; Vollero, A. A novel anomaly detection approach to identify intentional AIS on-off switching. *Expert Syst. Appl.* **2017**, *78*, 110–123. [CrossRef]

30. Tu, E.; Zhang, G.; Rachmawati, L.; Rajabally, E.; Huang, G.B. Exploiting AIS data for intelligent maritime navigation: A comprehensive survey from data to methodology. *IEEE Trans. Intell. Transp. Syst.* **2018**, *19*, 1559–1582. [CrossRef]

31. Liu, X.; Jin, Y. Design of Transfer Reinforcement Learning Mechanisms for Autonomous Collision Avoidance. In Proceedings of the 8th International Conference on Design Computing and Cognition, Politecnico di Milano, Italy, 2–4 July 2018; pp. 339–358.

32. Perera, L.P. Autonomous Ship Navigation Under Deep Learning and the Challenges in COLREGs. In Proceedings of the ASME 37th International Conference on Ocean, Offshore and Arctic Engineering, Madrid, Spain, 17–22 June 2018; pp. 1–10.

33. Cheng, Y.; Zhang, W. Concise deep reinforcement learning obstacle avoidance for underactuated unmanned marine vessels. *Neurocomputing* **2018**, *272*, 63–73. [CrossRef]

34. Xu, Z.; Yang, Y.; Hauptmann, A.G. A discriminative CNN video representation for event detection. In Proceedings of the IEEE Conference on Computer Vision and Pattern Recognition, Boston, MA, USA, 7–12 June 2015; pp. 1798–1807.

35. Yu, B.; Yin, H.; Zhu, Z. Spatio-Temporal Graph Convolutional Networks: A Deep Learning Framework for Traffic Forecasting. In Proceedings of the Twenty-Seventh International Joint Conference on Artificial Intelligence (IJCAI-18), Stockholm, Sweden, 13–19 July 2018, pp. 3634–3640.

36. Zonoozi, A.; Kim, J.J.; Li, X.L.; Cong, G. Periodic-CRN: A Convolutional Recurrent Model for Crowd Density Prediction with Recurring Periodic Patterns. In Proceedings of the Twenty-Seventh International Joint Conference on Artificial Intelligence (IJCAI-18), Stockholm, Sweden, 13–19 July 2018; pp. 3732–3738.

37. Greff, K.; Srivastava, R.K.; Koutník, J.; Steunebrink, B.R.; Schmidhuber, J. LSTM: A search space odyssey. *IEEE Trans. Neural Netw. Learn. Syst.* **2017**, *28*, 2222–2232. [CrossRef] [PubMed]

38. Chen, Z.; Zhang, L.; Jiang, C.; Cao, Z.; Cui, W. WiFi CSI Based Passive Human Activity Recognition Using Attention Based BLSTM. *IEEE Trans. Mob. Comput.* **2018**, *18*, 2714–2724. [CrossRef]

39. Yu, Z.; Ramanarayanan, V.; Suendermann-Oeft, D.; Wang, X.; Zechner, K.; Chen, L.; Tao, J.; Ivanou, A.; Qian, Y. Using bidirectional LSTM recurrent neural networks to learn high-level abstractions of sequential features for automated scoring of non-native spontaneous speech. In Proceedings of the IEEE Workshop on Automatic Speech Recognition and Understanding (ASRU), Scottsdale, AZ, USA, 13–17 December 2015; pp. 338–345.

40. Graves, A.; Jaitly, N.; Mohamed, A.R. Hybrid speech recognition with deep bidirectional LSTM. In Proceedings of the IEEE Workshop on Automatic Speech Recognition and Understanding (ASRU), Olomouc, Czech Republic, 8–12 December 2013; pp. 273–278.
41. Berrada, L.; Zisserman, A.; Kumar, M.P. Smooth Loss Functions for Deep Top-k Classification. In Proceedings of the International Conference on Learning Representations, Vancouver, BC, Canada, 30 Aprial–3 May 2018; pp. 1–25.
42. Liu, Y.; Zheng, H.; Feng, X.; Chen, Z. Short-term traffic flow prediction with Conv-LSTM. In Proceedings of the 2017 9th International Conference on Wireless Communications and Signal Processing (WCSP), Nanjing, China, 11–13 October 2017; pp. 1–6.
43. Duan, Z.; Yang, Y.; Zhang, K.; Ni, Y.; Bajgain, S. Improved deep hybrid networks for urban traffic flow prediction using trajectory data. *IEEE Access* **2018**, *6*, 31820–31827. [CrossRef]

Article

A Vision-Based Machine Learning Method for Barrier Access Control Using Vehicle License Plate Authentication

Kh Tohidul Islam [1,2], Ram Gopal Raj [1,*], Syed Mohammed Shamsul Islam [3,4], Sudanthi Wijewickrema [2], Md Sazzad Hossain [5], Tayla Razmovski [2] and Stephen O'Leary [2]

[1] Department of Artificial Intelligence, Faculty of Computer Science & Information Technology, University of Malaya, Kuala Lumpur 50603, Malaysia; kh.tohidulislam@gmail.com

[2] Department of Surgery (Otolaryngology), Faculty of Medicine, Dentistry and Health Sciences, University of Melbourne, Melbourne, VIC 3010, Australia; swijewickrem@unimelb.edu.au (S.W.); tayla.razmovski@gmail.com (T.R.); sjoleary@unimelb.edu.au (S.O.)

[3] Discipline of Computing and Security, School of Science, Edith Cowan University (ECU), Joondalup, WA 6027, Australia; syed.islam@ecu.edu.au

[4] Department of Computer Science and Software Engineering, The University of Western Australia, Crawley, WA 6009, Australia

[5] Faculty of Information Technology, Monash University, Melbourne, VIC 3800, Australia; sazzad.hossain@monash.edu

* Correspondence: ramdr@um.edu.my; Tel.: +60-12-307-0435

Received: 6 December 2019; Accepted: 6 January 2020; Published: 24 June 2020

Abstract: Automatic vehicle license plate recognition is an essential part of intelligent vehicle access control and monitoring systems. With the increasing number of vehicles, it is important that an effective real-time system for automated license plate recognition is developed. Computer vision techniques are typically used for this task. However, it remains a challenging problem, as both high accuracy and low processing time are required in such a system. Here, we propose a method for license plate recognition that seeks to find a balance between these two requirements. The proposed method consists of two stages: detection and recognition. In the detection stage, the image is processed so that a region of interest is identified. In the recognition stage, features are extracted from the region of interest using the histogram of oriented gradients method. These features are then used to train an artificial neural network to identify characters in the license plate. Experimental results show that the proposed method achieves a high level of accuracy as well as low processing time when compared to existing methods, indicating that it is suitable for real-time applications.

Keywords: automatic license plate recognition; intelligent vehicle access; histogram of oriented gradients; artificial neural networks

---

1. Introduction

Automatic vehicle license plate recognition (AVLPR) is used in a wide range of applications including automatic vehicle access control, traffic monitoring, and automatic toll and parking payment systems. Implementation of AVLPR systems is challenging due to the complexity of the natural images from which the license plates need to be extracted, and the real-time nature of the application. An AVLPR system depends on the quality of its physical components which acquire images and the algorithms that process the acquired images. In this paper, we focus on the algorithmic aspects of an AVLPR system, which includes the localization of a vehicle license plate, character extraction and character recognition. For the process of license plate localization, researchers have proposed various methods including, connected component analysis (CCA) [1], morphological analysis with edge statistics [2], edge point analysis [3], color processing [4], and deep learning [5]. The rate of accuracy for these localization methods varies from 80.00% to 99.80% [3,6,7]. The methods most commonly used for the recognition stage are optical character recognition (OCR) [8,9], template matching [10,11], feature extraction and classification [12,13], and deep learning based methods [5,14].

In recent years, several countries including the United Kingdom, United States of America, Australia, China, and Canada have successfully used real-time AVLPR in intelligent transport systems [3,15,16]. However, this is not yet widely used in Malaysia. The road transportation department of Malaysia has authorized the use of three types of license plates. The first contains white alphanumeric characters embossed or pasted on a black background. The second type is allocated to vehicles belonging to diplomatic personnel and also contains white alphanumeric characters, but the background on these plates is red. The third type is assigned to taxi cabs and hired vehicles, consisting of black alphanumeric characters on a white plate. There are also some rules the characters must satisfy, for example, there are no leading zeros in a license plate sequence. Additionally, the letters "I" and "O" are excluded from the sequences due to their similarities with the numbers "1" and "0" and only military vehicles have the letter "Z" included in their license plates. The objective of this study was to implement a fast and accurate method for automatic recognition of Malaysian license plates. Additionally, this method can be easily applied to similar datasets.

The paper is organized as follows. A discussion of the existing literature in the field is given in Section 2. In Section 3, we introduce the proposed method: localizing the license plate based on a deep learning method for object detection, image feature extraction through histogram of oriented gradients (HOG), and character recognition using an artificial neural network (ANN) [17]. In Section 4, we establish the suitability of the method for real-time license plate detection through experiments, including comparisons with similar methods. The paper concludes in Section 5 with a discussion of the findings.

2. Related Work

A typical AVLPR system consists of three stages: detection or localization of the region of interest (i.e., the license plate from the image), character extraction from the license plate, and recognition of those characters [18–20].

2.1. Detection or Localization

The precision of an AVLPR system is typically influenced by the license plate detection stage. As such, many researchers have focused on license detection as a priority. For instance, Suryanarayana et al. [21] and Mahini et al. [22] used the Sobel gradient operator, CCA, and morphological operations to extract the license plate region. They reported 95.00% and 96.5% correct localization, respectively. To monitor highway ticketing systems, a hybrid edge-detection-based method for segmenting the vehicle license plate region was introduced by Hongliang and Changping [2], achieving 99.60% detection accuracy. According to Zheng et al. [23], if the vertical edges of the vehicle image are extracted while the edges representing the background and noise are removed, the vehicle license plate can be easily segmented from the resultant image. In their findings, the overall segmentation accuracy reported was approximately 97%. Luo et al. [24] proposed a license plate detection system for Chinese vehicles, where a single-shot multi-box detector was used for the detection method [25] and achieved 96.5% detection accuracy on their database.

A wavelet transform based method was applied by Hsieh et al. [26] to detect the license plate region from a complex background. They successfully localized the vehicle license plate in three steps. Firstly, they used Haar scaling [27] as a function for wavelet transformation. Secondly, they roughly localized the vehicle license plate by finding the reference line with the maximum horizontal variation in the transformed image. Finally, they localized the license plate region below the reference line by calculating the total pixel values (the region with the maximum pixel value was considered as the plate region) followed by geometric verification using metrics such as the ratio of length and width of the region. They achieved 92.40% detection accuracy on average. A feature-based hybrid method for vehicle license plate detection was introduced by Niu et al. [28]. Initially, they used color processing (blue–white pairs) for possible localization of the license plate. They then used morphological processing such as open and close operations, followed by CCA. They used geometrical features (e.g., size) to remove unnecessary small regions and finally used HOG features in a support vector machine (SVM) to detect the vehicle license plate and achieved 98.06% detection accuracy with their database.

2.2. License Plate Character Segmentation

Correct segmentation of license plate characters is important, as the majority of incorrect recognition is due to incorrect segmentation, as opposed to issues in the recognition process [29]. Several methods have been introduced for character segmentation. For instance, Arafat et al. [9] proposed a license plate character segmentation method based on CCA. The detected license plate region was converted into a binary image and eight connected components were used for character region labeling. They achieved 95.40% character segmentation accuracy. A similar method was also introduced by Tabrizi et al. [13], where they achieved 95.24% character segmentation accuracy.

Chai and Zuo [30] used a similar process for segmenting vehicle license plate characters. To remove unnecessary small character regions from the detected license plate, a vertical and horizontal projection method, alongside morphological operations and CCA, was used. They achieved 97.00% character segmentation accuracy. Dhar et al. [31] proposed a vehicle license plate recognition system for Bangladeshi vehicles using edge detection and deep learning. Their method for character segmentation involved a combination of edge detection, morphological operations, and analysis of segmented region properties (e.g., ratio of height and width). Although they did not mention any segmentation results, they achieved 99.6% accuracy for license plate recognition. De Gaetano Ariel et al. [32] introduced an algorithm for Argentinian license plate character segmentation. They used horizontal and vertical edge projection to extract the characters, with a 96.49% accuracy level.

2.3. Recognition or Classification

Some researchers recognized license plates using adaptive boosting in conjunction with Haar-like features and training cascade classifiers on those features [33–35]. Several researchers have used template matching to recognize the license plate text [10,11]. Feature extraction based recognition has also proven to be accurate in vehicle license plate recognition [12,13,28]. Samma et al. [12] introduced fuzzy support vector machines (FSVM) with particle swarm optimization for Malaysian vehicle license plate recognition. They extracted image features using Haar-like wavelet functions and using a FSVM for classification and they achieved 98.36% recognition accuracy. A hybrid k-nearest neighbors and support vector machine (KNN-SVM) based vehicle license plate recognition system was proposed by Tabrizi et al. [13]. They used operations such as filling, filtering, dilation, and edge detection (using the Prewitt operator [36]) for license plate localization after color to grayscale conversion. For feature extraction, they used a structural and zoning feature extraction method. Initially, a KNN was trained with all possible classes including similar and dissimilar characters (whereas the SVM was trained only on similar character samples). Once the KNN ascertained which "similar character" class the target character belonged to, the SVM performed the next stage of classification to determine the actual class. They achieved 97.03% recognition accuracy.

Thakur et al. [37] introduced an approach that used a genetic algorithm (GA) for feature extraction and a neural network (NN) for classification in order to identify characters in vehicle license plates. They achieved 97.00% classification accuracy. Jin et al. [3] introduced a solution for license plate recognition in China. They used hand-crafted features on a fuzzy classifier to obtain 92.00% recognition accuracy. Another group of researchers proposed a radial wavelet neural network for vehicle license plate recognition [38]. They achieved 99.54% recognition accuracy.

Brillantes et al. [39] utilized fuzzy logic for Filipino vehicle license plate recognition. Their method was effective in identifying license plates from different issues which contained characters of different fonts and styles. They segmented the characters using CCA along with fuzzy clustering. They then used a template matching algorithm to recognize the segmented characters. The recognition accuracy of their methods was 95.00%. Another fuzzy based license plate region segmentation method was introduced by Mukherjee et al. [40]. They used fuzzy logic to identify edges in the license plate in conjunction with other edge detection algorithms such as Canny and Sobel [41,42]. A template matching algorithm was then used to recognize the license plate text from the segmented region and achieved a recognition accuracy of 79.30%. A hybrid segmentation method combining fuzzy logic and k-means clustering was proposed by Olmí et al. [43] for vehicle license plate region extraction. They developed SVM and ANN models to perform the classification task and achieved an accuracy level of 95.30%.

2.4. Recent Methods of AVLPR

Recently, deep learning based image classification approaches have received more attention from researchers as they can learn image features on their own, in addition to performing classification [44]. Therefore, no feature extraction is required for deep learning approaches. However, despite the advantages of using deep learning in image classification, it requires a large training image database and very high computational power. Li et al. [45] investigated a method of identifying similar characters on license plates based on convolution neural networks (CNN). They used CNNs as feature extractors and also as classifiers. They achieved 97.20% classification accuracy. Another deep learning method based on the AlexNet [46] was introduced by Lee et al. [47] for AVLPR, where they re-trained the AlexNet to perform their task on their database and achieved 95.24% correct recognition. Rizvi et al. [5] also proposed a deep learning based approach for Italian vehicle license plate recognition on a mobile platform. They utilized two deep learning models, one to detect and localize the license plate and the characters present, and another as a character classifier. They achieved 98.00% recognition accuracy with their database.

Another deep learning method called "you only look once" (YOLO) was developed for real-time object detection, which is now being used in AVLPR [48]. For example, Kessentini et al. [49] proposed a two-stage deep learning approach that first used YOLO version 2 (YOLO v2) for license plate detection [50]. Then, they used a convolutional recurrent neural network (CRNN) based segmentation-free approach for license plate character recognition. They achieved 95.31% and 99.49% character recognition accuracy in the two stages, respectively. Another YOLO based method was developed by Hendry and Chen [51] for vehicle license plate recognition in Taiwan. Here, for each character, detection and recognition was carried out using a YOLO model, totaling 36 YOLO models used for 36 classes. They achieved 98.22% and 78.00% accuracy for vehicle license plate detection and recognition, respectively.

Similarly, Yonetsu et al. [52] also introduced a two-stage YOLO v2 model for Japanese license plate detection. To increase accuracy, they initially detected the vehicle, followed by the detection of the license plate. In clear weather conditions, they achieved 99.00% and 87.00% accuracy for vehicle and license plate detection, respectively. A YOLO based three-stage Bangladeshi vehicle license plate detection and recognition method was implemented by Abdullah et al. [53]. Firstly, they used YOLO version 3 (YOLOv3) as their detection model [54]. In the second stage, they segmented the license plate region and character patches. Finally, they used a ResNet-20 deep learning model for the character recognition [55]. They achieved 95.00% and 92.70% accuracy for license plate detection character recognition, respectively. Laroca et al. [56] used YOLO for license plate detection and then another method proposed by Silva and Jung [57] for character segmentation and recognition. They tested their performance on their own database (UFPR-ALPR), which is now publicly available for research purposes. They achieved 98.33% and 93.53% accuracy for vehicle license plate detection and recognition, respectively.

3. Methodology

In the proposed system, a digital camera was placed at a fixed distance and height to be able to capture images of vehicle license plates. When a vehicle is at a predefined distance from the camera, it captures an image of the front of the vehicle, including the license plate. This image then went through several pre-processing steps to eliminate the unwanted background and localize the license plate region. Once this region was extracted from the original image, a character segmentation algorithm was used to segment the characters from the background of the license plate. The segmented characters were then identified using an ANN classifier trained on HOG features. Figure 1 illustrates the steps involved for the proposed AVLPR system.

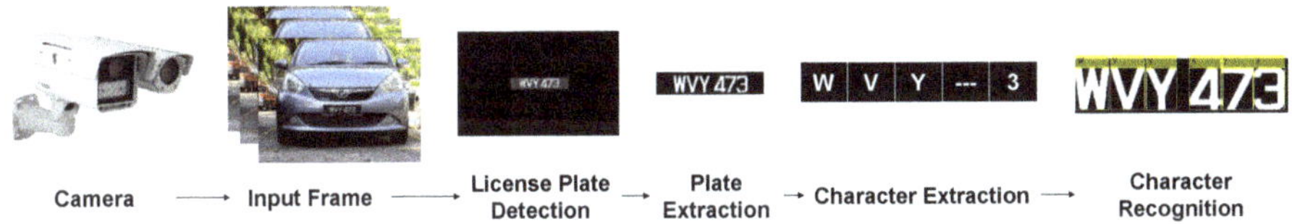

Figure 1. Outline of the proposed system for automatic license plate recognition.

3.1. Image Acquisition

A digital camera was used as an image acquisition device. This camera was placed at a height of 0.5 m from the ground. An ideal distance to capture images of an arriving vehicle was pre-defined. To detect whether a vehicle was within this pre-defined distance threshold, we subtracted the image of the background (with no vehicles) from each frame of the obtained video. If more than 70% of the background was obscured, it was considered that a vehicle was within this threshold. To avoid unnecessary background information, the camera lens was set to $5\times$ zoom. The speed of the vehicles when the images were captured was around 20 km/h. Camera specifications and image acquisition properties are shown in Table 1.

Table 1. Image acquisition properties.

Name	Description
Image acquisition device name	Canon® Power Shot SX530 HS
Zooming capabilities	$50\times$ Optical zoom
Camera zooming position	$5\times$ Optical zoom
Weather	Daylight, rainy, sunny, cloudy
Capturing period	Day and Night
Background	Complex; not fixed
Horizontal field-of-view	Approximately 75°
Image dimension	4608×3456
Vehicle speed limit	20 km/h; 5.56 m/s
Capturing distance	15 meter

3.2. Detection of the License Plate

Once the image was acquired, it was then processed to detect the license plate. First, the contrast of the RGB (red, green, and blue) images was improved using histogram equalization. As the location of the license plate in the acquired images was relatively consistent, we extracted a pre-defined rectangular region from the image to be used in the next stages of processing. Figure 2 shows the specifications of the region of interest (ROI). Therefore, we reduced the size of the image to be processed from 4608×3456 pixels in the original image to 2995×1891 pixels. The region to be extracted was defined using the x and y coordinates of the upper left corner (XOffset and YOffset) and the width and height of the rectangle.

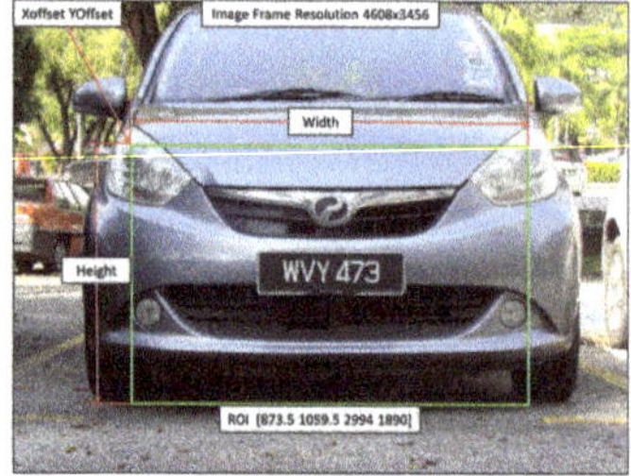

Figure 2. The initial extracted region of interest, defined by the green bounding box.

To detect the license plate, the ROI extracted in the previous stage was first resized to 128×128 pixels. Then, based on previous studies, a deep learning based approach (YOLO v2, as discussed in [50]) was used to extract the license plate region. The network accepts 128×128 pixels RGB images as an input and processes them in an end-to-end manner to produce a corresponding bounding box for the license plate region. This network has 25 layers: one Image Input layer, seven Convolution layers, six Batch Normalization layers, six Rectified Linear Unit (ReLU) layers, three Max Pooling layers, one YOLO v2 Transform Layer, and one YOLO v2 Output layer [58]. Figure 3 shows the YOLO v2 network architecture. We used this network to detect the license plate region only, not for license plate character recognition. The motivation behind this methodology is to reduce total processing time with low computational power without compromising the accuracy.

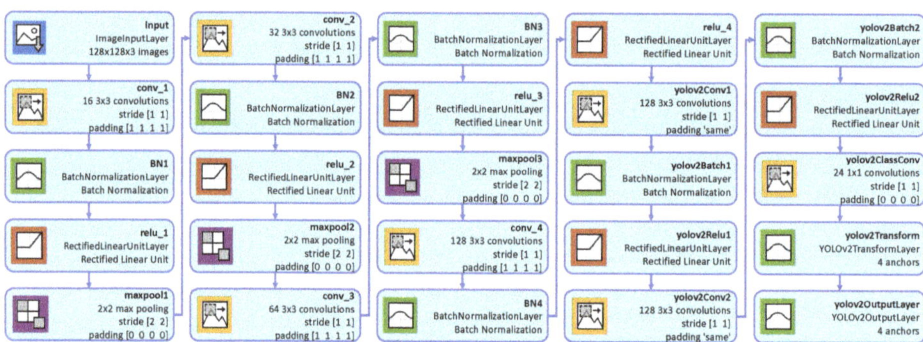

Figure 3. The architecture of YOLO V2 network.

For training, we used the stochastic gradient descent with momentum (SGDM) optimizer of 0.9, Initial Learn Rate of 0.001, and Max Epochs of 30 [59]. We chose these values as they provided the best performance with low computational power in our experiments. To improve the network accuracy, we used training image augmentation by randomly flipping the images during the training phase. By using this image augmentation, we increased the variations in the training images without actually having to increase the number of labeled images. Note that, to observe unbiased evaluation, we did not perform any augmentation to test images and preserved it as unmodified. An example detection result is shown in Figure 4.

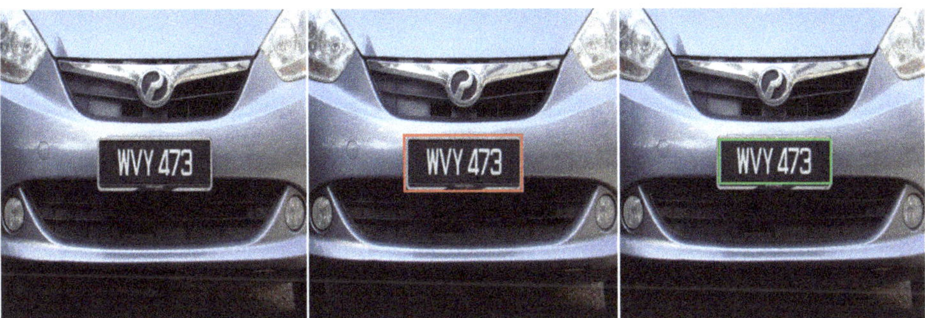

Figure 4. Detection result: from left to right, input image, ground truth image (red bounding box), and output image with detected license plate ROI (green bounding box).

3.3. Alphanumeric Character Segmentation

In the next stage, the alphanumeric characters that made up the license plate were extracted from the region resulting from the previous step. To this end, we first applied a gray level histogram equalization algorithm to increase the contrast level of the license plate region. Next, we converted the resulting image into a binary image using a global threshold of 0.5 (in the range [0, 1]). Then, the image was smoothed using a 3×3 median filter and salt-and-paper noise was removed from it. Any object that remained in the binary image after these operations was considered to represent a character. We segmented these characters using CCA. Each segmented character was resized to 56×56 pixels. The character segmentation process is illustrated in Figure 5 and Algorithm 1.

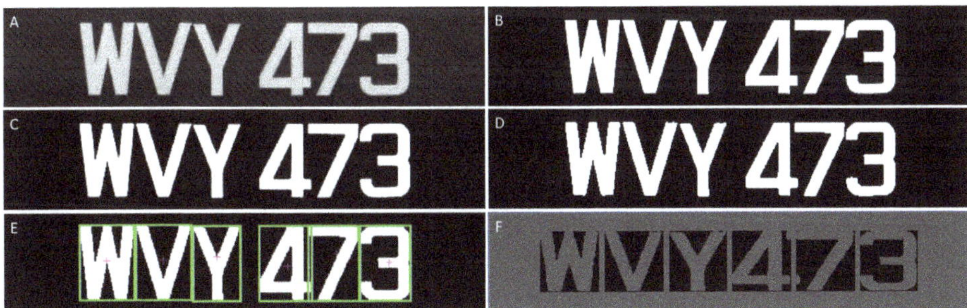

Figure 5. Alphanumeric character segmentation: (**A**) before gray level histogram equalization; (**B**) after gray level histogram equalization; (**C**) median filtered and noise removed image; (**D**) binary image; (**E**) character region segmentation; and (**F**) extracted and resized characters.

Algorithm 1: Alphanumeric character segmentation algorithm.

   Result: AlphanumericCharacter (AC)
   Input: LicensePlate (P);
   HEI = histogramEqualizationImage(CI);
   BI = binaryImage(HEI);
   NRI = noiseRemovedImage(BI);
   if *hasObject* **then**
       CO = countObject(NRI);
       BB = defineBoundingBoxUsingConnectedComponentAnalysis(CO);
       CC = cropEachCharacter(CO,BB);
       RC = Resize(CC,[53×53]);
   end

3.4. Feature Extraction

To be able to identify the characters of a license plate, we needed to first extract features from them that define their characteristics. For this purpose, we used HOG features [17] as it has successfully been used in many applications. To calculate the HoG features, the image was subdivided into smaller neighborhood regions (or "cells") [60]. Then, for each cell, at each pixel, the kernels $[-1, 0, +1]$ and $[-1, 0, +1]^T$ were applied to get the horizontal (G_x) and vertical (G_y) edge values, respectively. The magnitude and orientation of the gradient were calculated as $M(x, y) = \sqrt{(G_x^2 + G_y^2)}$ and $\theta(x, y) = tan^{-1}\left(\frac{G_y}{G_x}\right)$, respectively. Histograms of the unsigned angle (0°to 180°) weighted on the magnitude of each cell were then generated. Cells were combined into blocks and block normalization was performed on the concatenated histograms to account for variations in illumination. The length of the resulting feature vector depends on factors such as image size, cell size, and bin width of the histograms. For the proposed method, we used a cell size of 4×4 and each histogram was comprised of 9 bins, in order to achieve a balance between accuracy and efficiency. The resulting feature vector was of size 1×6084. Figure 6 gives a visualization of HOG features for different cell sizes.

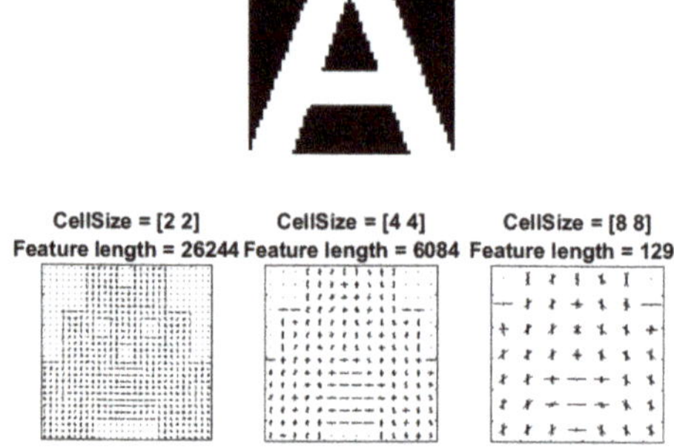

Figure 6. HOG feature vector visualization with different cell sizes.

3.5. Artificial Neural Network (ANN) Architecture

Once the features were extracted from the segmented characters, we trained an artificial neural network to identify them. As each character was encoded using a 1×6084 feature vector, the ANN had 6084 input neurons. The hidden layer comprised 40 neurons and the output layer had 36, equal to the number of different alphanumeric characters under consideration. Figure 7 shows the proposed recognition process along with the architecture of the ANN.

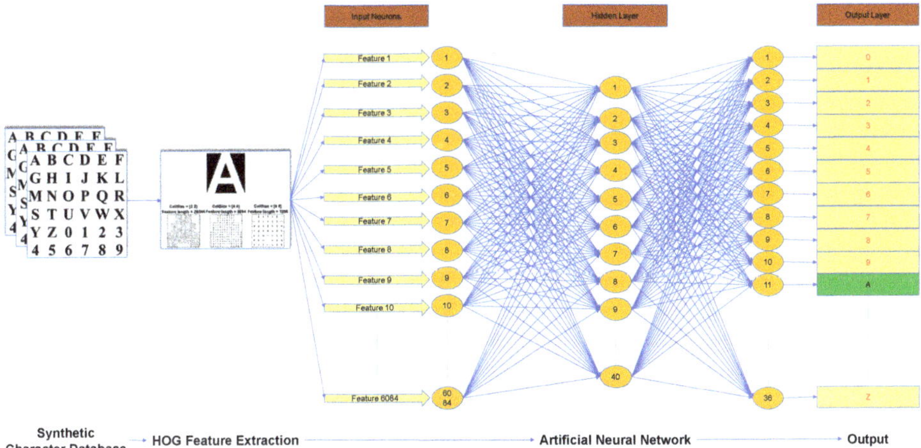

Figure 7. Proposed recognition process along with the artificial neural network architecture.

4. Experimental Results

Initially, we created a synthetic character image database to train and test the classification method. In addition, we created a database of real images, using the image acquisition method discussed above, to test the proposed method. We justified the selection of HOG as the feature extraction method and ANN as the classifier by comparing its performance with other feature extraction and classification methods. We also compared the performance of our method to similar existing methods. To conduct these experiments, an Intel®Core™i5 computer with 8 Gigabytes of RAM, running Windows Professional 64-bit operating system was used. MATLAB® framework (MathWorks, Natick, MA, USA) was used for the implementation of the proposed method as well as the experiments.

4.1. Generation of a Synthetic Data for Training

Using synthetic images is convenient as it enables the creation of a variety of training samples without having to manually collect them. The database was created by generating characters using random fonts and sizes (12—72 in steps of 2). In addition, one of four styles (normal, bold, italic, or bold with italic) was also used in the font generation process. To align with the type of characters available in license plates, we included the 10 numerals (0–9) as well as the characters of the English alphabet (A–Z). We also rotated the characters in the database by a randomly chosen angle between $\pm 0°$ and $\pm 15°$. Each class contained 200 samples consisting of an equal number of rotated and unrotated images. In total, $36 \times 200 = 7200$ images were used to train the ANN. Some characters from the synthetic training database are shown in Figure 8.

Figure 8. Synthetic samples for training purposes.

4.2. Performance on Synthetic Data

The ANN, discussed in Section 3, was trained on 70% of the synthetic database (5040 randomly selected samples). The remaining images were split evenly for validation and testing (1080 samples each). To determine the ideal number of hidden neurons, ANNs with different numbers of neurons (10, 20, and 40) were trained on these data five times each (see Table 2). The network with the best performance (with 40 hidden neurons) was selected as our trained network. We further increased the number of neurons to 60 and recorded comparably high processing time without reducing the error (%). The overall accuracy of the classification was 99.90% and the only misclassifications were between the classes **0** and **O**.

Table 2. Training performance with respect to hidden neuron size. The best performance is highlighted in bold.

Hidden Neurons	Repetition Number	Iterations	Time	Performance	Gradient	Error (%)
	1	160	0:00:45	1.47×10^{-3}	9.03×10^{-4}	1.50×10^{-0}
	2	179	0:00:51	8.10×10^{-4}	9.09×10^{-3}	6.11×10^{-1}
10	3	273	0:01:17	6.00×10^{-4}	9.62×10^{-4}	8.06×10^{-1}
	4	189	0:00:53	2.02×10^{-3}	1.34×10^{-3}	1.86×10^{-0}
	5	214	0:01:00	1.34×10^{-3}	6.66×10^{-4}	1.19×10^{-0}
	1	167	0:01:07	2.00×10^{-4}	1.04×10^{-3}	2.50×10^{-1}
	2	186	0:01:18	4.86×10^{-6}	1.97×10^{-4}	2.50×10^{-1}
20	3	140	0:00:59	1.51×10^{-4}	6.95×10^{-4}	3.33×10^{-1}
	4	165	0:01:10	6.12×10^{-5}	2.72×10^{-4}	2.22×10^{-1}
	5	137	0:00:58	1.04×10^{-4}	5.59×10^{-4}	3.61×10^{-1}
	1	124	0:01:26	3.65×10^{-5}	2.35×10^{-4}	2.78×10^{-1}
	2	124	0:01:25	2.78×10^{-5}	2.12×10^{-4}	1.67×10^{-1}
40	3	**151**	**0:01:44**	$\mathbf{1.29 \times 10^{-5}}$	$\mathbf{7.51 \times 10^{-4}}$	$\mathbf{1.11 \times 10^{-1}}$
	4	142	0:02:03	7.50×10^{-6}	3.67×10^{-4}	1.67×10^{-1}
	5	105	0:01:18	1.22×10^{-4}	1.81×10^{-3}	3.33×10^{-1}

4.3. Performance on Real Data

To test the performance on real-data, we acquired images at several locations in Kuala Lumpur, Malaysia using the process discussed in Section 3.1. In total, 100 vehicle license plates were used in this experiment, where each plate contained 5–8 alphanumeric characters. Then, 671 characters were extracted from the license plate images using the process discussed above. Classes **I**, **O**, and **Z** were not used here as they are not present in the license plates, as discussed above. The number of characters in each class is shown in Table 3. An accuracy level of 99.70% was achieved. The misclassifications in this experiment occurred among the classes **S**, **5**, and **9**, likely due to similarities in the characters. Real-time classification results for a sample license plate image are shown in Figure 9.

Table 3. The number of characters extracted from license plate images in each class.

Characters	0	1	2	3	4	5	6	7	8	9	A
Quantity	29	37	36	40	40	41	39	51	35	33	13
Characters	B	C	D	E	F	G	H	J	K	L	M
Quantity	16	7	7	10	5	7	7	13	12	12	7
Characters	N	P	Q	R	S	T	U	V	W	X	Y
Quantity	10	15	9	12	9	13	3	15	73	6	9

Figure 9. Real-time experimental results for license plate character recognition.

4.4. Comparison of Different Feature Extraction and Classification Methods

We also compared the proposed method with combinations of feature extraction and classification methods with respect to accuracy and processing time. Bag of words (BoF) [61], scale-invariant feature transform (SIFT) [62], and HOG were the feature extraction methods used. The classifiers used were: stacked auto-encoders (SAE), k-nearest neighbors (KNN), support vector machines (SVM), and ANN [63]. Processing time was calculated as the average time taken for the procedure to complete, as discussed in Section 3 (license plate extraction, characters extraction, feature extraction, and classification). The same 100 images used in the previous experiment were used here. The results are shown in Table 4.

Table 4. Comparison of performance for different combinations of feature extraction and classification methods. The best performance per metric is highlighted in bold.

Method	Processing Time (s)					Accuracy (%)
	Plate Extraction	Character Extraction	Feature Extraction	Classification	Total	
BoF+SAE	0.15	0.25	0.21	0.015	0.625	95.73
BoF+KNN	0.15	0.25	0.21	0.024	0.634	88.25
BoF+SVM	0.15	0.25	0.21	0.020	0.630	89.78
BoF+ANN	0.15	0.25	0.21	0.021	0.631	98.33
SIFT+SAE	0.15	0.25	0.28	0.019	0.699	93.75
SIFT+KNN	0.15	0.25	0.28	0.030	0.710	87.38
SIFT+SVM	0.15	0.25	0.28	0.027	0.707	88.94
SIFT+ANN	0.15	0.25	0.28	0.026	0.706	96.18
HOG+SAE	0.15	0.25	0.27	0.018	0.688	94.30
HOG+KNN	0.15	0.25	0.27	0.028	0.698	97.60
HOG+SVM	0.15	0.25	0.27	0.025	0.695	98.90
Proposed (HOG+ANN)	**0.15**	**0.25**	**0.27**	**0.010**	**0.690**	**99.70**

4.5. Performance Comparison with Other Similar Methods

Table 5 compares the performance of the proposed algorithm with other similar AVLPR methods in the literature. We reimplemented these methods on our system and trained and tested them on the same datasets to ensure an unbiased comparison. The training and testing was performed on our synthetic and real image databases, respectively. Note that the processing time only reports the time taken for the feature extraction and classification stages of the process. Accuracy denotes the classification accuracy. Since the training dataset was balanced, we did not consider bias performance metrics such as sensitivity and precision. As can be seen in Table 5, the proposed method outperformed the other compared methods.

Table 5. The classification performance for the proposed method compared to similar existing methods. The best performance per metric is highlighted in bold.

Method	Feature Extraction Method	Classifier	Total Time (s)	Accuracy (%)
Jin et al. [3]	Hand-Crafted	Fuzzy	0.432	92.00
Arafat et al. [9]	OCR	OCR	0.681	97.86
Samma et al. [12]	Haar-like	FSVM	0.649	98.36
Tabrizi et al. [13]	KNN+SVM	KNN+SVM	0.721	97.03
Niu et al. [28]	HOG	SVM	0.645	96.60
Li et al. [45]	CNN	CNN	0.825	99.20
Thakur et al. [37]	GA	ANN	0.532	97.00
Cheng et al. [38]	SCDCS-LS	RWNN	0.659	99.54
Lee et al. [47]	AlexNet	AlexNet	0.983	99.58
Proposed	**HOG**	**ANN**	**0.280**	**99.70**

4.6. Comparison of Methods with Respect to the Medialab Database

We compared the performance of our method to the methods discussed above, on a publicly available database (Medialab LPR (License Plate Recognition) database [64]) to investigate the transferability of results. This database contains still images and video sequences captured at various times of the day, under different weather conditions. We included all still images from this database except for ones that contained more than one vehicle. As our image capture system was specifically designed to capture only one vehicle at a time in order to simplify the subsequent image processing steps, images with multiple vehicles are beyond our scope.

The methods were compared with respect to the different stages of a typical AVLPR system (detection, character segmentation, and classification). Table 6 shows the comparison results (detection, character segmentation, and classification accuracy show the percentages of vehicle number plate regions detected, characters accurately extracted, and accurate classifications, respectively). As our method of pre-defined rectangular region selection (shown in Figure 2) was specifically designed for our image acquisition system, we did not consider this step in the comparison as different image acquisition systems were used in obtaining images for this database. Instead, the full image was used for detecting the license plate.

Table 6. Performance comparison on the Medialab LPR database. The stages that were not addressed in the original papers are denoted by "—". The best performance per metric is highlighted in bold.

Method	Accuracy (%)		
	Detection	Segmentation	Classification
Jin et al. [3]	95.73	98.87	91.25
Arafat et al. [9]	98.30	99.30	96.57
Samma et al. [12]	96.25	—	98.05
Tabrizi et al. [13]	96.98	96.85	96.54
Niu et al. [28]	98.45	—	96.38
Li et al. [45]	—	—	98.52
Thakur et al. [37]	97.85	98.37	97.35
Cheng et al. [38]	—	—	99.38
Lee et al. [47]	—	—	97.38
Proposed	**99.30**	**99.45**	**99.50**

Note that some of the methods in Table 6 do not address all three stages of the process. This is due to the fact that some papers only performed the latter parts of the process (for example, using pre-segmented characters for classification) and others did not clearly mention the methods they used. All methods were trained on our synthetic database as discussed above (no retraining was performed on the Medialab LPR database).

As can be seen in Table 6, the proposed method outperformed the other methods compared. However, the overall classification performance for all methods was slightly lower than that on our database (Table 5). We hypothesize that it could be due to factors such as differences in resolution and capture conditions (for example, weather, and time of day) of the images in the two databases.

4.7. Comparison of Methods with Respect to the UFPR-ALPR Database

We further compared the performance of our method with other existing methods (discussed above) on another publicly available database (UFPR-ALPR), which includes images that are more challenging [56]. This database contains multiple images of 150 vehicles (including motorcycles and cars) from real-world scenarios. The images were collected from different devices such as mobile cameras (iPhone® 7 Plus and Huawei® P9 Lite) and GoPro® HERO®4 Silver.

Since we used a digital camera to capture images in our method, we only considered those images in the UFPR-ALPR database that were captured by a similar device (GoPro® HERO®4 Silver). The database is split into three sets: 40%, 40%, and 20% for training, testing, and validation, respectively. We only considered the testing portion of the database to perform this comparison. In addition, we did not consider motorcycle images here, as our method was developed for cars. First, we converted the original image from 1920×1080 to 1498×946 pixels (to keep it consistent with the images of our database). Then, we performed the detection and recognition processes on the resized images.

As can be seen in Table 7, the proposed method outperformed the other methods with respect to accuracy of detection, segmentation, and classification. However, the overall classification performance for all methods were lower than on ours and Medialab LPR databases (Tables 5 and 6). We hypothesize that this could be due to factors such as the uncontrolled capture conditions (i.e, speed of vehicle, weather, and time of day) of the images in the three databases.

Table 7. Performance comparison on the UFPR-ALPR database. The stages that were not addressed in the original papers are denoted by "—". The best performance per metric is highlighted in bold.

Method	Accuracy (%)		
	Detection	Segmentation	Classification
Jin et al. [3]	85.48	91.75	85.35
Arafat et al. [9]	85.45	93.45	90.37
Samma et al. [12]	80.35	—	91.70
Tabrizi et al. [13]	84.45	90.50	92.86
Niu et al. [28]	85.80	—	89.32
Li et al. [45]	—	—	92.71
Thakur et al. [37]	82.35	91.22	90.85
Cheng et al. [38]	—	—	92.50
Lee et al. [47]	—	—	92.75
Proposed	**98.45**	**93.85**	**95.80**

5. Conclusions

In this paper, we propose a methodology for automatic vehicle license plate detection and recognition. This process consists of the following steps: image acquisition, license plate extraction, character extraction, and recognition. We demonstrated through experiments on synthetic and real license plate data that the proposed system is not only highly accurate but is also efficient. We also compared this method to similar existing methods and showed that it achieved a balance between accuracy and efficiency, and as such is suitable for real-time detection of license plates. A limitation of this work is that our method was only tested on a database of Malaysian license plate images captured by the researchers and the publicly available Medialab LPR and UFPR-ALPR databases. In the future, we will explore how it performs on other license plate databases. We will also investigate the use of multi-stage deep learning architectures (detection and recognition) in this domain. Furthermore, since we proposed barrier access control for the controlled environment, the current image acquisition system was set up so that only one vehicle was visible in the field of view. As a result, the image only captured one vehicle per frame, simplifying the process of license plate detection. In future work, we will extend our methodology to identify license plates using more complex images containing multiple vehicles. In addition, we will increase the size of the training database in an effort to minimise misclassification of similar classes.

Author Contributions: Conceptualization, K.T.I.; Data curation, K.T.I. and R.G.R.; Formal analysis, K.T.I., R.G.R., S.M.S.I. and S.W.; Funding acquisition, R.G.R., S.M.S.I., S.W. and S.O.; Investigation, K.T.I., S.M.S.I., S.W., M.S.H. and T.R.; Methodology, K.T.I., S.M.S.I. and S.W.; Project administration, R.G.R., S.M.S.I., S.W. and S.O.; Resources, R.G.R., S.M.S.I., S.W. and S.O.; Software, K.T.I., S.M.S.I. and S.O.; Supervision, R.G.R., S.M.S.I., S.W. and S.O.; Validation, K.T.I., R.G.R., S.M.S.I., S.W., M.S.H. and T.R.; Visualization, K.T.I., S.M.S.I., M.S.H. and T.R.; and Writing—original draft, K.T.I.; Writing—review and editing, K.T.I., R.G.R., S.M.S.I., S.W., M.S.H., T.R. and S.O. All authors have read and agreed to the published version of the manuscript.

Funding: This Research was funded by The University of Malaya Grant Number-BKS082-2017, IIRG012C-2019 and UMRG RP059C 17SBS, Melbourne Research Scholarship (MRS), and Edith Cowan University School of Science Collaborative Research Grant Scheme 2019.

Acknowledgments: The authors would like to thank Md Yeasir Arafat of the Department of Electrical Engineering, Faculty of Engineering, University of Malaya, Kuala Lumpur, Malaysia, for his input on image pre-processing techniques.

Conflicts of Interest: The authors declare no conflict of interest.

Data Availability: The datasets used in this study are available from the corresponding author upon request.

Abbreviations

The following abbreviations are used in this manuscript:

ANN	artificial Neural Network
AVLPR	automatic vehicle license plate recognition
BoF	bag of words
CCA	connected component analysis
CNN	convolution neural network
FSVM	fuzzy support vector machines
GA	genetic algorithm
HOG	histogram of oriented gradients
KNN	k-nearest neighbors
NN	neural network
OCR	optical character recognition
ReLU	rectified linear unit
RGB	red, green, and blue
ROC	receiver operating characteristic
ROI	region of interest
SAE	stacked auto-encoders
SIFT	scale-invariant feature transform
SGDM	stochastic gradient descent with momentum
SVM	support Vector Machine
YOLO	you only look once

References

1. Saha, S.; Basu, S.; Nasipuri, M.; Basu, D.K. Localization of License Plates from Surveillance Camera Images: A Color Feature Based ANN Approach. *Int. J. Comput. Appl.* **2010**, *1*, 27–31. [CrossRef]
2. Hongliang, B.; Changping, L. A hybrid license plate extraction method based on edge statistics and morphology. In Proceedings of the 17th International Conference on Pattern Recognition, Cambridge, UK, 23–26 August 2004. [CrossRef]
3. Jin, L.; Xian, H.; Bie, J.; Sun, Y.; Hou, H.; Niu, Q. License Plate Recognition Algorithm for Passenger Cars in Chinese Residential Areas. *Sensors* **2012**, *12*, 8355–8370. [CrossRef] [PubMed]
4. Shi, X.; Zhao, W.; Shen, Y. Automatic License Plate Recognition System Based on Color Image Processing. In Proceedings of the Computational Science and Its Applications, Singapore, 9–12 May 2005; Gervasi, O., Gavrilova, M.L., Kumar, V., Laganá, A., Lee, H.P., Mun, Y., Taniar, D., Tan, C.J.K., Eds.; Springer: Berlin/Heidelberg, Germany, 2005; pp. 1159–1168.
5. Rizvi, S.; Patti, D.; Björklund, T.; Cabodi, G.; Francini, G. Deep Classifiers-Based License Plate Detection, Localization and Recognition on GPU-Powered Mobile Platform. *Future Internet* **2017**, *9*, 66. [CrossRef]
6. Rafique, M.A.; Pedrycz, W.; Jeon, M. Vehicle license plate detection using region-based convolutional neural networks. *Soft Comput.* **2018**, *22*, 6429–6440. [CrossRef]
7. Salau, A.O.; Yesufu, T.K.; Ogundare, B.S. Vehicle plate number localization using a modified GrabCut algorithm. *J. King Saud Univ. Comput. Inf. Sci.* **2019**. [CrossRef]
8. Kakani, B.V.; Gandhi, D.; Jani, S. Improved OCR based automatic vehicle number plate recognition using features trained neural network. In Proceedings of the 2017 8th International Conference on Computing, Communication and Networking Technologies (ICCCNT), Delhi, India, 3–5 July 2017. [CrossRef]
9. Arafat, M.Y.; Khairuddin, A.S.M.; Paramesran, R. A Vehicular License Plate Recognition Framework For Skewed Images. *KSII Trans. Internet Inf. Syst.* **2018**, *12*. [CrossRef]
10. Yogheedha, K.; Nasir, A.; Jaafar, H.; Mamduh, S. Automatic Vehicle License Plate Recognition System Based on Image Processing and Template Matching Approach. In Proceedings of the 2018 International Conference on Computational Approach in Smart Systems Design and Applications (ICASSDA), Kuching, Malaysia, 15–17 August 2018. [CrossRef]
11. Ansari, N.N.; Singh, A.K.; Student, M.T. License Number Plate Recognition using Template Matching. *Int. J. Comput. Trends Technol.* **2016**, *35*, 175–178. [CrossRef]

12. Samma, H.; Lim, C.P.; Saleh, J.M.; Suandi, S.A. A memetic-based fuzzy support vector machine model and its application to license plate recognition. *Memetic Comput.* **2016**, *8*, 235–251. [CrossRef]
13. Tabrizi, S.S.; Cavus, N. A Hybrid KNN-SVM Model for Iranian License Plate Recognition. *Procedia Comput. Sci.* **2016**, *102*, 588–594. [CrossRef]
14. Gao, Y.; Lee, H. Local Tiled Deep Networks for Recognition of Vehicle Make and Model. *Sensors* **2016**, *16*, 226. [CrossRef]
15. Leeds City Council. Automatic Number Plate Recognition (ANPR) Project. 2018. Available online: http://data.gov.uk/dataset/f90db76e-e72f-4ab6-9927-765101b7d997 (accessed on 6 February 2019).
16. Dynamics, S. Australia's Leading ANPR—Automatic Number Plate Recognition Provider. 2015. Available online: http://www.sensordynamics.com.au/ (accessed on 6 February 2019).
17. Dalal, N.; Triggs, B. Histograms of Oriented Gradients for Human Detection. In Proceedings of the 2005 IEEE Computer Society Conference on Computer Vision and Pattern Recognition (CVPR'05), San Diego, CA, USA, 20–25 June 2005. [CrossRef]
18. Silva, S.M.; Jung, C.R. License Plate Detection and Recognition in Unconstrained Scenarios. Available online: http://www.inf.ufrgs.br/~smsilva/alpr-unconstrained/ (accessed on 7 February 2019).
19. Resende Gonçalves, G.; Alves Diniz, M.; Laroca, R.; Menotti, D.; Robson Schwartz, W. Real-Time Automatic License Plate Recognition through Deep Multi-Task Networks. In Proceedings of the 2018 31st SIBGRAPI Conference on Graphics, Patterns and Images (SIBGRAPI), Parana, Brazil, 29 October–1 November 2018; pp. 110–117. [CrossRef]
20. Ullah, F.; Anwar, H.; Shahzadi, I.; Ur Rehman, A.; Mehmood, S.; Niaz, S.; Mahmood Awan, K.; Khan, A.; Kwak, D. Barrier Access Control Using Sensors Platform and Vehicle License Plate Characters Recognition. *Sensors* **2019**, *19*, 3015. [CrossRef]
21. Suryanarayana, P.; Mitra, S.; Banerjee, A.; Roy, A. A Morphology Based Approach for Car License Plate Extraction. In Proceedings of the 2005 Annual IEEE India Conference-Indicon, Chennai, India, 11–13 December 2005. [CrossRef]
22. Mahini, H.; Kasaei, S.; Dorri, F.; Dorri, F. An Efficient Features—Based License Plate Localization Method. In Proceedings of the 18th International Conference on Pattern Recognition (ICPR'06), Hong Kong, China, 20–24 August 2006. [CrossRef]
23. Zheng, D.; Zhao, Y.; Wang, J. An efficient method of license plate location. *Pattern Recognit. Lett.* **2005**, *26*, 2431–2438. [CrossRef]
24. Luo, Y.; Li, Y.; Huang, S.; Han, F. Multiple Chinese Vehicle License Plate Localization in Complex Scenes. In Proceedings of the 2018 IEEE 3rd International Conference on Image, Vision and Computing (ICIVC), Chongqing, China, 27–29 June 2018; pp. 745–749. [CrossRef]
25. Liu, W.; Anguelov, D.; Erhan, D.; Szegedy, C.; Reed, S.; Fu, C.Y.; Berg, A.C. SSD: Single Shot MultiBox Detector. In *Proceedings of the Computer Vision—ECCV 2016, Amsterdam, The Netherlands, 11–14 October 2016*; Leibe, B., Matas, J., Sebe, N., Welling, M., Eds.; Springer International Publishing: Cham, Switzerland, 2016; pp. 21–37.
26. Hsieh, C.T.; Juan, Y.S.; Hung, K.M. Multiple License Plate Detection for Complex Background. In Proceedings of the 19th International Conference on Advanced Information Networking and Applications (AINA'05) Volume 1 (AINA papers), Taipei, Taiwan, 28–30 March 2005. [CrossRef]
27. Haar, A. Zur Theorie der orthogonalen Funktionensysteme. *Math. Ann.* **1910**, *69*, 331–371. [CrossRef]
28. Niu, B.; Huang, L.; Hu, J. Hybrid Method for License Plate Detection from Natural Scene Images. In Proceedings of the First International Conference on Information Science and Electronic Technology, Wuhan, China, 21–22 March 2015; Atlantis Press: Paris, France, 2015. [CrossRef]
29. Arafat, M.Y.; Khairuddin, A.S.M.; Khairuddin, U.; Paramesran, R. Systematic review on vehicular licence plate recognition framework in intelligent transport systems. *IET Intell. Transp. Syst.* **2019**. [CrossRef]
30. Chai, D.; Zuo, Y. Extraction, Segmentation and Recognition of Vehicle's License Plate Numbers. In *Advances in Information and Communication Networks*; Arai, K., Kapoor, S., Bhatia, R., Eds.; Springer International Publishing: Cham, Switzerland, 2019; pp. 724–732.
31. Dhar, P.; Guha, S.; Biswas, T.; Abedin, M.Z. A System Design for License Plate Recognition by Using Edge Detection and Convolution Neural Network. In Proceedings of the 2018 International Conference on Computer, Communication, Chemical, Material and Electronic Engineering (IC4ME2), Rajshahi, Bangladesh, 8–9 February 2018; pp. 1–4. [CrossRef]

32. De Gaetano Ariel, O.; Martín, D.F.; Ariel, A. ALPR character segmentation algorithm. In Proceedings of the 2018 IEEE 9th Latin American Symposium on Circuits Systems (LASCAS), Puerto Vallarta, Mexico, 25–28 February 2018; pp. 1–4. [CrossRef]

33. Kahraman, F.; Kurt, B.; Gökmen, M. License Plate Character Segmentation Based on the Gabor Transform and Vector Quantization. In *Proceedings of the Computer and Information Sciences—ISCIS 2003, Antalya, Turkey, 3–5 November 2003*; Yazıcı, A.; Şener, C., Eds.; Springer: Berlin/Heidelberg, Germany, 2003; pp. 381–388.

34. Wu, Q.; Zhang, H.; Jia, W.; He, X.; Yang, J.; Hintz, T. Car Plate Detection Using Cascaded Tree-Style Learner Based on Hybrid Object Features. In Proceedings of the 2006 IEEE International Conference on Video and Signal Based Surveillance, Sydney, Australia, 22–24 November 2006. [CrossRef]

35. Zhang, H.; Jia, W.; He, X.; Wu, Q. Learning-Based License Plate Detection Using Global and Local Features. In Proceedings of the 18th International Conference on Pattern Recognition (ICPR'06), Hong Kong, China, 20–24 August 2006. [CrossRef]

36. Prewitt, J.M. Object enhancement and extraction. *Pict. Process. Psychopictorics* **1970**, *10*, 15–19.

37. Thakur, M.; Raj, I.; P, G. The cooperative approach of genetic algorithm and neural network for the identification of vehicle License Plate number. In Proceedings of the 2015 International Conference on Innovations in Information, Embedded and Communication Systems (ICIIECS), Coimbatore, India, 19–20 March 2015. [CrossRef]

38. Cheng, R.; Bai, Y.; Hu, H.; Tan, X. Radial Wavelet Neural Network with a Novel Self-Creating Disk-Cell-Splitting Algorithm for License Plate Character Recognition. *Entropy* **2015**, *17*, 3857–3876. [CrossRef]

39. Brillantes, A.K.M.; Bandala, A.A.; Dadios, E.P.; Jose, J.A. Detection of Fonts and Characters with Hybrid Graphic-Text Plate Numbers. In Proceedings of the TENCON 2018—2018 IEEE Region 10 Conference, Jeju, South Korea, 28–31 October 2018; pp. 629–633. [CrossRef]

40. Mukherjee, R.; Pundir, A.; Mahato, D.; Bhandari, G.; Saxena, G.J. A robust algorithm for morphological, spatial image-filtering and character feature extraction and mapping employed for vehicle number plate recognition. In Proceedings of the 2017 International Conference on Wireless Communications, Signal Processing and Networking (WiSPNET), Chennai, India, 22–24 March 2017; pp. 864–869. [CrossRef]

41. Canny, J. A Computational Approach to Edge Detection. *IEEE Trans. Pattern Anal. Mach. Intell.* **1986**, *PAMI-8*, 679–698. [CrossRef]

42. Sobel, I. An isotropic 3×3 image gradient operater. In *Machine Vision for Three-Dimensional Scenes*; ResearchGate: Berlin, Germany, 1990; pp. 376–379.

43. Olmí, H.; Urrea, C.; Jamett, M. Numeric Character Recognition System for Chilean License Plates in semicontrolled scenarios. *Int. J. Comput. Intell. Syst.* **2017**, *10*, 405. [CrossRef]

44. Han, J.; Yao, J.; Zhao, J.; Tu, J.; Liu, Y. Multi-Oriented and Scale-Invariant License Plate Detection Based on Convolutional Neural Networks. *Sensors* **2019**, *19*, 1175. [CrossRef]

45. Li, S.; Li, Y. A Recognition Algorithm for Similar Characters on License Plates Based on Improved CNN. In Proceedings of the 2015 11th International Conference on Computational Intelligence and Security (CIS), Shenzhen, China, 19–20 December 2015; doi:10.1109/cis.2015.9. [CrossRef]

46. Krizhevsky, A.; Sutskever, I.; Hinton, G.E. ImageNet Classification with Deep Convolutional Neural Networks. In Proceedings of the 25th International Conference on Neural Information Processing Systems, Lake Tahoe, NV, USA, 3–6 December 2012; Curran Associates Inc.: Dutchess County, NY, USA, 2012; Volume 1, pp. 1097–1105.

47. Lee, S.; Son, K.; Kim, H.; Park, J. Car plate recognition based on CNN using embedded system with GPU. In Proceedings of the 2017 10th International Conference on Human System Interactions (HSI), Ulsan, South Korea, 17–19 July 2017. [CrossRef]

48. Redmon, J.; Divvala, S.; Girshick, R.; Farhadi, A. You Only Look Once: Unified, Real-Time Object Detection. In Proceedings of the IEEE Conference on Computer Vision and Pattern Recognition (CVPR), Las Vegas, NV, USA, 27–30 June 2016. [CrossRef]

49. Kessentini, Y.; Besbes, M.D.; Ammar, S.; Chabbouh, A. A two-stage deep neural network for multi-norm license plate detection and recognition. *Expert Syst. Appl.* **2019**, *136*, 159–170. [CrossRef]

50. Redmon, J.; Farhadi, A. YOLO9000: Better, Faster, Stronger. In Proceedings of the IEEE Conference on Computer Vision and Pattern Recognition (CVPR), Honolulu, HI, USA, 21–26 July 2017. [CrossRef]

51. Chen, R.C. Automatic License Plate Recognition via sliding-window darknet-YOLO deep learning. *Image Vis. Comput.* **2019**, *87*, 47 – 56. [CrossRef]
52. Yonetsu, S.; Iwamoto, Y.; Chen, Y.W. Two-Stage YOLOv2 for Accurate License-Plate Detection in Complex Scenes. In Proceedings of the 2019 IEEE International Conference on Consumer Electronics (ICCE), Las Vegas, NV, USA, USA, 11–13 January 2019; pp. 1–4. [CrossRef]
53. Abdullah, S.; Mahedi Hasan, M.; Muhammad Saiful Islam, S. YOLO-Based Three-Stage Network for Bangla License Plate Recognition in Dhaka Metropolitan City. In Proceedings of the 2018 International Conference on Bangla Speech and Language Processing (ICBSLP), Sylhet, Bangladesh, 21–22 September 2018; pp. 1–6. [CrossRef]
54. Redmon, J.; Farhadi, A. Yolov3: An incremental improvement. *arXiv* **2018**, arXiv:1804.02767.
55. He, K.; Zhang, X.; Ren, S.; Sun, J. Deep Residual Learning for Image Recognition. In Proceedings of the 2016 IEEE Conference on Computer Vision and Pattern Recognition (CVPR), Las Vegas, NV, USA, 27–30 June 2016; doi:10.1109/cvpr.2016.90. [CrossRef]
56. Laroca, R.; Severo, E.; Zanlorensi, L.A.; Oliveira, L.S.; Gonçalves, G.R.; Schwartz, W.R.; Menotti, D. A Robust Real-Time Automatic License Plate Recognition Based on the YOLO Detector. In Proceedings of the 2018 International Joint Conference on Neural Networks (IJCNN), Rio de Janeiro, Brazil, 8–13 July 2018; pp. 1–10. [CrossRef]
57. Silva, S.M.; Jung, C.R. Real-Time Brazilian License Plate Detection and Recognition Using Deep Convolutional Neural Networks. In Proceedings of the 2017 30th SIBGRAPI Conference on Graphics, Patterns and Images (SIBGRAPI), Niteroi, Brazil, 17–20 October 2017; pp. 55–62. [CrossRef]
58. Nair, V.; Hinton, G.E. Rectified Linear Units Improve Restricted Boltzmann Machines. In Proceedings of the 27th International Conference on International Conference on Machine Learning, Haifa, Israel, 21–24 June 2010; Omnipress: Madison, WI, USA, 2010; pp. 807–814.
59. Robbins, H.; Monro, S. A Stochastic Approximation Method. *Ann. Math. Stat.* **1951**, *22*, 400–407. [CrossRef]
60. Takahashi, K.; Takahashi, S.; Cui, Y.; Hashimoto, M. Remarks on Computational Facial Expression Recognition from HOG Features Using Quaternion Multi-layer Neural Network. In *Engineering Applications of Neural Networks*; Mladenov, V., Jayne, C., Iliadis, L., Eds.; Springer International Publishing: Cham, Switzerland, 2014; pp. 15–24.
61. Sivic, J.; Zisserman, A. Efficient Visual Search of Videos Cast as Text Retrieval. *IEEE Trans. Pattern Anal. Mach. Intell.* **2009**, *31*, 591–606. [CrossRef]
62. Lowe, D. Object recognition from local scale-invariant features. In Proceedings of the Seventh IEEE International Conference on Computer Vision, Kerkyra, Greece, 20–27 September 1999. [CrossRef]
63. Altman, N.S. An Introduction to Kernel and Nearest-Neighbor Nonparametric Regression. *Am. Stat.* **1992**, *46*, 175–185. [CrossRef]
64. Anagnostopoulos, I.; Psoroulas, I.; Loumos, V.; Kayafas, E.; Anagnostopoulos, C. Medialab LPR Database. Multimedia Technology Laboratory, National Technical University of Athens. Available online: http://www.medialab.ntua.gr/research/LPRdatabase.html (accessed on 7 November 2019).

Article

On the Application of Time Frequency Convolutional Neural Networks to Road Anomalies' Identification with Accelerometers and Gyroscopes

Gianmarco Baldini [1,*], Raimondo Giuliani [1] and Filip Geib [2]

[1] European Commission, Joint Research Centre, 21027 Ispra, Italy; raimondo.giuliani@ec.europa.eu
[2] Faculty of Electrical Engineering, Czech Technical University in Prague, 160 00 Prague, Czech Republic; filip9geib@gmail.com
* Correspondence: gianmarco.baldini@ec.europa.com; Tel.: +39-0332-78-6618

Received: 29 September 2020; Accepted: 4 November 2020; Published: 10 November 2020

Abstract: The detection and identification of road anomalies and obstacles in the road infrastructure has been investigated by the research community using different types of sensors. This paper evaluates the detection and identification of road anomalies/obstacles using the data collected from the Inertial Measurement Unit (IMU) installed in a vehicle and in particular from the data generated by the accelerometers' and gyroscopes' components. Inspired by the successes of the application of deep learning to various identification problems, this paper investigates the application of Convolutional Neural Network (CNN) to this specific problem. In particular, we propose a novel approach in this context where the time-frequency representation (i.e., spectrogram) is used as an input to the CNN rather than the original time domain data. This approach is evaluated on an experimental dataset collected using 12 different vehicles driving for more than 40 km of road. The results show that the proposed approach outperforms significantly and across different sampling rates both the application of CNN to the original time domain representation and the application of shallow machine learning algorithms. The approach achieves an identification accuracy of 97.2%. The results presented in this paper are based on an extensive optimization both of the CNN algorithm and the spectrogram implementation in terms of window size, type of window, and overlapping ratio. The accurate detection of road anomalies/obstacles could be useful to road infrastructure managers to monitor the quality of the road surface and to improve the accurate positioning of autonomous vehicles because road anomalies/obstacles could be used as landmarks.

Keywords: convolutional neural networks; deep learning; time-frequency; Inertial Measurement Unit (IMU); road anomalies

---

1. Introduction

Detection of road anomalies like potholes, road cracks, and road safety features/obstacles like speed bumps (also called speed humps) has been investigated by the research community in recent years using a variety of sensors including cameras, Light Detection And Ranging (LiDAR)s, and Inertial Measurement Units (IMU)s. One of the main applications for the detection of road anomalies is the monitoring of the road conditions, which can be used to repair the road surface once a road anomaly (e.g., pothole) is detected or to improve the comfort and safety of the vehicle in Advanced Driver Assistance Systems (ADAS). An additional aspect is to use the identification of road anomalies as landmarks to enhance maps and improve localization to a high degree of accuracy. The upcoming evolution of modern vehicles to autonomous vehicles could benefit from this information and use it for a variety of purposes. It could be used to improve the position of the autonomous vehicle. It could be used improve its travel plan for the comfort of the passengers (e.g., the autonomous vehicle could slow down before a known pothole). It could even be used to mitigate cybersecurity attacks on the positioning information (e.g., the position information would be correlated with the detected road anomaly to mitigate Global Navigation Satellite Systems (GNSS) spoofing attacks). The use of the position of the road anomaly to support localization algorithms was already mentioned in [1,2]. In addition, studies like [3] suggest that artificial position measurements obtained by detecting road anomalies associated with map locations will be increasingly used in the automotive sector and automated vehicles also in synergy with Simultaneous Localization And Mapping (SLAM) techniques by the robotics community.

Detection of road anomalies can be performed using different techniques including machine learning algorithms [4,5], and in recent times, Deep Learning (DL) algorithms have been adopted with significant success. In many cases, DL is used to detect and identify road anomalies on the basis of the images collected by the camera installed on the vehicle [6,7]. Processing of data images can be a time consuming and error prone task [5], and it is limited by the lighting (e.g., darkness) or environmental conditions (e.g., fog, rain) [8]. An alternative approach, which is not affected by these issues, is to use the data provided by the IMU, and DL was applied to such data in recent studies [9]. Then, the application of DL for road anomaly detection using data from accelerometers and gyroscopes is a recent research area, which is gaining significant momentum in the research community, and it is further explored in this paper.

This paper proposes the combination of time-frequency transform and CNN (called CNN-SP) for the detection and identification of road anomalies in the road infrastructure, which has not been proposed in the literature yet (to the knowledge of the authors) for this specific problem. As shown in the results provided in this paper, CNN-SP provides a superior classification performance to the direct application of CNN to the original time domain signal. This paper uses a relatively large set of vehicles (12) in the data collection phase to improve generalization of the results. The application of CNN-SP is evaluated using an experimental dataset collected by the authors with many hours of driving on a realistic road path with various types of road anomalies and obstacles. The approach is evaluated using data both from the accelerometer and the gyroscope. Finally, this study provides an extensive evaluation of the different sampling rates on the identification accuracy.

The structure of this paper is as follows: Section 2 provides a literature review on the detection of road anomalies using sensors installed on the vehicles with a particular focus on accelerometers and gyroscopes. Section 3 describes the materials and methods used for the analysis including the description of the adopted machine learning algorithms and the related evaluation metrics. Section 4 provides the results of the analysis including the optimization of the CNN-SP approach and a comparison among the results provided by the different machine learning algorithms. Finally, Section 5 provides the conclusions.

2. Literature Review

Detection of road anomalies and road surface conditions using the sensors installed on the vehicle has been investigated by the research community in recent years. Detection of road anomalies through cameras can provide high accuracy especially with the recent application of deep learning. Examples of the application of deep learning (and convolutional neural networks in particular) in combination with camera images to assess road surface condition were provided in [6,7]. The use of data images can be a time consuming and error prone task [5], and it is limited by the lighting (e.g., darkness) or environmental conditions (e.g., fog, rain) [8]. For this reason, this paper does not use an approach based on camera images, and other sensors are used.

The application of accelerometers, gyroscopes, and magnetometers to detect road anomalies like potholes and obstacles (e.g., speed humps) has been investigated by the research community in recent years [10]. This is also due to the decreasing costs of IMUs and their increasing use in the automotive sector either because they are inserted and used in the vehicle systems or because smartphones (which are equipped with IMUs) can be deployed and installed in vehicles. A very recent and detailed survey on the use of accelerometers and gyroscopes (as inertial sensing sources of information) was presented in [4] where an analysis focused on identifying methods that capture signals provided by inertial sensors such as accelerometers and gyroscopes to recognize transient or persistent events associated with the vehicle's movement was presented. The results of the survey show that a limited amount of the reviewed papers used time-frequency analysis, and no paper in the review used the combination of the time-frequency and convolutional neural networks, as proposed in this paper.

Detection of road anomalies like potholes using accelerometers was implemented by the authors in [11,12] where a high detection rate was achieved using the data collected by the accelerometers in the Z direction, but no deep learning approach was used. In a similar way, the authors in [13] used a generic smartphone application reading data from built-in accelerometers sensors to map and measure the locations of potholes and speed bumps, which can be used to evaluate the road conditions. The data were collected in a cloud system, where the analysis was performed. The authors in [5] proposed the Android application RoadSense, which automatically predicts the quality of the road based on a triaxial accelerometer and a gyroscope. The study used frequency domain features for the classification combined with different machine learning algorithms. As in previous papers, the focus was on pothole detection to monitor the smoothness of the road surface, but the identification of the road anomalies was not attempted.

Four recent works are very similar to the study presented in this paper. In [14], different road anomalies like bump, pothole, and normal conditions (flat road) were detected and identified using an improved Gaussian background model and the K Nearest Neighbor (KNN) algorithm applied to the accelerometers' recording. The described approach provides a high accuracy (96.03%) of recognition of the road surface pothole and a high accuracy of the road surface bump of 94.12%. In comparison to [14], this paper applies a more sophisticated deep learning approach in combination with a time-frequency transform, which is shown to provide a higher identification accuracy than the time domain representation (i.e., original accelerometer signal). In addition, gyroscope data were also used in addition to the accelerometer data. Finally, a larger set of 12 vehicles is used than the two vehicles used in [14], which improves the generalizations of the results.

The authors of [8] proposed a novel approach to identify the profile of a pothole, which is a more challenging task than the detection alone. The depth of the pothole was used as a profile metric, and the accelerometer data collected by a smartphone were used to conduct the analysis together with the GPS data. The authors of [8] used four different vehicles to collect the data with two samplings rate of 100 and 200 Hz. Twenty-three different types of potholes were considered on an overall set of 2760 segments. In comparison to [8], this paper uses a larger set of vehicles (12), a similar set of road anomalies/obstacles (19), and different sampling rates (50, 100, 150, 200 and 250 Hz). In addition, this paper proposes a deep learning approach, while the analysis in [8] was done with Cumulative Distribution Functions (CDFs). On the other hand, the authors of [8] also investigated different placements of the smartphone in the vehicle with the result that the placement on the dashboard provided the best identification accuracy. Based on the results of [8], this paper adopts the position of the smartphone on the dashboard as well.

The authors of [15] used a DL approach based on CNN (as in this paper) to perform the detection of road anomalies. The results provided in [15] confirmed the superior performance of the adoption of CNN in comparison to shallow machine learning algorithms, which prompted the authors of this paper to apply a CNN based approach as well. In comparison to the CNN approach used by the authors of [15], where the CNN was applied directly to the data collected from the accelerometers, this paper first transforms the accelerometer data in the spectral domain using the spectrogram defined in Section 3.4, then the spectral representation is fed to a CNN (called CNN-SP in the rest of this paper). The results presented in this paper show that CNN-SP outperforms the application of CNN directly on the source data (called CNN-1D in the rest of this paper) both for accelerometer and gyroscope data and across different sampling rates. The application of CNN-SP is evaluated using an experimental dataset collected by the authors through many hours of driving on a realistic road path with various types of road anomalies. The study results presented in this paper use the dataset from [16], where it was used for the different goal of automotive vehicle authentication. Additional details on the dataset are presented in Section 3.

In [17], the authors used both accelerometers and gyroscopes like the study done in this paper to detect speed bumps. The authors applied a number of features to the data recorded by the accelerometer and gyroscope mounted on the vehicle. Then, a genetic algorithm was used to find a logistic model that accurately detects road abnormalities. The results were quite encouraging as the approach was able to achieve an accuracy of 0.9714 in a blind evaluation.

In [9], the authors proposed a road anomaly detection approach based on the application of three different DL algorithms: Deep Feedforward Network (DFN), Convolutional Neural Network (CNN), and Recurrent Neural Network (RNN), and the results were compared. The results showed that the application of DL algorithms was very effective in detecting road anomalies, thus proving the approach also used in this paper. In comparison to this paper, the authors in [9] used a larger set of signals beyond the accelerometers and gyroscopes used in this paper since they also included the shock responses of the four absorbers and the rotation speed. From this point of view, reference [9] represents a significant progress in comparison to the literature. The authors of [9] also compared three different DL algorithms. On the other hand, this paper uses a set of 12 vehicles in comparison to the single vehicle used in [9]. This paper also evaluates the impact of different samples rates on the identification accuracy, while [9] used only the sampling rate of 100 Hz. Finally, this paper uses a time-frequency CNN approach rather than the direct application of DL to the signals in the time domain.

Next, we summarize the progress introduced by this paper in comparison to the references identified in the previous paragraphs. As shown in the results provided in this paper, the time-frequency CNNs provide a superior classification performance to the direct application of CNN to the original time domain signal, as proposed in the literature [9,15]. In comparison to the literature [8,9,13–15,17], which used only one or two vehicles (with the exception of [8], where four vehicles were used), this paper uses a relatively large set of vehicles (12) to improve the generalization of the results. Then, the focus of the paper is to investigate how accelerometer and gyroscope readings collected from different vehicles are used to detect road anomalies and obstacles. The impact of various speeds by different vehicles is also mitigated to support the application of time-frequency CNN. In comparison to the literature, this study uses the data both from the accelerometers and the gyroscopes, which was adopted only by a few studies [9,17], as the accelerometer data were generally used [15]. Finally, this study provides a more extensive, to the knowledge of the authors, evaluation of the different sampling rates on the identification accuracy.

3. Materials and Methods

3.1. Materials

As mentioned before, the dataset used for this experiment consisted of 12 different cars. The specifications of the vehicles are listed in Table 1. The same driver and co-driver were present in every car during the experimental data collection to minimize the potential bias of the driving behavior. Even if it is acknowledged that in practical operations, the co-driver may be different or may be absent, we wanted to limit the number of variables in the study. Future developments will investigate the presence of different passengers and their number (see Section 5). For the IMU, we used the microelectromechanical system based motion tracker supplied by Xsens (Enschede, The Netherlands) with Model Number $MTi - 100 - 2A8G4$. The technical specification of the Xsens sensor are reported in Table 2.

Table 1. Order and specifications(brand and model) of the cars used in the data collection.

Car	Manufacturer	Model	Generation	Version
1	Fiat Automobiles (Turin, Italy)	Panda	2nd	Active
2	Fiat Automobiles (Turin, Italy)	Panda	2nd	Active
3	Fiat Automobiles (Turin, Italy)	Panda	2nd	Active
4	Fiat Automobiles (Turin, Italy)	Panda	2nd	Active
5	Fiat Automobiles (Turin, Italy)	Panda	2nd	Active
6	Fiat Automobiles (Turin, Italy)	Panda	2nd	Active
7	Fiat Automobiles (Turin, Italy)	Punto	2nd	3-door
8	Fiat Automobiles (Turin, Italy)	Doblo	1st	Facelift
9	Fiat Automobiles (Turin, Italy)	Tipo	3rd	Hatchback
10	Mitsubishi (Tokyo, Japan)	Colt	6th	CZ3
11	Škoda Auto (Mladá Boleslav, Czech Republic)	Octavia	3rd	Estate
12	Mazda Motor Corp. (Hiroshima, Japan)	Mazda3	4th	Hatchback

Table 2. Technical specificationsof the sensor used in the data collection: Xsens with Model Number $MTi - 100 - 2A8G4$.

Accelerometer		
Parameter	Measurement unit	Value
Standard full range	m/s^2	200
Initial bias error	m/s^2	0.05
In-run bias stability	μg	15
Bandwidth (-3 dB)	Hz	375
Non-linearity	%	0.01
Gyroscope		
Parameter	Measurement unit	Value
Standard full range	$°/s$	450
Initial bias error	$°/s$	0.2
In-run bias stability	μg	10
Bandwidth (-3 dB)	Hz	415
Non-linearity	%	0.1

The Xsens $MTi - 100$ sensor used for data collection is designed to measure the three axis acceleration and rate of turn at a 2000 Hz sampling rate and the three axis magnetic field at a 100 Hz sampling rate. The actual sampling rate used in the analysis was actually smaller to emulate the sampling rate from a smartphone. In particular, increasing sampling rates of 50, 100, 150, 200 and 250 Hz were used, as this is the range from low cost phones to more sophisticated phones. The IMU was mounted using a strong double sided foam tape at the same spot and orientation for every car. We decided to place the sensor on the top of car's dashboard in the middle of the car, because it is a common placement position for the IMU in the literature [12]. In the recent review by Menegazzo et al. [4], the placement of the IMU on the top of car's dashboard was the one mostly adopted in the literature. The results from [8] also showed that the placement on the dashboard provided the smallest classification error in comparison to the placement of the IMU in other locations on the vehicles (e.g., left or right side of the car). The image of the placement of the IMU in three vehicles is shown in Figure 1.

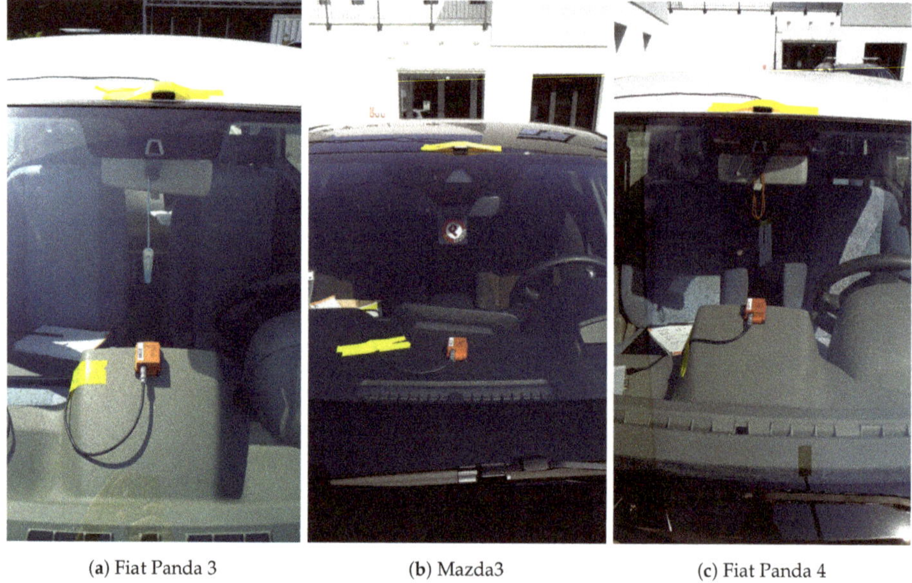

(**a**) Fiat Panda 3   (**b**) Mazda3   (**c**) Fiat Panda 4

Figure 1. Placement of the sensor used for the data collection in the vehicles for three vehicle models.

The data collection was performed using the controlled positioning strategy. As described in [4], this strategy consists of a simple technique where the sensor is placed on the vehicle so that the axes in both reference frames coincide, i.e., the sensor axes are aligned with the vehicle axes, and it is not necessary to apply preprocessing for reorientation. This technique was used for both accelerometer and gyroscope data. Then, the x-axis of the IMU was always pointing towards the driving direction and the z-axis in the vertical direction. A description of the reference frames used in this paper is shown in Figure 2 with a depiction of the body frame and the world frame. The data collection and analysis were performed on all three axes' data both for the accelerometers and gyroscopes, but the accelerometer Z and the gyroscope Y provided the optimal classification accuracy.

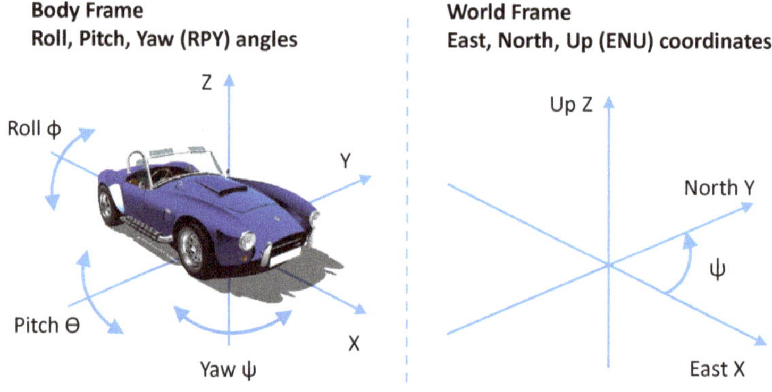

Figure 2. Reference frames.

The description of the brand and model of each of the 12 cars is shown in Table 1. The models were chosen to have a significant number of cars of the same model (the Fiat Panda), but to include as well car models (Mazda3, Octavia) that are different from the Fiat Panda regarding the weight and engine power. The use of 12 different vehicles mitigated the problem of defining a model for pothole detection based on a single vehicle because the data collected by each vehicle were shuffled in the classification process.

The path where the vehicles were driving was a loop in the European Commission Joint Research Centre (JRC) premises. The path is slightly longer than 2 km. Since the data collection was performed for each of the 12 vehicles driving 20 times on the loop, the entire driving data collection was performed on more than 480 km of road. The JRC campus was built in 1960–1965, and the creation of the road infrastructure was more than 50 years old at the time of writing this paper. The road infrastructure is well maintained, and it is all asphalted, even if the asphalt is relatively old and presents various road anomalies like potholes, cracks, transverse cracks, and patches. Different parts of the loop also have different maintenance records because the loop crosses different sectors of the JRC campus. Then, the road surface condition is uneven across the different parts of the loop. The loop includes 15 main road anomalies and 4 obstacles of two different types, which were selected for the data analysis. With reference to the classification of exteroceptions identified in [4], the road anomalies included potholes, cracks, transverse cracks, and patches, while the obstacles were of types rumble strips and speed humps/speed bumps. A picture of some examples of the speed bumps and the road anomalies (potholes in the road) are shown in Figure 3 with the related recordings with the accelerometer in the Z direction in Figure 4.

Figure 3. Examples of obstacles and road anomalies (picture).

The road anomalies are identified in the rest of the paper with the identifier RFX with X = 01, ..., 15 while the obstacles are identified with SBY with Y = 01, ..., 04. The segments of the loop, which are outside RFX and SBY, are identified as NORMZ with Z = 01, ..., 22. See Section 3.2 for further details. Note that even the NORM road segments are not exempt from the presence of small road anomalies or uneven surfaces. The map with the pictorial description of the driving path and the position of the road anomalies/obstacles is shown in Figure 5.

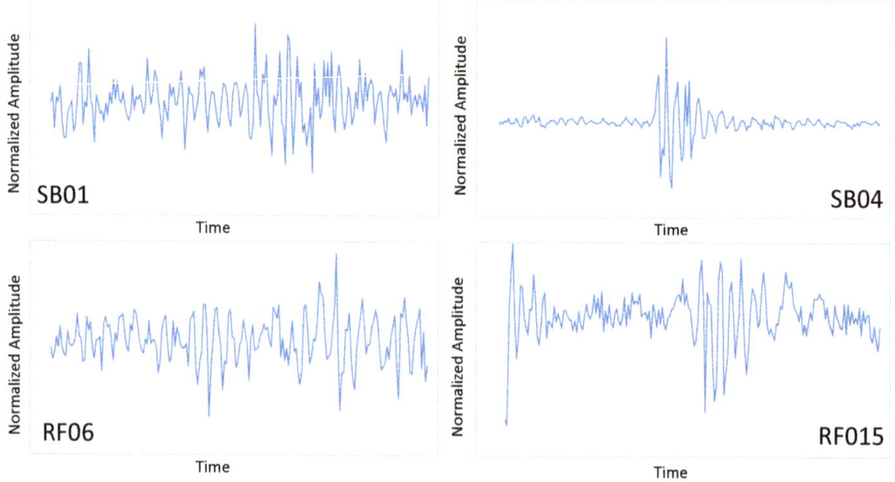

Figure 4. Examples of obstacles and road anomalies (recordings with the accelerometer in the Z vertical direction).

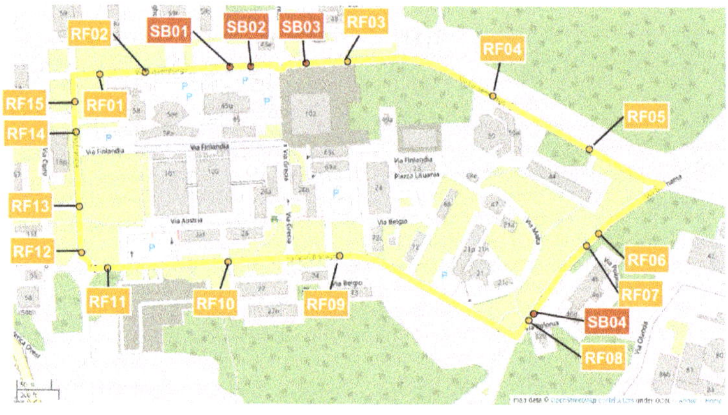

Figure 5. Map of the driving path/loop with the position of the road anomalies/obstacles.

3.2. Methodology

The overall methodology is described in Figure 6, and each step is described in the following paragraphs:

- Normalization and synchronization: As a first step, the data collected from the IMU in each of the 12 vehicles were synchronized among the 20 laps and normalized. A GNSS receiver was also installed in the vehicle and synchronized with the IMU. This process was repeated for different sampling rates: 50, 100, 150, 200 and 250 Hz. All these sampling rates were obtained by downsampling by the related factor (e.g., a factor of 10 for 200 Hz) the initial data collected at 2000 Hz from the IMU. In this paper, we consider only the analysis of the accelerometer data in the Z direction (the vertical direction) and the gyroscope in the Y direction (in the direction of the vehicle). The reason for this choice was to minimize the degrees of freedoms in the analysis and because a heuristic analysis of the data related to the other axes of the accelerometers and gyroscopes showed that the obtained accuracy was inferior to the one obtained using the accelerometer in the Z direction and the gyroscope in the Y direction. This is to expected because the IMUs were mostly stimulated for those axes by the roughness of the road surface, and this is also consistent with literature [11]. In the rest of this paper, the Accelerometer data in the Z direction is called AccZ, and the Gyroscope data in the Y direction is called GyroY. The relationship between the RPY angles and the ENU coordinates was the same as described in [16]. The synchronization was performed by applying the moving variance to the data from the accelerometer and by correlating the results across the laps and the vehicle. As is well known in the literature, full synchronization is not always possible because the vehicles move at different speeds over the road anomalies, and each vehicle has its response to the stimulus created by the road anomaly. On the other side, these are problems derived from data collection in a realistic environment, and such processing issues are likely to be present in any real scenario.

- The entire loop was divided into three different types of segments, and each segment was labeled with type normal, road anomaly identifier (e.g., an asphalt break is RF15), and obstacle identifier (e.g., SB01). There were 4 obstacles (identified with SB01, SB02, SB03 and SB04), 15 road anomalies or road features (identified with RFX and X= 01, ..., 15), and 22 normal segments (identified with NORM), which were obviously the majority in the entire loop. Each segment had a time duration of 4 s. This time duration was chosen because it was long enough to include the driving time of a vehicle over each of the road anomalies/obstacles considered in the study. The approach was tested on different driving speeds since each vehicle (of the dataset of 12 vehicles) was driving at a different speed and the speed was different in each loop (20 loops), even for the same vehicle. Then, the data collection was representative of the real-world conditions when vehicle speeds may be different. Indeed, this was the focus of the study to evaluate if the proposed approach was able to compensate the data taken with different speeds. As the segment duration was fixed, the sample length of each segment was obviously longer for higher sampling rates (e.g., a segment was 200 samples long for a sampling rate of 50 Hz). The labels were generated by using the GNSS position and by manually checking for each segment that the road anomaly and the obstacle were correctly assigned to each segment. This manual step was needed because the GNSS accuracy may not be precise enough to identify the precise location of the road anomaly/feature.

As we had 12 vehicles for 20 loops, the analysis took into consideration a total of $12 \times 20 \times 41$ segments = 9840 segments based on $(4 + 15 + 1) = 20$ different classes.

- The spectrogram was applied to the time domain digital output from the IMU (AccZ and GyroY) for the considered sampling rates and by using different values of the hyperparameter. The definition of the spectrogram and its hyperparameters is provided in Section 3.4.
- The segments of the different representations were used as an input to a CNN, which is described in Section 3.3. The initial time segment was also used as the input to three machine learning algorithms, also described in Section 3.3. We note that all the data measurements from the 12 vehicles were used for classification. Then, the classification using CNN was performed on a model based on the data from all 12 vehicles, which enhanced the generalization of the proposed approach in comparison to models based on the data collected by a single vehicle.

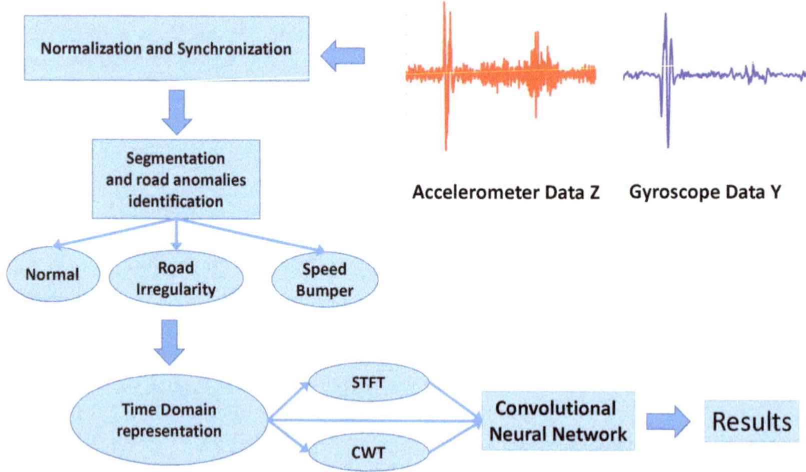

Figure 6. Overall methodology.

3.3. Machine Learning

As mentioned before, three different machine learning algorithms were used, and the results were compared: a CNN using different representations, Support Vector Machine (SVM), and K Nearest Neighbor (KNN). Each machine learning algorithm was based on a set of hyperparameters and their optimal values, which are summarized in Table 3. The following paragraphs describe each specific algorithm and how the hyperparameter values were identified. All the machine learning algorithms were implemented in MATLAB. The three machine learning algorithms were chosen on the basis of the following considerations.

Table 3. Hyperparameters values and ranges of the adopted machine learning algorithms.

Machine Learning Algorithm	Hyperparameter Name	Optimal Hyperparameter Value	Optimal Hyperparameter Range
CNN	Number of filters (Nf)	$Nf = 30$	20, ..., 40
CNN	Solver	RMSProp	Choice between RMSProp, stochastic gradient descent and stochastic gradient descent with momentum
CNN	$L2$	0.001	0.0001, ..., 0.01
CNN	Batch size	128	64, ..., 256
CNN	Number of Convolutional layers (NC)	$NC = 3$	2, ..., 5
SVM	Kernel	RBF	Choice between linear, RBF, and polynomial (orders 2 and 3) kernels
SVM	RBF scaling factor	$\gamma = 2^5$	2^2, ..., 2^{10}
SVM	C factor	$C = 2^7$	2^2, ..., 2^{10}
KNN	K factor	5	1, ..., 20
KNN	Distance	Euclidean distance	Choice between Euclidean, Manhattan, Chebyschev, and Mahalanobis distances

The main objective of this study was to investigate the performance of Deep Learning (DL) for this particular problem of the identification of road anomalies and obstacles. Among other DL algorithms, CNN has been widely applied to image processing since the excellent results from the ImageNet Large Scale Visual Recognition Competition (ILSVRC) in 2012 [18]. The approach proposed in this paper is based on the transformation to images of the accelerometer/gyroscope readings in the time domain collected on the vehicles while driving (see Figures 7 and 8). The intimate relationship between the layers and spatial information in CNNs renders them well suited for image processing and for extracting the discriminating characteristics of the road anomalies and obstacles [19].

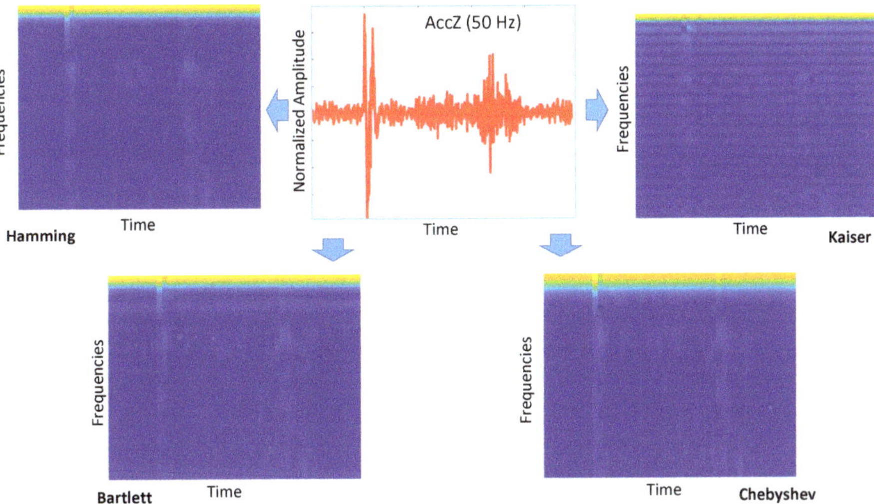

Figure 7. Application of the spectrogram to the Accelerometer Z (AccZ) output from the IMU.

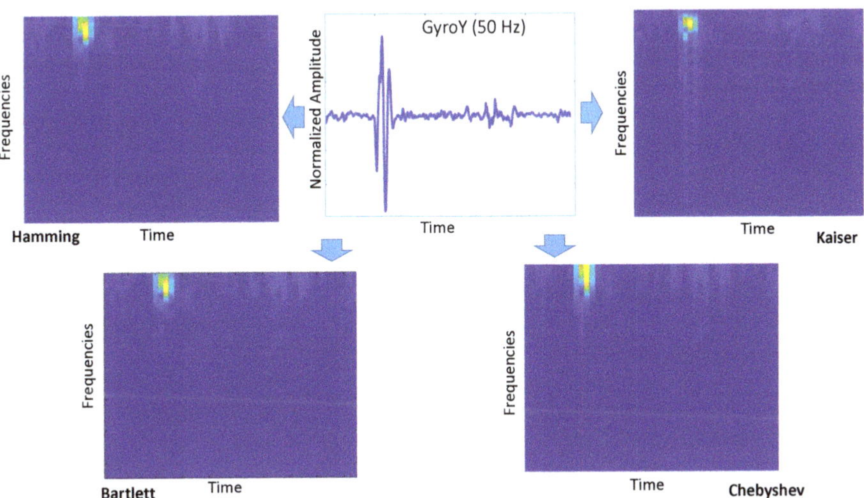

Figure 8. Application of the spectrogram to the Gyroscope Y (GyroY) output from the IMU.

The relatively small size of the dataset (9600 segments of road) for deep learning may generate the risk of overfitting. This risk was mitigated by tuning the hyperparameters (e.g., *L2*) to improve generalization, by using a dropout layer (see Figure 9) and by using cross-validation using 12 folds, as described in the subsequent paragraphs. We also highlight that the number of segments was comparable to the ones used in the literature where deep learning was also applied [9,15].

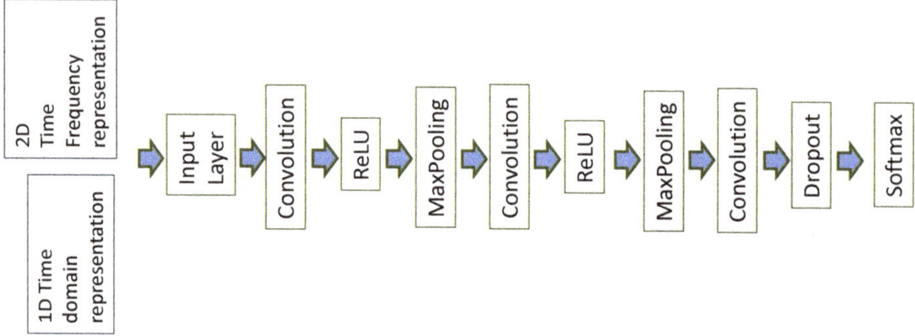

Figure 9. CNN architectureused for the classification.

Then, two shallow machine learning algorithms were used to compare their classification performance to the deep learning CNN algorithm described above. In particular, SVM and KNN were used. The SVM is a computational learning method based on statistical learning theory, whose algorithm constructs and then searches the separating hyperplanes with the maximum margin by transforming the problem description into the dual space by means of the Lagrangian. SVM is reported to be successfully applied in many classification problems (object detection and recognition, information and image retrieval) with a good generalization ability [20]. SVM can deliver a unique solution, since the optimality problem is convex, while other algorithms (like neural networks) may have multiple solutions associated with local minima. In addition, SVM is well known for its effectiveness in high-dimensional spaces, as in the dataset used in this study.

The SVM was also used since it is widely adopted in the literature for road anomaly detection when a machine learning approach is used. The survey presented in [4] for the machine learning algorithms showed that SVM is the most used machine learning algorithm and the second in the overall ranking (the first approach in the ranking is the threshold based approach, which is not a machine learning algorithm and, therefore, is not adopted in this paper).

The KNN was chosen as a baseline for comparison with the other two algorithms as it is relatively simple and naturally lends itself to multi-class problems, as in this case.

The architecture of the CNN is shown in Figure 9, and a brief description is provided here. The input layer of the CNN depended on the type of adopted transformation (e.g., in the 1D time domain, the representation is 1×200). The convolutional layers' parameters (e.g., stride) were optimized according to the specific input, as this input changes for the different representations (e.g., time or spectrogram). Padding was used. The number of filters was set to 30 (an analysis range between 20 and 40 was considered), and the max pooling was set to 4. The solver RMSProp was used as it provided a superior performance to other solvers (stochastic gradient descent and stochastic gradient descent with momentum) for this specific dataset with a learning rate of 0.001 (optimization range between 0.0001 and 0.01). The batch size was set to 128. The number of epochs was set to 160. To mitigate overfitting, the *L2* parameter was set to 0.0005 (optimization range between 0.0001 and 0.01), and a dropout layer was used. Cross-entropy was used as the loss function. The Number of Convolutional layers (*NC*) was also optimized in the range *NC* = 2 to *NC* = 5. The optimal value for the identification accuracy of *NC* was determined to be *NC* = 3, which defines the architecture described in Figure 9. Because the number of hyperparameters to optimize was significant (number of filters, solver, L2, batch size, and number of convolutional layers), each parameter was optimized in a sequential way by keeping fixed the value of some parameters and performing the optimization on other parameters. The first parameter, to be optimized, was the number of convolutional layers NC (i.e., the CNN architecture) with $Nf = 20$, solver = SGD, L2 = 0.001, and batch size = 64. Then, the solver parameter was optimized with the identified number of convolutional layers (i.e., NC = 3). Finally, a three-dimensional grid approach was used to select the optimal values of the parameters Nf, L2, and batch size. The resulting values from the optimization process are presented in Table 3.

SVM is a supervised learning model that classifies data by creating a hyperplane or set of hyperplanes in a high- or infinite-dimensional space, to distinguish the samples belonging to different classes. As this paper addresses a multi-class machine learning problem, a multi-class SVM was used, which was based on an Error-Correcting Output Coding (ECOC) classifier for multiclass learning, where the classifier consists of multiple binary learners. In particular, we used the OneVsOne approach where for each binary learner, one class is positive, another is negative, and the algorithm ignores the rest. This design exhausts all combinations of class pair assignments. Various kernels (e.g., linear, polynomial) were tried, and the one providing the best performance was the Radial Basis Function (RBF) kernel, where the values of the scaling factor γ must be optimized together with the parameter C [21]. In this paper, the SVM was applied with the RBF with a scaling factor $\gamma = 2^5$ and a $C_{factor} = 2^7$. The optimization was performed on a range of values for γ and C between 2^2 and 2^{10} using a grid search approach. The optimal values of the parameters are presented in Table 3.

K Nearest Neighbor (KNN) is an approach to data classification that estimates how likely a data point is to be a member of one class or another depending in which group the data points nearest to it are. KNN is an example of a lazy learner algorithm, meaning that it does not build a model using the training set until a query of the dataset is performed. The main hyperparameter in KNN is the K factor, which must be optimized for the specific classification problem. The type of distance metric used to calculate the "nearest" must also be chosen carefully [22]. The optimization was performed on a range of values of K from 1 and 10 for each of the different distance: Euclidean, Manhattan, Chebyschev, and Mahalanobis. The optimal values of the hyperparameters, which provided the highest accuracy, are listed in Table 3.

All the optimized values presented in this section were calculated using a linear or grid approach for each sampling rate and then averaged for the analysis of the time domain data.

For all three machine learning algorithms, a 12-fold cross-validationwas used with the partition of the entire set into 12 exclusive folds. For each fold, eleven of twelve parts of the initial dataset were used for the training plus validation, while the remaining 1/12 part was used for testing. The validation part was 1/4 of the training plus validation portion (thus, the validation part was 11/48 of the entire dataset in each fold). Then, the results were averaged.

The evaluation metrics were the accuracy, precision, and recall, and they are described in the following equations:

$$Accuracy = \frac{(TP + TN)}{(TP + TN + FP + FN)} \tag{1}$$

$$Precision = \frac{(TP)}{(TP + FP)} \tag{2}$$

$$Recall = \frac{(TP)}{(TP + FN)} \tag{3}$$

where *TP* is the number of True Positives, *FP* is the number of False Positives, *FN* is the number of False Negatives, and *TN* is the number of True Negatives.

3.4. Time Frequency Transforms

This section is focused on the description of the spectrogram and the related hyperparameters. In addition, because Section 4 provides a comparison with another time-frequency representation based on Continuous Wavelet Transform (CWT), the CWT definition and hyperparameters are also described here.

Consider a signal $s(\tau)$ of length N (the original signal in the time domain, called 1D-T in the rest of this paper) and a window $w(\tau)$ of length W_{size} (where $N >> W_{size}$), whose Fourier transforms are respectively $S(f)$ and $W(f)$, as shown in the following equations:

$$S(f) = \int_{-\infty}^{\infty} s(\tau)e^{-j2\pi f\tau}d\tau \tag{4}$$

$$W(f) = \int_{-\infty}^{\infty} w(\tau)e^{-j2\pi f\tau}d\tau \tag{5}$$

The localized spectrum representation of $s(\tau)$ at time $\tau = t$ is obtained by multiplying the signal by the window $w(\tau)$ centered in time $\tau = t$, as shown in the following equation:

$$s_w(t,\tau) = s(\tau)w(\tau - t), \tag{6}$$

through the application of the FT, the Short Time Fourier Transform (STFT) $F_s^w(t,f)$ is obtained:

$$F_s^w(t,f) = \mathcal{F}_{\tau \to f}\{s(\tau)w(\tau - t)\} \tag{7}$$

The squared magnitude of the STFT, denoted by $S_s^w(t,f)$, is called the spectrogram and is expressed by the following equation:

$$S_s^w(t,f) = |F_s^w(t,f)|^2 \tag{8}$$

As described in Section 4, the spectrogram was adopted for the analysis presented in this paper because it provided a superior identification performance in comparison to the phase component of the STFT, as shown in Table 4.

In the spectrogram definition, we evaluated the impact of three main hyperparameters: (1) the type of window $w(\tau)$, (2) the length of the window or window size W_{size}, and (3) how much overlapping was between different windows on the overall length N of the signal $x(\tau)$. The overlapping is defined as O_{lap}, and it is defined as a percentage of the window size W_{size} while the window is sliding across $x(\tau)$. Because the parameter W_{size} is dependent on the sample ratio at which the data are collected by the IMU and it would not be the same across different sampling rates, the parameter W_R is instead used in the rest of this paper. W_R is defined as the ratio of N/W_{size}. For example, a value of $W_R = 10$ at a sampling rate of 50 Hz results in $W_{size} = 20$.

Four different window types were used: Hamming, Bartlett, Chebyschev, and Kaiser [23], and they were chosen because of their different filtering behaviors. Three different window sizes with different values of W_R were identified: $W_R = 4, 5$, and 10. Finally, three different overlapping factors O_{lap} were evaluated: 33%, 50%, and 66%. The results of the evaluation are presented in the next Section 4.

The other time-frequency representation was based on the wavelets. The CWT provides an overcomplete representation of a signal by letting the translation and scale parameter of the wavelets vary continuously [24].

The CWT is expressed by:

$$C(a,b) = \frac{1}{|a|^{1/2}} \int_{-\infty}^{\infty} s(\tau)\psi\left(\frac{(\tau - b)}{a}\right) d\tau \tag{9}$$

where ψ is the mother wavelet, a is the scale, and b is the translational value. In this analysis, we chose the Morse mother wavelet [25] and the CWT implementation in MATLAB from MathWorks (i.e., the cwt function).

4. Results

In an initial step, the impact of the hyperparameters of the spectrogram definitions on the identification accuracy was evaluated.

4.1. Optimization of the Spectrogram Hyperparameters

This section is focused on the optimization of the hyperparameters in the spectrogram.

Figure 10 presents the results for the calculated accuracy using the CNN-SP approach using AccZ. The results are presented as a bar graph with different bars related to the hyperparameters of the spectrogram definition including the size of the window (based on the W_R factor) and the different values of the overlapping factor O_{lap}. Four different graphs are presented for the different sampling rates: 50 Hz (Figure 10a), 100 Hz (Figure 10b), 200 Hz (Figure 10c), and 250 Hz (Figure 10d). The sampling rate of 150 Hz is not presented because of the lack of space and because it is quite similar to other graphs. The graphs in Figure 10 shows that for $W_R = 4$ and $W_R = 5$, the highest accuracy was obtained with $O_{lap} = 66\%$, while for $W_R = 10$, the highest accuracy was obtained with $W_R = 10$. On the other side, it is noted that for most of the shown sampling rates (e.g., 50, 100 and 200 Hz), the accuracy obtained with $W_R = 10$ was the lowest of all the values of O_{lap}, while this was different for the sampling rate of 250 Hz. The use of higher sampling rates also increased the classification time; then, if there is no loss of classification accuracy (as shown in Figure 10), it is preferable to use the optimal parameters for the lower sampling rate (i.e., 50 Hz). In this case, such an optimal value was obtained with $W_R = 5$ and $O_{lap} = 66\%$. These results were obtained with the Hamming window. The choice of this window is supported by the results presented in Figure 11 for different windows and overlap values O_{lap} with $W_R = 5$. In fact, the results depicted in the figures below show that the optimal identification accuracy is obtained with the Hamming window across all the sampling rates and the O_{lap} values, and this is the type of window chosen in the subsequent results.

The results obtained for GyroY are shown in Figure 12. In general, the results for GyroY confirm the previous results obtained with AccZ, but with greater variability. In fact, $O_{lap} = 66\%$ provides better results for the sampling rate of 100 Hz (Figure 12b) and the sampling rate of 200 Hz (Figure 12c) for $W_R = 4$ and $W_R = 5$, but in other cases, $O_{lap} = 50\%$ provides better results. We highlight that the highest classification accuracy (i.e., reaching the threshold of 0.97) is quite similar to the one obtained with AccZ. These results indicate that the use of AccZ or GyroY provides an equivalent classification accuracy. All these results are obtained with the Hamming window. As shown in Figure 13, the Hamming window provides better results than the other windows across all the different sampling rates. From this point of view, these results obtained for the gyroscope are consistent with the results obtained with the accelerometer.

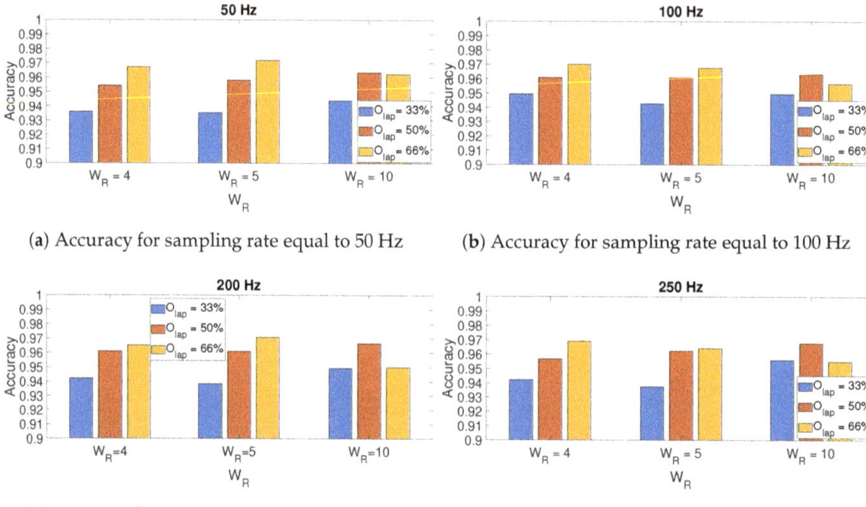

(**a**) Accuracy for sampling rate equal to 50 Hz

(**b**) Accuracy for sampling rate equal to 100 Hz

(**c**) Accuracy for sampling rate equal to 200 Hz

(**d**) Accuracy for sampling rate equal to 250 Hz

Figure 10. Accuracy for AccZ in relation to the size of window (based on the W_R factor) used in the spectrogram for different sampling rates and different values of the overlapping factor O_{lap}. The Hamming window is used to obtain these results.

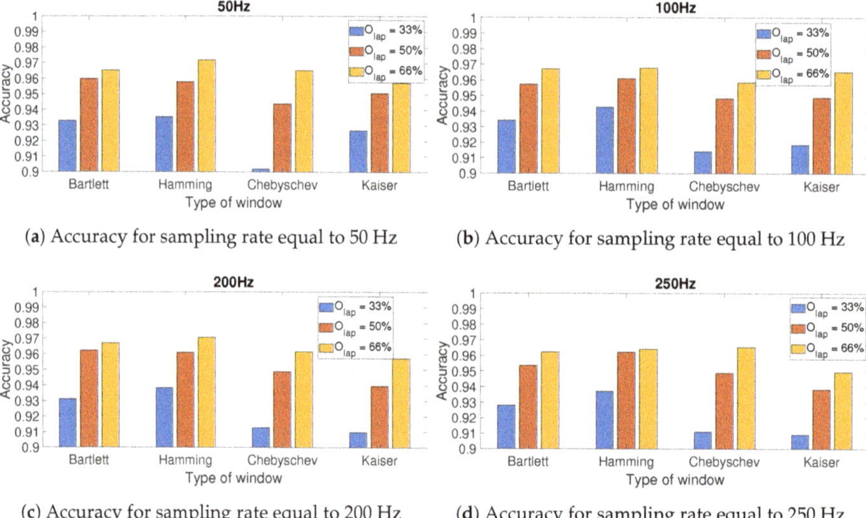

(**a**) Accuracy for sampling rate equal to 50 Hz

(**b**) Accuracy for sampling rate equal to 100 Hz

(**c**) Accuracy for sampling rate equal to 200 Hz

(**d**) Accuracy for sampling rate equal to 250 Hz

Figure 11. Accuracy for AccZ in relation to the type of window used in the spectrogram for different sampling rates and different values of the overlapping factor O_{lap} and $W_R = 5$.

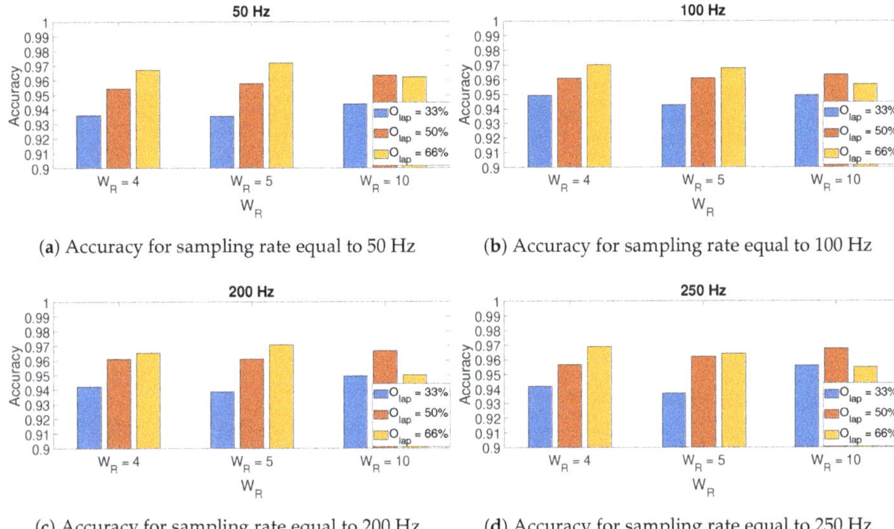

(**a**) Accuracy for sampling rate equal to 50 Hz (**b**) Accuracy for sampling rate equal to 100 Hz

(**c**) Accuracy for sampling rate equal to 200 Hz (**d**) Accuracy for sampling rate equal to 250 Hz

Figure 12. Accuracy for GyroY in relation to the size of window (based on the W_R factor) used in the spectrogram for different sampling rates and different values of the overlapping factor O_{lap}. The Hamming window is used to obtain these results.

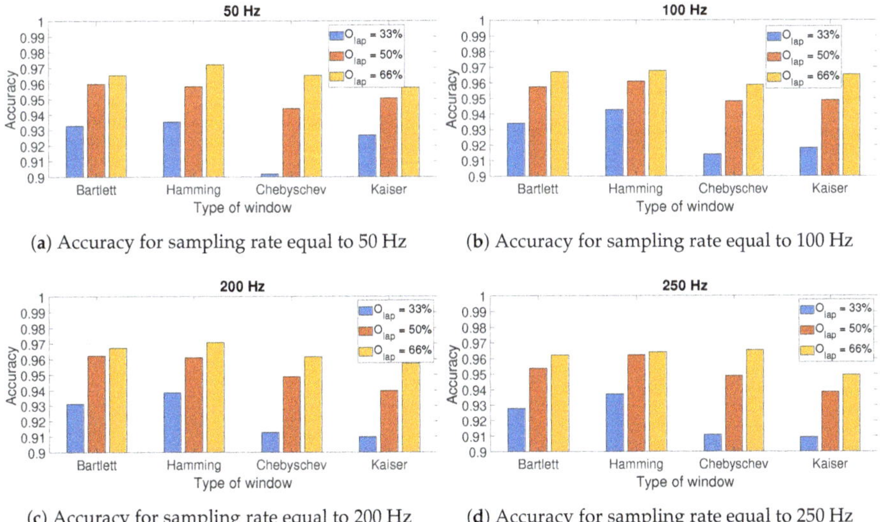

(**a**) Accuracy for sampling rate equal to 50 Hz (**b**) Accuracy for sampling rate equal to 100 Hz

(**c**) Accuracy for sampling rate equal to 200 Hz (**d**) Accuracy for sampling rate equal to 250 Hz

Figure 13. Accuracy for GyroY in relation to the type of window used in the spectrogram for different sample rates and different values of the overlapping factor O_{lap} and $W_R = 5$.

4.2. Analysis and Comparison of the Approaches

This section presents the results of the comparison of the proposed approach with the shallow machine learning algorithms and CNN-1D. Figure 14 provides the comparison of all the different approaches evaluated in the analysis for the accelerometer in the Z direction (AccZ), while Figure 14 provides the comparison for the gyroscope in the Y direction (GyroY). As mentioned before, CNN-SP is the application of CNN to the spectrogram of the original signal in the time domain; CNN-1D is the application of the CNN to 1D-T (the original signal in the time domain). SVM and KNN represent the direct application of the respective shallow machine learning algorithm to 1D-T. The presented results are based on the mean (average) of the results calculated for each fold across the 12 folds. The results clearly show that all the CNN based approaches significantly outperform the shallow machine learning approach in a consistent way across all the different sampling rates, which confirms the results obtained by the authors in [15] for a different dataset. In addition, the approach proposed in this paper based on the combination of the CNN with the spectrogram (CNN-SP) outperforms in a consistent way the direct application of CNN on the time domain (CNN-1D), which is the novel finding of this paper. The other interesting results provided by Figures 14 and 15 are that the CNN-SP approach provides a relatively uniform value of the accuracy across the different sampling rates, while CNN-1D drops significantly with higher sampling rates. A potential reason for this behavior is that the CNN algorithm is able to extract similar discriminating features across the different sampling rates when applied in combination with the spectrogram, while lower sampling rates in 1D-T introduce a smoothing effect in CNN-1D, which benefits the classification accuracy. A similar consideration can be applied to the application of the SVM and KNN algorithms, whose classification accuracy also degrades with increasing sampling rates.

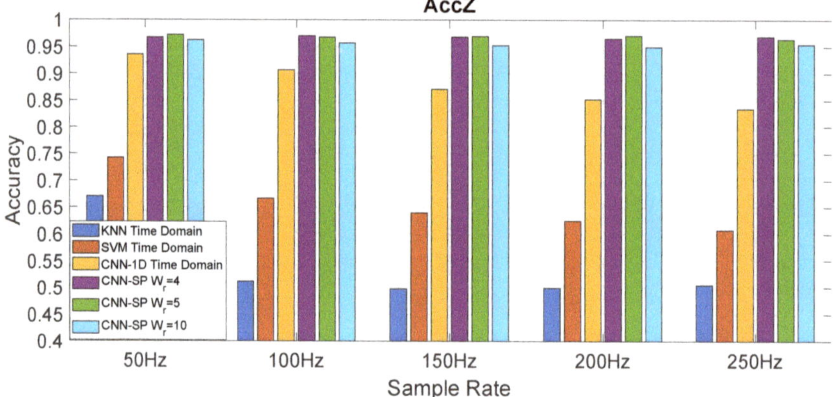

Figure 14. Comparison of the accuracy among the spectrogram CNN approach (CNN-SP) proposed in this paper (for different window sizes), the CNN directly applied to the time domain representation (CNN-1D), and the shallow machine learning techniques (SVM, KNN) applied to the data of the Accelerometer in the Z direction (AccZ).

For completeness, we also report the accuracy obtained for another time-frequency representation as mentioned previously (i.e., the CWT) and for the application of the STFT in phase (the complex amplitude of the STFT is the spectrogram already considered before). The results are shown in Table 4. Note that the results presented for the STFT phase and the CWT are the result of an optimization process similar to what was done for the spectrogram in Section 4.1. For the CWT, this was done on the type of the wavelet and the number of octaves for frequency. For the STFT (phase), this was based on the same parameters identified in Section 4.1. The results clearly confirm that CNN-SP provides a superior classification performance in comparison to the other approaches. In particular, the use of the magnitude-only components of CNN-SP provides a better identification accuracy than the phase-only component (and the combination of amplitude plus phase, which is provided here). The other spectral approach based on the use of CWT also provides an inferior performance, which indicates that the wavelets may not be appropriate for this context (note that a similar result for vehicle authentication was also obtained in [16]).

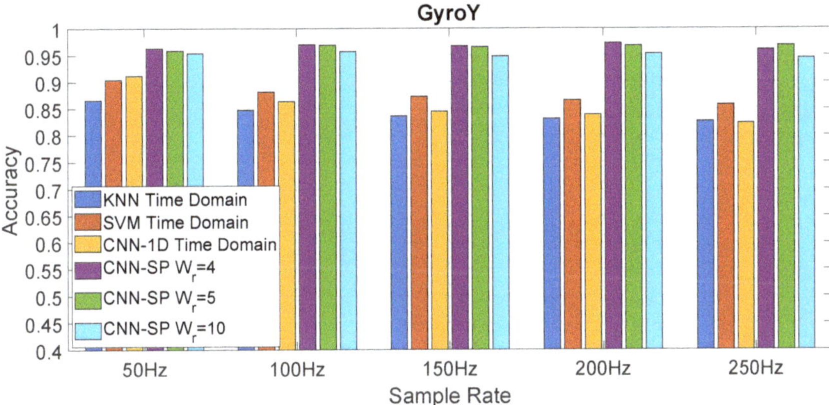

Figure 15. Comparison of the accuracy among the spectrogram CNN approach (CNN-SP) proposed in this paper (for different window sizes), the CNN directly applied to the time domain representation (CNN-1D), and the shallow machine learning techniques (SVM, KNN) applied to the data of the Gyroscope in the Y direction (GyroY).

Table 4. Accuracy obtained for the different representations at 50 Hz. Bold represents the highest value.

Machine Learning Algorithm and Representation	Accuracy
CNN-SP (magnitude) (Hamming window $O_{lap} = 66\%$, $W_R = 5$)	**0.9720**
CNN with STFT (phase) (Hamming window $O_{lap} = 66\%$, $W_R = 5$)	0.8047
CWT-CNN (magnitude and Morlet wavelet)	0.9421
CWT-CNN (phase and Morlet wavelet)	0.7812
CNN-1D	0.9354
SVM	0.7422
KNN	0.6728

The accuracy metric only provides a limited view of the classification performance. For this reason, we also provide in the following paragraphs and figures the estimate of the recall and precision obtained for the specific set of parameters and for different machine learning algorithms. The following figures show the recall and precision for each specific road anomaly when compared to the other anomalies, and they give a better understanding of the areas where the machine learning failed to identify the correct samples (false positives and false negatives).

Figure 16a,b provides respectively the precision and recall for each road anomaly as a percentage using the CNN-SP approach with a sampling rate of 50 Hz, $O_{lap} = 66\%$, and $W_R = 5$. Similar results were obtained for the other values of the CNN-SP hyperparameters, but they are provided here for space reasons. These figures confirm the previous findings on the accuracy, which shows that it is possible to obtain a great identification accuracy using the CNN-SP approach. On the other hand, the figures show that there can be great variability for the identification of each road anomaly/obstacle. Of course, this depends on the type of road anomaly/obstacle and how much different from the other road anomalies/obstacles it is. One relevant aspect is the relatively low value of recall and precision for the class NORM. The reason is that the machine learning algorithm confuses in some cases the road anomaly/obstacle with the normal segment (identified as NORM in the figures), because the digital output from the accelerometers/gyroscopes is similar. In fact, even a NORM segment includes some small irregularities (e.g., cracks, uneven condition of the road), which may be difficult to distinguish. This result is to be expected, and it is consistent with the findings in the literature [15].

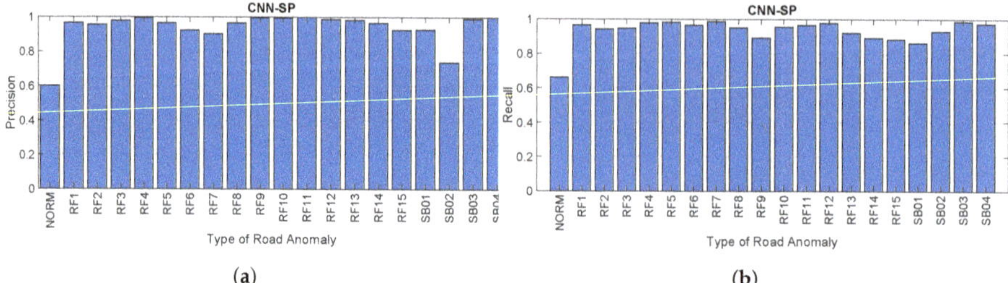

Figure 16. Precision and Recall using the $CNN - SP$ approach. (**a**) Precision for each road anomaly for CNN-SP (sampling rate of 50 Hz, $O_{lap} = 66\%, W_R = 5$) and (**b**) recall for each road anomaly for CNN-SP (sampling rate of 50 Hz, $O_{lap} = 66\%, W_R = 5$).

On the other side, such low values of precision and recall may be a significant problem when used in real-world applications since most real-world road segments are normal. A potential reason for these results is that no thresholds were applied for the detection of road anomalies/obstacles, but the machine learning algorithms were directly applied to the data recordings. Different approaches can be used to mitigate this issue. Some of these approaches are based on a similar analysis from previous studies in the literature, and they can be classified into different phases of the overall methodology: data collection, data pre-processing, or data processing/classification (e.g., based on the ML/DL algorithm). One potential approach in the data collection phase would be to widen the sensor inputs using the shock responses of the four absorbers or the wheel speed from the four wheels, as was done in [9] to enhance the discriminating power of the classification algorithm. Another approach in the data pre-processing is to introduce a smoothing filter in the data pre-processing phase. We note that the evaluation of the identification performance with different sampling rates described in this study is also a crude form of smoothing for the lower sampling rates. A review of the application of filters in the literature for road anomalies'/obstacles' detection was provided in Section 2.2 of [10]. Another possibility would be to set a threshold to analyze only the most significant road anomalies/obstacles and remove the smaller ones, which are present in the normal segment (see Section 2.3 of [10] for a review on threshold techniques). Note that the thresholds' definitions do not need to be static, but they can also be dynamic to adapt themselves to the road conditions, as proposed in [26]. The application of both techniques would require significant effort to identify the most effective smoothing filter or the appropriate values of the threshold. For this reason, this analysis is postponed to future developments (see Section 5). In the data processing/classification phase, the existing dataset could be widened using data augmentation techniques (e.g., adding noise) to improve generalization and robustness. Another approach would be to use other time-frequency transforms (e.g., Stockwell transform) in combination with CNN or with other deep learning algorithms.

Figure 17a,b provides respectively the precision and recall for each road anomaly as a percentage using the CNN-1D approach at a sampling rate of 50 Hz. The results are similar to what was achieved for the CNN-SP approach, but the recall and precision are slightly lower, as already foreseen from the results presented in Figure 14. In particular, the same phenomenon of a lower value of precision and recall is also noted here.

Figure 18a,b provides respectively the precision and recall for each road anomaly as a percentage using the SVM algorithm at a sampling rate of 50 Hz. The figures show the drastic drop in the classification performance in comparison to the CNN based approach. In addition, we note a greater variability in the identification of the specific road anomalies among them. In general, the precision is negatively impacted as the number of False Positives (FP) is quite significant. In particular, the algorithm confuses specific road features (e.g., RF14) or obstacles (e.g., SB01 and SB02) as normal road segments (i.e., NORM). In comparison, recall is usually higher for the same road anomalies, which shows that the False Negatives (FN) are more limited in number than the False Positives (FP).

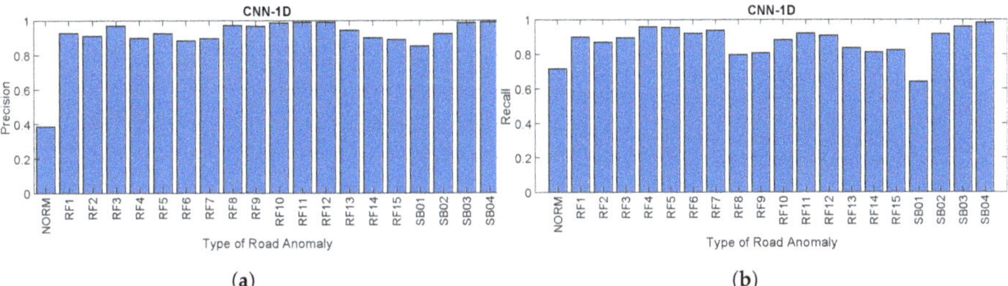

(**a**)   (**b**)

Figure 17. Precision and recall using the CNN-1D approach. (**a**) Precision for each road anomaly for CNN-1D (sampling rate of 50 Hz) and (**b**) recall for each road anomaly for CNN-1D (sampling rate of 50 Hz).

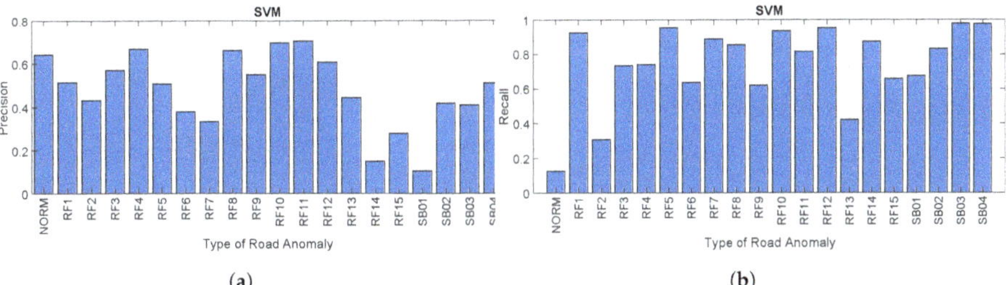

(**a**)   (**b**)

Figure 18. Precision and recall using the SVM algorithm. (**a**) Precision for each road anomaly for SVM (sampling rate of 50 Hz) and (**b**) recall for each road anomaly using SVM (sampling rate of 50 Hz).

Figure 19a,b provides respectively the precision and recall for each road anomaly as a percentage using the KNN algorithm at a sampling rate of 50 Hz. These results confirm what was already obtained for the accuracy. The variability is even increased in comparison to the SVM approach. In addition, the recall for the NORM segment reaches quite low values, which highlights the difficulty of the algorithm to distinguish the NORM segments from the other segments. It is also noted that the lower values of precision and recall are usually for specific road anomalies (e.g., RF14 or SB01), in a similar way to what was obtained for the previous figures and algorithms.

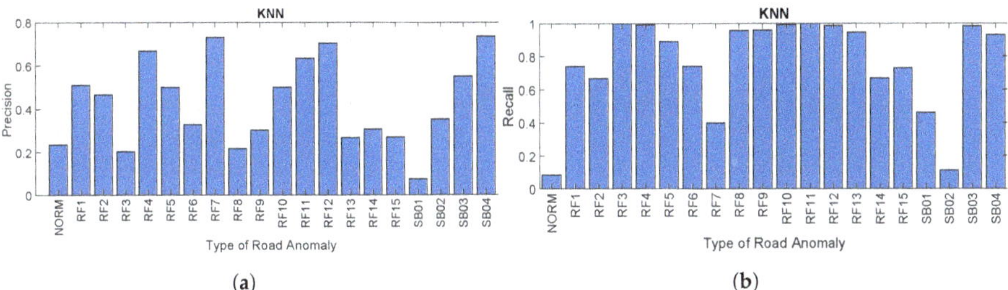

(**a**)   (**b**)

Figure 19. Precision and recall using the KNN algorithm. (**a**) Precision for each road anomaly using KNN (sampling rate of 50 Hz) and (**b**) recall for each road anomaly using KNN (sampling rate of 50 Hz).

4.3. Discussion

The results shown in the previous subsections prove the superiority of the time-frequency CNN (TF-CNN) for road anomaly identification in comparison to shallow machine learning algorithms (SVM, KNN) and to the application of CNN to the initial time domain representation (accelerometer and gyroscope). The difference in classification performance appears for all the considered sampling rates (i.e., 50, 100, 150, 200 and 250 Hz) and both for accelerometer and gyroscope data. With accelerometer data, the improvement in the identification accuracy of TF-CNN is even more visible than with gyroscope data. The final identification accuracy taking into consideration all the road anomalies and the vehicles used in the data collection is slightly more than 97%, which is consistent with the findings in the literature. We note that this value of identification accuracy was obtained by using a set of 12 different vehicles rather than the one or two vehicles commonly used in literature. Then, the high accuracy obtained with TF-CNN is even more remarkable because it manages to mitigate the differences of the mechanical configuration of the vehicles, and it can be used to support crowd-sourcing approaches (with many different vehicles on the road) for the detection and mapping of the road anomalies in the road infrastructure.

Regarding the choice of the time-frequency transform, the results in Table 4 show that the spectrogram in combination with CNN (CNN-SP) provides a slightly better identification accuracy than the application of the complex magnitude of the CWT (CNN-CWT). Additional time-frequency transforms could also be applied, and this will be the scope of future studies (see Section 5).

Another important result from Figures 14 and 15 is that CNN-SP is able to provide a high identification accuracy (higher than 97%) even at lower sampling rates (e.g., 50 or 100 Hz), which supports the deployment of the proposed approach even with cost effective mass-market smartphones, which are now equipped with accelerometers and gyroscopes with a sampling rate of 100 Hz.

We note that the study presented in this paper conducted an extensive analysis of the hyperparameters present in the definition of the time-frequency transforms (e.g., window size, type of window, overlapping ratio) to evaluate their impact on the identification accuracy. The results show that although some hyperparameters do not have a significant impact, other hyperparameters like the overlapping ratio must be carefully tuned in a training phase to achieve the optimal identification accuracy.

One aspect that must still be improved is the relatively low precision and recall of the NORM segments. Potential approaches to mitigate this issue would be to introduce a smoothing filter in the data pre-processing phase, but this would entail an extensive analysis of the choice of the most appropriate filter, so for this reason, this analysis is postponed to future developments (see Section 5).

5. Conclusions and Future Developments

This paper proposes a novel approach for the detection and identification of road anomalies using data collected from accelerometers and gyroscopes installed on the vehicles. The approach is based on the transformation of the collected data into the spectral domain (via a spectrogram), which is then given as an input to CNN. This approach is compared against the direct application of CNN on the samples collected in the time domain and on the application of shallow machine learning algorithms like SVM and KNN, as is usually proposed in the existing research literature. This approach is evaluated on a dataset, created by the authors, by collecting IMU (e.g., accelerometers, gyroscopes) recordings on many km of road infrastructure using 12 different automotive vehicles. A comprehensive analysis of the influence of the hyperparameters on the classification performance is presented for the data collected from the accelerometers and gyroscopes. The results show that this approach is able to obtain a very high identification accuracy for each road anomaly, and it is able to distinguish accurately between obstacles intentionally created by road traffic authorities against road anomalies (potholes), which are the consequences of road degradation and roadworks. Beyond the analysis of the road roughness, such high accuracy can also be used to correctly identify specific road anomalies and obstacles on the road infrastructure, which can be used as landmarks in maps to improve the positions of vehicles (including autonomous vehicles) in the future.

Future developments will extend the scope of the study presented in this paper in various directions. One direction in the data collection phase is to use additional sensors' inputs (e.g., different placements of the accelerometers/gyroscopes in the vehicle, shock absorber responses of the four absorbers, or the wheels' speed). Another direction in the data collection phase would be to collect the data with different passengers and drivers in the vehicles. One direction in the processing phase is to investigate the application of additional time-frequency transformations in combination with other deep learning architectures and algorithms. Future developments will also investigate the combination of thresholds or smoothing filters with CNN-SP to enhance the precision and recall results for the classification of the NORM segments. In particular, we will investigate adaptive filters whose definition and parameters are modified depending on the conditions of the road or the vehicle model.

Author Contributions: Conceptualization, G.B. and R.G.; methodology, G.B., R.G., and F.G.; software, G.B., R.G., and F.G.; validation, G.B., R.G., and F.G.; investigation, G.B. and F.G.; writing, original draft preparation, G.B.; writing, review and editing, G.B., R.G., and F.G.; supervision, G.B. All authors have read and agreed to the published version of the manuscript.

Funding: No specific funding was used to conduct the study presented in this paper.

Acknowledgments: We acknowledge the directorates I and C of the European Commission Joint Research Centre for providing most of the vehicles used in the study.

Conflicts of Interest: The authors declare no conflict of interest.

Abbreviations

The following abbreviations are used in this manuscript:

1D-T	One-Dimensional Time domain
CNN	Convolutional Neural Network
CNN-1D	Convolutional Neural Network-One-Dimensional
CNN-SP	Convolutional Neural Network-Spectrogram
CWT	Continuous Wavelet Transform
ECOC	Error-Correcting Output Codes
FP	False Positive
FN	False Negative
ENU	East-North-Up
GNSS	Global Navigation Satellite System
IMU	Inertial Measurement Unit
KNN	K Nearest Neighbor
LiDAR	Light Detection And Ranging
RPY	Role-Pitch-Yaw
RBF	Radial Basis Function
RF	Road Feature
SB	Speed Bump
STFT	Short Time Fourier Transform
SVM	Support Vector Machine
TN	True Negative
TP	True Positive

References

1. Luettel, T.; Himmelsbach, M.; Wuensche, H.J. Autonomous ground vehicles—Concepts and a path to the future. *Proc. IEEE* **2012**, *100*, 1831–1839. [CrossRef]
2. Baldini, G.; Giuliani, R.; Dimc, F. Road safety features identification using the inertial measurement unit. *IEEE Sens. Lett.* **2018**, *2*, 1–4. [CrossRef]
3. Wahlström, J.; Skog, I.; Händel, P. Smartphone-based vehicle telematics: A ten-year anniversary. *IEEE Trans. Intell. Transp. Syst.* **2017**, *18*, 2802–2825. [CrossRef]
4. Menegazzo, J.; von Wangenheim, A. Vehicular Perception Based on Inertial Sensing: A Structured Mapping of Approaches and Methods. *SN Comput. Sci.* **2020**, *1*, 1–24. [CrossRef]
5. Allouch, A.; Koubâa, A.; Abbes, T.; Ammar, A. Roadsense: Smartphone application to estimate road conditions using accelerometer and gyroscope. *IEEE Sens. J.* **2017**, *17*, 4231–4238. [CrossRef]
6. Yang, X.; Li, H.; Yu, Y.; Luo, X.; Huang, T.; Yang, X. Automatic pixel-level crack detection and measurement using fully convolutional network. *Comput.-Aided Civ. Infrastruct. Eng.* **2018**, *33*, 1090–1109. [CrossRef]

7. Maeda, H.; Sekimoto, Y.; Seto, T.; Kashiyama, T.; Omata, H. Road damage detection and classification using deep neural networks with smartphone images. *Comput.-Aided Civ. Infrastruct. Eng.* **2018**, *33*, 1127–1141. [CrossRef]

8. Xue, G.; Zhu, H.; Hu, Z.; Yu, J.; Zhu, Y.; Luo, Y. Pothole in the dark: Perceiving pothole profiles with participatory urban vehicles. *IEEE Trans. Mob. Comput.* **2016**, *16*, 1408–1419. [CrossRef]

9. Luo, D.; Lu, J.; Guo, G. Road Anomaly Detection Through Deep Learning Approaches. *IEEE Access* **2020**, *8*, 117390–117404. [CrossRef]

10. Sattar, S.; Li, S.; Chapman, M. Road surface monitoring using smartphone sensors: A review. *Sensors* **2018**, *18*, 3845. [CrossRef]

11. Mednis, A.; Strazdins, G.; Zviedris, R.; Kanonirs, G.; Selavo, L. Real time pothole detection using android smartphones with accelerometers. In Proceedings of the 2011 International Conference on Distributed Computing in Sensor Systems and Workshops (DCOSS), Barcelona, Spain, 27–29 June 2011; pp. 1–6.

12. Eriksson, J.; Girod, L.; Hull, B.; Newton, R.; Madden, S.; Balakrishnan, H. The pothole patrol: Using a mobile sensor network for road surface monitoring. In Proceedings of the 6th International Conference on Mobile Systems, Applications, and Services, Breckenridge, CO, USA, 17–20 June 2008; pp. 29–39.

13. Kalim, F.; Jeong, J.P.; Ilyas, M.U. CRATER: A crowd sensing application to estimate road conditions. *IEEE Access* **2016**, *4*, 8317–8326. [CrossRef]

14. Du, R.; Qiu, G.; Gao, K.; Hu, L.; Liu, L. Abnormal road surface recognition based on smartphone acceleration sensor. *Sensors* **2020**, *20*, 451. [CrossRef] [PubMed]

15. Varona, B.; Monteserin, A.; Teyseyre, A. A deep learning approach to automatic road surface monitoring and pothole detection. *Pers. Ubiquitous Comput.* **2019**, *24*, 519–534. [CrossRef]

16. Baldini, G.; Geib, F.; Giuliani, R. Continuous authentication of automotive vehicles using inertial measurement units. *Sensors* **2019**, *19*, 5283. [CrossRef]

17. Celaya-Padilla, J.M.; Galván-Tejada, C.E.; López-Monteagudo, F.E.; Alonso-González, O.; Moreno-Báez, A.; Martínez-Torteya, A.; Galván-Tejada, J.I.; Arceo-Olague, J.G.; Luna-García, H.; Gamboa-Rosales, H. Speed bump detection using accelerometric features: A genetic algorithm approach. *Sensors* **2018**, *18*, 443. [CrossRef]

18. Russakovsky, O.; Deng, J.; Su, H.; Krause, J.; Satheesh, S.; Ma, S.; Huang, Z.; Karpathy, A.; Khosla, A.; Bernstein, M.; et al. Imagenet large scale visual recognition challenge. *Int. J. Comput. Vis.* **2015**, *115*, 211–252. [CrossRef]

19. Arel, I.; Rose, D.C.; Karnowski, T.P. Deep machine learning-a new frontier in artificial intelligence research [research frontier]. *IEEE Comput. Intell. Mag.* **2010**, *5*, 13–18. [CrossRef]

20. Cortes, C.; Vapnik, V. Support-vector networks. *Mach. Learn.* **1995**, *20*, 273–297. [CrossRef]

21. Cristianini, N.; Shawe-Taylor, J. *An Introduction to Support Vector Machines and Other Kernel-Based Learning Methods*; Cambridge University Press: Cambridge, UK, 2000.

22. Zhang, M.L.; Zhou, Z.H. A k-nearest neighbor based algorithm for multi-label classification. In Proceedings of the 2005 IEEE International Conference on Granular Computing, Beijing, China, 25–27 July 2005; Volume 2, pp. 718–721.

23. Oppenheim, A.V. *Discrete-Time Signal Processing*; Pearson Education India: Delhi, India, 1999.

24. Lilly, J.M.; Olhede, S.C. Generalized Morse wavelets as a superfamily of analytic wavelets. *IEEE Trans. Signal Process.* **2012**, *60*, 6036–6041. [CrossRef]

25. Olhede, S.C.; Walden, A.T. Generalized morse wavelets. *IEEE Trans. Signal Process.* **2002**, *50*, 2661–2670. [CrossRef]

26. Harikrishnan, P.; Gopi, V.P. Vehicle vibration signal processing for road surface monitoring. *IEEE Sens. J.* **2017**, *17*, 5192–5197. [CrossRef]

Publisher's Note: MDPI stays neutral with regard to jurisdictional claims in published maps and institutional affiliations.

MDPI

St. Alban-Anlage 66

4052 Basel

Switzerland

Tel. +41 61 683 77 34

Fax +41 61 302 89 18

www.mdpi.com

Sensors Editorial Office

E-mail: sensors@mdpi.com

www.mdpi.com/journal/sensors